2184

QD
142
C 4.3
1982

CRC SERIES IN ANALYSIS FOR ENVIRONMENTAL CONTROL

Editor-in-Chief
James W. Robinson
Professor
College of Chemistry and Physics
Louisiana State University
Baton Rouge, Louisiana

ANALYSIS OF PESTICIDES IN WATER

Volume I: Significance, Principles, Techniques, and Chemistry of Pesticides
Volume II: Chlorine- and Phosphorus-Containing Pesticides
Volume III: Nitrogen-Containing Pesticides

Alfred S. Y. Chau, Senior Editor
B. K. Afghan, Co-Editor
Canada Centre for Inland Waters
Burlington, Ontario
Canada

CHEMICAL ANALYSIS OF INORGANIC CONSTITUENTS OF WATER

Editor
Jon C. Van Loon
Department of Geology and Chemistry
University of Toronto
Canada

ORGANIC ANALYSIS OF WATER POLLUTION: CHEMICAL ANALYSIS

B. K. Afghan, Senior Editor
Alfred S. Y. Chau, Co-Editor
Canada Centre for Inland Waters
Burlington, Ontario
Canada

Chemical Analysis
of
Inorganic Constituents
of Water

Editor

Jon C. Van Loon

Departments of Geology and Chemistry
and
the Institute for Environmental Studies
University of Toronto
Canada

CRC Press, Inc.
Boca Raton, Florida

Library of Congress Cataloging in Publication Data

Main entry under title:

Chemical analysis of inorganic constituents
    of water.

    (CRC series in analysis of environmental
control)
    Includes index.
    1. Water—Analysis. 2. Chemistry,
Inorganic. I. Van Loon, J. C. (Jon Clement),
1937- . II. Series.
QD142.C43   628.1'61   81-12209
ISBN 0-8493-5209-6   AACR2

Direct all inquiries to CRC Press, Inc., 2000 Corporate Blvd., N.W., Boca Raton, Florida, 33431.

© 1982 by CRC Press, Inc.

International Standard Book Number 0-8493-5209-6

Library of Congress Card Number 81-12209
Printed in the United States

FOREWORD

The assessment of the environmental impact of man's endeavors must be made as rapidly and painlessly as possible. It is not only aesthetically rewarding to retrieve clean lakes, rivers, air, and countrysides; it is possibly vital to man's continued existence. Emotional reaction serves no useful long-range purpose, but in fact eventually boomerangs and disenchants the public. Nevertheless, it is incumbent upon man to safeguard his environment. Clearly, the forces involved are quite beyond our comprehension at this time. When we are operating from a position of ignorance as we are today, it is most important to be sure that the risks we take are minimal, although in many cases a 'fail safe' posture has been adopted which can be extremely expensive. Discussion on the possible short-term and long-term impacts of various pollutants is valuable but probably endless. The truth ultimately emerges from reliable data interpreted with wisdom and understanding. Analytical chemistry provides the invaluable bridge between speculation and firm data. We can generate firm data only with reliable analytical techniques, skilled scientists, and clear minds.

In an effort to provide means of collecting and dispersing this information to all interested parties, we have invited a number of scientists of stature to produce monographs in their field of expertise. The objective of these monographs is to document analytical procedures and techniques that are useful to the environmentalists.

In general, three groups of people will be interested in these monographs: industrial engineers and scientists who are monitoring both liquid and gaseous effluents from industrial effluents; and environmentalists who are trying to assess pollution levels and amass data on long-term health effects and other effects of pollution.

J. W. Robinson
Editor-in-Chief
June 1977

INTRODUCTION

It would seem at first glance that the analysis of water would be the simplest of all analytical tasks. While this may be correct for distilled water solutions containing an easily detectable amount of a single inorganic salt, it is far from the truth for natural waters. The latter are dynamic systems containing living as well as non-living, organic together with inorganic, and dissolved as well as insoluble substances. Problems can occur during or after sampling which may change the sample composition drastically from its true form. These are due to a variety of causes, principal among which are contamination and loss of a substance due to precipitation complexation adsorption or ion exchange on the container wall.

Natural waters vary greatly in matrix constituents. A laboratory will normally be required to analyze samples ranging from precipitation and soft water, rivers and lakes which are generally low in impurities, to highly polluted effluents. A single analytical method for a substance will not necessarily be applicable to such a wide range of compositions.

Compounding these difficulties is the question of sampling. Little attention is generally given to the problem; hence, it is common for sampling to be the largest source of error in the final result.

Sampling

No agreement can be found in the literature on sampling methods. Generally three types of sampling will be done on water bodies: grab, composite, or continuous. Precipitation samples may be short interval, event, or composite. Samples are taken with a fixed volume sampling device or by pumping.

Natural waters can be very heterogeneous vertically, horizontally, and with time. This is due not only to man-made pollution but also can be caused by natural phenomena such as erosion, currents, thermoclines, and precipitation washout of dust.

Grab, short interval, and event sampling can thus lead to erroneous estimates of water composition. A series of such samples taken with time and, in the case of rivers and lakes, vertically and horizontally, may give a useful result.

Composite samples are obtained by combining a large number of samples taken at different intervals. The intervals may either be temporal or spacial or both. In this case an average composition can be obtained. A distinct disadvantage of this approach is that any important compositional variations within the water system become undetectable.

In designing a sampling strategy, the objectives of the work must be borne in mind, together with the problems mentioned above. A detailed discussion of sampling is beyond the scope of this book. The reader will, however, find more details here and there throughout the following chapters. There are a number of publications from which further information can be obtained, e.g., Erickson,[1] Wagner,[2] Kleiber and Erlebach,[3] and Cullen.[4]

General Precautions
Contamination

In the case of many parameters which must be measured, contamination of the water is an ever-present danger. Thiers[5] classifies the problem as follows:

1. Positive contamination results from the addition of contaminants to the sample.
2. Negative contamination occurs when losses of the substance in question (analyte) occur.
3. Pseudo-contamination is an error introduced by the presence of a substance other than the analyte.

Positive and negative contamination pose the most serious problems. Causes of the former are more obvious and many. Negative contamination arises frequently due to precipitation, adsorption, or ion exchange on the surface of containers. Adsorption of analyte on particulate matter, if the latter is present, may also pose a problem.

Sample Containers

The parameter(s) being measured on a particular sample will dictate the type of suitable sampling and storage container for a water sample. In some cases, e.g. heavy metals, the constraints are very severe; in others, e.g. major ions, almost any clean, leak-proof container will suffice.

There are two basic choices in containers, glass or plastic. In both cases a tightly fitting screw-cap top is essential.

Glass bottles are suitable for many purposes. The notable exception is storage of water to be analyzed for heavy metals. In this latter case, adsorption and/or ion exchange on the glass surface can invalidate their use. If a glass container is chosen, borosilicate glass is best. Soft glass (soda-lime glass) should most definitely not be used when sodium is to be determined. Presently, plastic bottles have almost entirely superceded glass for the storage of waters for inorganic analysis. Compared to borosilicate glass most plastic bottles are inexpensive, less fragile, and have minimal problems due to ion exchange. On the negative side most plastics have a porosity which allows samples to evaporate slightly over long storage periods. This problem is particularly serious in the storage of standard solutions.

Second choice among the plastics is high-density linear polyethylene. Compared to other softer plastic containers, these are rugged and suffer less from evaporation and adsorption/ion exchange problems.

Bottle Cleaning

Plastic bottles, including Teflon®, may contain impurities which will result in sample contamination. Metallic impurities, present as a result of the fabrication process, can cause serious positive contamination in trace metal analysis. Other additives used as stabilizers may leach out, causing negative or pseudo-contamination problems.

The method of cleaning is dependent on the use to which the bottles will be put. At the very least, however, most inorganic analysis applications require that the containers be washed with detergent, rinsed thoroughly with tap and finally distilled or deionized water. In most cases, a non-ionic detergent used in hot water is best.

When heavy metals are to be analyzed, an acid soaking following the detergent wash is essential. When an acid treatment is needed, there is no general agreement in the literature as to the type or strength of acid and the method of treatment. The present author uses warm 1:1 nitric or hydrochloric acid. This mixture is added, followed by a period of vigorous shaking. Subsequently the container is rinsed thoroughly with distilled or deionized water.

There is a considerable disagreement on the nature and treatment of storage containers for waters to be analyzed for mercury. Some researchers recommend glass bottles and other plastics. An acid wash as described above is recommended; however, several workers suggest rinsing with oxidizing reagents such as permanganate or persulfate as an additional precaution.

Preservation

Depending on the parameter to be measured it may be necessary to add a preservative to the sample to prevent adverse changes during storage. Such a procedure is generally recommended for waters to be analyzed for heavy metals.

For all the heavy metals except mercury, acid only is added as a preservative. General agreement does not exist on the nature and final concentration of the acid. The present author recommends that the water be brought to pH 1 with high-purity nitric acid. This procedure will result in metals being leached from particulate material. A prior filtration of the sample through pre-washed membranes of 0.45 μm will minimize this effect.

Filtration

Depending upon the nature of the sample, the parameter to be measured, and the data desired, it may be necessary to perform a filtration. This is often essential in the case of trace and minor elements.

This treatment may aid in maintaining sample integrity due to the resulting reduced bacterial activity. As indicated in the preceding section, prior to the acidification of waters for heavy metal analysis, a filtration through 0.45 μm porosity membrane filters is commonly done. In this way, a distinction can be made between dissolved and total metal in a water. Although it is a generally accepted procedure to use 0.45 μm porosity size to make this distinction, the present author has evidence that, in some cases, particulate passes through such a filter. Most commonly, membrane-type filters are employed. These must be acid washed and distilled water rinsed to eliminate metal contaminants.

Reagents

It is crucial, particularly in the case of trace element analysis, to use reagents of high purity. In many cases, analytical reagent grade chemicals will suffice. However, when working at very low concentration it may be necessary to use specially purified materials. The latter are now generally available, but at very high cost, from most of the larger chemical suppliers. A blank must be run with any trace analysis. The magnitude of the blank will control the detection limit which can be obtained.

Water used for dilutions and for the preparation of standard solutions should be of high quality. Distillation and/or ion exchange procedures are generally employed for water purification. A good quality borosilicate glass still will often produce water of the required purity. In some cases a double distillation or a combination of distillation and deionization may be required. Distillation performed at a temperature just below the boiling point of water is extremely slow, but produces very high quality water. In areas where water has a very high salt content, a mixed-bed ion exchange column can be used prior to distillation. If organic matter is a problem, distillation in the presence of permanganate can be done.

There are commercially available deionizing units which can be used in place of distillation. This equipment removes particulate, ions, and organic matter. The present author has a personal bias favoring the distillation approach to purification of water.

Standard Solutions

Great care is essential in the preparation of standard solutions. Only reagents of highest purity should be employed. In the case of organic reagents this may necessitate distillation or recrystallization of commercial material. Reagent-grade inorganic chemicals are normally satisfactory, but special high-purity material may be necessary in some instances. The latter are available from specialty chemical companies in North America and Europe.

Quality Control

Recently there has been a general awakening to the necessity of a good quality control program. Despite this, it is still common to find reports of poor agreement between different laboratories analyzing the same parameter in the same sample.

It is important here to stress that good precision (repeatability) does not necessarily signify that an accurate result has been obtained. To maximize the possibility of the latter, testing of both natural and synthetic reference samples is essential on a routine basis.

Jon C. VanLoon

REFERENCES

1. Erickson, P., Marine Trace Metal Sampling and Storage, National Research Council of Canada Report 1, NRCC No. 16472, Ottawa, Canada, 1977.
2. Wagner, R., Z. Anal. Chem. 282, 315, 1976.
3. Kleiber, P. and Erlebach, W. E., Limitations in Single Water Samples in Representing Mean Water Quality, Tech. Bull. No. 103, Inland Waters Directorate, Water Quality Branch, Vancouver, British Columbia, Canada, 1977.
4. Cullen, D. H., Special Technical Publication 573, Am. Soc. for Testing Materials, 1975.
5. Thiers, R. E., Methods Biochem. Anal., 5, 273, 1975.

EDITOR-IN-CHIEF

J. W. Robinson is Professor of Chemistry and Chairman of the Analytical Division at Louisiana State University, Baton Rouge, Louisiana. He earned his degrees at The University of Birmingham, England (B.Sc., 1949; Ph.D., 1952; D.Sc., 1977) and obtained American citizenship in 1965.

He worked at both Exxon Research Company and Ethyl Research Corporation for a number of years before returning to Louisiana State University to join its Chemistry Faculty.

Dr. Robinson has written more than 130 publications, as well as two texts: *Undergraduate Instrumental Analysis*, the 3rd edition of which is currently in press, and *Atomic Absorption Spectroscopy*, the 2nd edition of which was published in 1975. He is Editor of two international journals, *Spectroscopy Letters* and *Environmental Science and Engineering*. He is also assistant editor of *Applied Spectroscopy Reviews*. He is a former chairman of the Gordon Research Conference on Analytical Chemistry and of the L.S.U. International Symposium on Analytical Chemistry. He is also director of the Saul Gordon Workshop on Atomic Absorption Spectroscopy.

Dr. Robinson is a Guggenheim Fellow and an Awardee of the Honor Scroll of the American Institute of Chemists.

THE EDITOR

Jon C. Van Loon, Ph. D., is a Professor at the University of Toronto and is cross appointed to the Departments of Geology and Chemistry and the Institute for Environmental Studies. His principal research interest is the determination of the chemical forms of the elements in complex samples using the techniques of valence electron atomic spectrometry as detectors for chromatography.

Jon Van Loon spent a research leave at CSIRO, Division of Chemical Physics, Melbourne, Australia, in 1975, doing research into developing high-intensity radiation sources for atomic fluorescence spectrometry. In 1981 he was a visiting scientist at the University of Tokyo in the Department of Chemistry.

He is a member of the Spectroscopy Society of Canada and the Chemical Institute of Canada. He served as the General Chairman of the 4th International Conference on Atomic Spectroscopy held in Toronto, Canada, in October, 1974.

Jon Van Loon has given 45 invited and plenary lectures at North American and International conferences. He has published over 100 papers in International Journals and has co-authored 2 books, edited 1 book, and authored a book on analytical atomic absorption spectrometry.

CONTRIBUTORS

B. K. Afghan
Analytical Methods Division
National Water Research Institute
Canada Centre for Inland Waters
Burlington, Ontario, Canada

F. J. Philbert
Water Quality Branch
Inland Waters Directorate, Ontario Region
Environment Canada
Canada Centre for Inland Waters
Burlington, Ontario, Canada

TABLE OF CONTENTS

Chapter 1

PHYSICAL PARAMETERS

F. J. Philbert

TABLE OF CONTENTS

I. INTRODUCTION*

Specific conductance, pH, temperature, color, turbidity, odor, taste, dissolved solids, and suspended solids are among the physical parameters that are most commonly measured for water quality monitoring, surveillance, and process control purposes and for the physical characterization of waters. Most of these parameters can be measured with a high degree of accuracy and precision in a relatively short time by means of a variety of instruments and procedures now available. On the other hand, some water types may be characterized in a very broad sense, for certain specific uses and, from an aesthetic standpoint, simply through sensory perception. For example, it is a very simple matter for almost anyone having absolutely no knowledge of, or experience in, analytical chemistry to tell when a water body is unsightly or its taste, color, or odor is unacceptable for normal human consumption.

Some tests (pH, odor, color, temperature, specific conductance, turbidity, residue) are sometimes denoted by the designation "immediates," implying that measurements should be made as soon as possible after sample collection and preferably at the site if practicable. This has been made possible in recent years through the availability of a variety of measuring devices for the automatic and/or manual measurement of single or groups of physical parameters including pH, dissolved oxygen, temperature, specific conductance, turbidity, and color. Nevertheless, there still remains a need for the development of improved, less expensive, and more reliable multiparameter sensing devices for automated water quality monitoring purposes. Odor, temperature, and pH are particularly susceptible to changes during the time interval between sample collection and analysis. For example, pH and possibly conductivity may change as a result of absorption of CO_2 from the air. The color and turbidity of some samples may change also during sample storage due to interactions that may result in sample coloration or discoloration, precipitate formation, and sorption. Probably the most common example of this occurrence is the precipitation of iron as the brown-colored ferric hydroxide when the lower valence salts are oxidized upon exposure of some samples to the air. The conversion of bicarbonates into the precipitable carbonates is also another example. It is thus highly desirable to avoid unnecessary contact of the sample with the air, whether trapped within the container or through direct sample exposure. It is desirable also that the time lapse between sample collection and analysis be kept at the absolute minimum. Above all, in cases where it is impractical to do the testing

---

* The use of specific names of equipment and material is not to be construed as an endorsement or promotion of that product by the author or Environment Canada.

Table 1

RECOMMENDED HANDLING PROCEDURE FOR SAMPLES FOR PHYSICAL
TESTING WHEN IMMEDIATE ON-SITE ANALYSIS IS IMPRACTICABLE

Parameter	Container	Preservation/Storage
Odor	Glass	Fill container completely and make air-tight by capping with a ground-glass or Teflon®-lined stopper. Store at approximately 4°C. Test as soon as possible and preferably within 24 hr of collection.
Taste	Glass	Store at approximately 4°C. Test as soon as possible and preferably within 24 hr of collection.
Color	Glass preferred to plastic	Fill container completely, store in dark at approximately 4°C. Test within 24 hr if feasible.
Temperature	Plastic or glass	Test *in situ* or immediately after collection.
pH	Plastic or borosilicate glass	Fill container completely. Maintain at approximately 4°C.[a] Test within 24 hr.
Turbidity	Plastic or glass	Fill container completely. Store in dark. Test within 24 hr if feasible.
Specific conductance	Plastic or borosilicate glass	Fill container completely and maintain at approximately 4°C. Test within 24 hr if feasible.
Dissolved solids	Plastic or borosilicate glass	Filter on-site if feasible. Do not delay analysis longer than is necessary.
Suspended solids	Plastic or glass	Make determinations as soon as possible.

[a]   Sample pH varies with temperature, and readings must be corrected appropriately if sample has been stored and measured at a temperature significantly different from its initial temperature.

in situ or on-site, it is essential that appropriate measures be taken to maintain sample integrity while they are being transported and/or awaiting tests.

The choice of sample container and the application of the appropriate preservation techniques are the two prime factors that determine whether sample integrity would be maintained or not. There have been numerous studies done and a great deal written on the subject of sample preservation. The fact that study results and recommended procedures are not always concordant tends to complicate matters. It is impossible to provide a set of absolute and precise instructions on sample preservation to cover every possible situation. Several factors, including the nature of the sample, constituents to be measured, and the measuring technique to be used, must be taken into account. In general however, the task of preserving a water sample for physical tests is a relatively simple one. Seldom if ever, is it necessary to add a preservative to the sample, and once the appropriate container is used, the desired results may be achieved by one or more of the following procedures:

1.   Fill container completely to exclude any trapped air.
2.   Store away from light.
3.   Keep refrigerated at approximately 4°C (care should be taken not to freeze the sample).

Table 1 lists the recommended handling procedures for the parameters being considered.

II. ODOR

The application of the sense of smell is probably the simplest means available to an individual to characterize in a general way, or to discriminate between, water types for certain specific uses. For instance, a water that may otherwise appear quite suitable for drinking or swimming, will, under normal circumstances, be considered to be unacceptable if it contains a disagreeable odor. Odor is invariably attributed to the presence of some volatile substance in the water. It is produced primarily by organic and also by some inorganic compounds. It is not unusual to find water containing significant amounts of decomposed organic materials and/or excrement having a characteristic smell of ammonia and/or hydrogen sulfide. Under the right conditions, some constituents may act synergistically with the resulting increase in the odor intensity of water. It is well known, for example, that some phenolic compounds in the presence of chlorine exude smells much stronger than either of the single compounds alone. Odor may originate from natural sources such as some algae. Whatever the source, however, whether it be plankton, contamination emanating from municipal or industrial waste, organic matter decomposition, microbial activity, or simply dissolved gases, the presence of water odor is indicative of a water quality problem of some kind.

Simple and subjective though it may be, the odor test is in fact a very valuable indicator of water contamination and it provides a means of assessing the suitability of water for the various applications for which it is intended. Odor is also a very useful parameter in monitoring activities in process control and water and waste treatment plants.

Apart from the health aspects there are also cosmetic or aesthetic and economic factors to be considered when dealing with water odor. An adequate supply of odor-free water is essential for domestic consumption and is in fact a prerequisite for the food, beverage, and pharmaceutical industries to name but a few.

In measuring odor, the most that can be derived from the test is some form of qualitative and, at best, an approximate quantitative measure of the odor intensity. Because the test is done solely by the application of the tester's sense of smell and due to the complexity of most odors, it is impracticable if not impossible to eliminate the element of subjectivity associated with the test or to obtain accurate and precise test results. The fact that power of odor testers is affected by factors such as fatigue and adaptation and that it may differ among individual testers by as much as some 2 to 3 orders of magnitude, is noteworthy. The following expressions typify the terms most commonly used to describe water odor: aromatic or spicy, balsamic, chlorinated, sulfurous, phenolic, estery, alcoholic, earthy, fishy, grassy, moldy, musty, peaty, marshy, and sweetish.

Odor is detected when the odor-causing substance is volatile and is released as a vapor from the sample. Therefore, the degree to which odor is detected depends on the vapor pressure and solubility of the substance in the sample. In reporting odor, it is generally useful to state the measuring temperature since temperature has an effect on odor. When conducting the test it is important to ensure the absence of any extraneous odors, such as from the tester's person or clothes and testing room, that are likely to affect the accuracy of the test.

Odor-causing substances in a sample could be determined and identified by certain physio-chemical analytical procedures. These methods however, are generally time-consuming and require sophisticated and expensive equipment. The sensory methods are the most common and convenient odor-testing procedures available.

A. Qualitative Test

The following is a procedure for a simple qualitative odor test.

Place about 100 to 150 mℓ of the sample in a glass Erlenmeyer flask or beaker of about 250 to 500 mℓ volume. Cover with a watch glass and smell at the mouth of the flask or beaker for any discernable odor. If odor is detected, describe it by using the descriptive terms that best depict it. The test may be repeated by first heating the sample to 60°C before shaking and smelling it.

B. Quantitative Determination

Qualitative odor results are not always enough, and it is often desirable to obtain some kind of quantitative measure of the determinand. The threshold odor test is considered to be the best procedure available for the quantitative measurement of odor. The unit of measurement is the "threshold odor number" (TON) which is the maximum dilution ratio at which odor is just detectable. The TON of a sample is determined by progressively diluting the sample with odor-free water until any distinctive odor is just barely discernible.

As pointed out above, different testers have different sensitivity to odor and the test is best conducted in one of two ways. The first one is to use a calibrating odor standard with known threshold odor values for screening the testers and for standardization purposes. A dilute solution containing about 0.05 to 1.0 mg/ℓ *n*-butanol is the standard generally used. The other approach is to have the same sample tested by at least two, and preferably more, people. From a practical standpoint, the former approach is preferred. When conducting the odor test, the accuracy and preciseness may be improved if:

1.   The tester has no knowledge of the extent of dilution of the samples being tested. This would normally require at least two persons to carry out the required test procedures. Thus the individual doing the actual odor-sensing should not be involved in the preparation of the samples for testing.
2.   Sample dilutions are done precisely and consistently. It may be necessary to use 10 mℓ pipets graduated in 0.1 mℓ when measuring sample aliquots for dilution.
3.   The tests are performed by a group of two or more qualified testers under identical conditions.
4.   No single tester performs the test continuously for periods in excess of about 15 min at a time. This limit may be extended up to about 1 hr if the odor intensity of the samples under test is relatively low.
5.   Opaque or dark-colored glass flasks such as the red actinic Erlenmeyer flasks are used for the test in order to eliminate possible bias due to the presence of color or turbidity in the sample.

1. Equipment and Reagents

Odor-free Erlenmeyer flasks — The Erlenmeyer flasks are made odor-free by cleaning with an odorless detergent followed by an acid solution such as chromic acid and then rinsing thoroughly. Odor-free water is then boiled in each flask for about 4 to 6 min shortly before use.

The use of odor-free dilution water is a key element for the proper conduct of this test. Odor-free water may be prepared by passing distilled water or tap water through a bed of activated carbon at a rate of about 6 ℓ/hr, or by boiling distilled water for a few minutes and allowing it to cool in an odor-free environment.

Table 2
RECOMMENDED DILUTIONS FOR VARIOUS ODOR INTENSITIES

Volume (ml) of Sample in Which Odor is First Detected	Volume (ml) of Sample Aliquots to be Diluted to 200 ml
200	200, 140, 100, 70, and 50
50	50, 35, 25, 17, and 12
12	12, 8.3, 5.7, 4.0, 2.8
2.8	Prepare a series of more diluted samples from an intermediate dilution.

2. Procedure

It is advisable to do a preliminary odor test on the sample in the following manner. Measure a series of four or five sample aliquots, ranging in volumes from about 2 ml to 200 ml, into 500 ml odor-free glass Erlenmeyer flasks equipped with ground-glass stoppers. The sample should be added to suitable volumes of odor-free dilution water contained in the flask to make the final volume exactly 200 ml. Include a flask containing 200 ml of odor-free dilution water only as a reference blank. This is to be used for comparison each time a sample aliquot is tested.

Stopper the flasks and heat the contents to 60°C±1°C by immersing the flasks in a thermostated water bath or on an electric hot plate. Shake the flask with the reference blank vigorously. Remove stopper and immediately sniff the vapors given off. Repeat the test in the flask (Flask A) containing the smallest sample aliquot in the dilution water. If an odor is detected, prepare another series of more diluted samples from an intermediate dilution prepared by making an exact tenfold dilution of the intermediate and repeat the above procedure. The order of testing should be from the highest to the least diluted sample fraction. However, the sequence should be broken by inserting a few reference blanks and some more-diluted sample aliquots in the series of test samples so as to avoid guessing or anticipation by the individual(s) determining the odor.

If no odor is detected in Flask A, repeat the test for odor in the flask containing the next larger sample aliquot and so on until the first flask (Flask B), in which there is a distinctive perceptible odor, is identified. The approximate threshold odor number (TON) of the sample could then be calculated. A more accurate TON measurement can now be made by the following procedure.

Prepare a new series of at least five, and preferably eight, sample dilutions so that the highest in the series is equivalent to the dilution used in Flask B for the preliminary test. The dilution guide given in Table 2 would be useful for this purpose. It is advisable to include two or more reference blanks in the series in a random fashion near the expected threshold. The odor test is then performed starting with the most diluted sample aliquot and progressing upwards until the first flask in which odor is just detectable with certainty, is identified.

3. Calculation

Knowing the volume of sample used in that particular flask, calculate the TON from the formula:

$$TON_T = \frac{V_1 + V_2}{V_1}$$

Table 3
DILUTION SCHEDULE AND EQUIVALENT TON/OII DATA[a]

Dilution	ml Sample and Dilutions A to E Brought to 200 ml with Dilution Water	Threshold Odor Number (TON)	Odor Intensity Index (OII)
Sample	200	1	0
	100	2	1
	50	4	2
	25	8	3
	12.5	16	4
Dilution A	50	32	5
(25 ml of sample	25	64	6
made up to 200	12.5	128	7
ml)			
Dilution B	50	256	8
(25 ml dilution A	25	512	9
further diluted	12.5	1024	10
to 200 ml)			
Dilution C	50	2050	11
(25 ml dilution B	25	4100	12
further diluted	12.5	8200	13
to 200 ml)			
Dilution D	50	16400	14
(25 ml dilution C	25	32800	15
further diluted	12.5	65500	16
to 200 ml)			
Dilution E	50	131000	17
(25 ml dilution D	25	262000	18
further diluted	12.5	524000	19
to 200 ml)	6.25	1050000	20

[a] Data taken from Leithe[1]

where V_1 = ml of sample used, V_2 = ml of dilution water, and T = temperature at which test is conducted.

4. Notes

1. While a 60°C sample temperature (T) is considered as standard for this test, there may be instances where a T value of 40°C or lower would be applicable as deemed appropriate according to the nature of the sample. When determined by two or more odor testers, report the median and the range of TON_T values obtained.

2. In addition to reporting the TON_T value, it is advisable to include a qualitative description of the odor sensed wherever possible.

3. Samples such as some types of wastewaters may have extremely strong odors such that the TON may exceed a million. In these instances, the use of odor intensity index (OII) units instead of the TON may be a more convenient way to report odor results. The relationship between TON and OII is given by $TON = 2^{OII}$. The dilution schedule and equivalent TON/OII data given in Table 3 aptly illustrate this relationship.

4. Water containing free chlorine must be pretreated by adding just enough sodium thiosulfate ($Na_2S_2O_3 \cdot 5H_2O$), sodium sulfide (Na_2SO_3) or sodium arsenite ($NaAsO_2$) to get rid of the excess chlorine.

III. TASTE

The taste and odor tests are closely related, and there may well be cases where what may be considered to be a gustatory sensory perception may actually be the result of an olfactory sensation. In fact, this parameter is sometimes called flavor which is essentially a combination of the sensory effects of taste, odor, temperature, and the feel of the sample. Taste may be manifested as a result of the presence of inorganic salts of some metals including iron, zinc, copper and sodium, but more so from organic compounds such as phenols, e.g., chlorophenols. The types of compounds contributing to the taste of a sample are varied. If one were to take a gas chromatographic profile of certain water samples, it may reveal the presence of numerous organic compounds, including some chlorinated compounds, which may affect to varying degrees both the odor and taste of the water. Therefore, as in the case of odor, it is extremely difficult, it not impossible, to determine all of the individual taste-contributing components in a water sample. The presence of algae and certain of their metabolic products may cause water to have a somewhat earthy, grassy, or even rotten taste, depending on the algae species present. Also, water containing significant amounts of detergents may have a soapy taste imparted to it.

Most water quality standards set limits on taste as one of the acceptance criteria for specific uses. Taste is a useful indicator of water pollution, and, although highly subjective, testing could be easily conducted in a matter of seconds. Care should be taken, however, to ensure that the test is not done on water such as wastewaters, untreated effluents, or waters from other unaesthetic sources that are likely to contain pathogenic organisms or significant amounts of toxins. In testing the quality of drinking water, it is safe to say that the taste test will be a matter-of-course.

Taste may be described in a general way by such adjectives as insipid, sweet, sour, bitter, or salty as determined by the taste sensory perception. It may be possible at times to describe the taste of a sample more definitively, e.g., when certain distinctive components, for example sulfur or sodium chloride, are present in predominantly large amounts. Waters with very low mineral content are likely to have an insipid taste. Water with high chloride content may have a brackish to salty taste, whereas the presence of excessive amounts of magnesium may result in a somewhat bitterish taste. Excessive amounts of iron or aluminum may impart a metallic or earthy taste, respectively, to water.

The taste intensity is measured by the threshold test in a manner similar to the odor threshold test. Tasteless water is used to dilute the sample until the dilution is reached at which taste is just barely detectable. The test result will depend on the sensitivity of the testers' taste buds, of course. Because of the similarities between the odor and taste tests, the precautions to be taken in conducting the odor test generally apply in the determination of taste as well. In particular, the following measures should be taken:

1. Keep samples out of contact with all taste-producing materials including plastics.
2. All glassware used in the test must be scrupulously clean.
3. All extraneous taste and/or odor sources must be removed from the test areas.
4. Smokers and drinkers should, preferably, not be designated as taste testers.

A. Reagents

Tasteless water may be prepared by following the same procedure described for the preparation of odor-free water. It should be noted however, that the recommended use of water filtered through activated carbon for this test is not unanimous because of the insipid taste that may result. Some authors have expressed a preference for the use of spring or well water.

B. Procedure

The procedure is similar to that described above for the odor test but for a few exceptions.

Wash several 50 mℓ glass beakers with odorless detergent followed by a chromic acid wash. Rinse thoroughly and then boil about 30 mℓ of distilled water in each beaker for about 5 min. Following the same procedure described for the odor test, prepare a series of sample dilutions duly interspersed with blanks. Place the beakers containing about 30 mℓ of each dilution and blanks in a water bath maintained at 40°C. After the temperature is stabilized, rinse mouth with a little tasteless water and then collect and hold a reasonable amount of the sample with the highest dilution in the mouth for approximately 10 sec. Rinse mouth with more tasteless water and repeat test using one of the blanks. Determine and record whether or not taste is present in the diluted sample relative to the blank. Repeat the preceding step using successively less diluted samples until the dilution at which a distinctive taste is just discernible is determined.

C. Calculation

Calculate the taste threshold number (TTN_t) from the formula

$$TTN_t = \frac{V_1 + V_2}{V_1}$$

where V_1 = mℓ of sample used V_2 = mℓ of dilution water, and t = temperature of sample at testing. Several temperatures ranging from about 8°C to 40°C have been recommended for the taste test. Some analysts have shown a preference for 30°C which is believed to be the optimum temperature for taste sensitivity. Since the taste test is intended primarily for determining the potability of water, it is probably best to do the test at the standard 15°C as recommended by Standard Methods.[2] As in the case of the odor test, a more reliable TTN value may be obtained if a panel of testers is used.

IV. COLOR

The color of water results from its optical properties causing changes in the spectral composition of transmitted visible light. So-called "pure" water, when viewed under the right conditions, exhibits a bluish color. Generally speaking, however, most people would consider "pure" water to be colorless. The real impact of color in water lies not so much in the fact that it is likely to have an adverse effect on the aesthetic characteristics of that water, but more so in its origin. The presence of color in water is an indication of pollution of some sort. Color may be of natural origin, such as from humus and peat materials, it may result from turbidity, it may stem from the presence of inorganic compounds such as ferric and manganese compounds, or it may result from the presence of certain classes of organic compounds; for example, the strong brown color sometimes observed in lake waters is generally attributed to certain organic matter derived from peaty material in lake sediments or tributary bogs. Some kinds of residential, agricultural, or industrial waste discharges usually impart a color ranging from yellow to brown to the receiving water.

Color is an important factor to be considered in determining the suitability of water for purposes such as for drinking and domestic use, recreation, as well as for industrial applications such as in the brewing and soft drink industries, pharmaceuticals manufacturing, and the food processing industry; for example, highly colored water, if used for domestic purposes, is likely to have undesirable effects such as staining of laundry, foods, etc. Color is essentially light seen at some specified wavelength and, like turbidity, determines to a great extent, the amount of light energy that a water body is likely to absorb when exposed to radiation. It affects the transparency of water and governs the amount of light penetrating the water. Thus the color of a water body is an important factor controlling aquatic plant and animal life. Nearly all inland waters contain varying amounts of dissolved yellow substances comprising a complex mixture of plant breakdown products consisting mainly of polymerized oxidized phenolics, which contribute considerably to the light-absorbing properties of water.

Color is highly pH dependent and is described as true or apparent color depending on whether the test is done on a turbid sample or not. Apparent color governs to some extent the temperature of a water body. In determining the true color of a sample, any suspended material present must be first removed by centrifugation or filtration. Otherwise, the resulting value represents apparent color. Centrifugation or filtration does not normally remove colloidal particles which may contribute to some significant extent to the sample color. There could also be problems associated with the preparation of the samples for true color measurement; for example, iron and manganese compounds may be precipitated or may be transformed to different oxidation states. Removal of suspended material by centrifugation is preferred to filtration, because in the latter instance, there is a possibility of some of the color being lost on the filter paper.

There are a number of methods available for color measurement. The most widely used methods are the visual comparison method, the instrumental methods, and the Forel-Ule Scale/Secchi disc method. The first two methods provide a means for making direct measurements of color. The Forel-Ule Scale/Secchi disc method enables one to make approximate *in-situ* measurements of apparent color of water systems.

A. The Visual Comparison Method

The method is relatively simple. The sample is compared with a series of platinum-cobalt standard solutions contained in matched color comparison tubes having shadowless bottoms (Nessler tubes, for example) or against colored glass discs. It is not recommended, for obvious reasons, that the test be performed by analysts who are color blind.

1. Equipment and Reagents

Matched color comparison tubes must be used. Prepare a stock solution of a platinum-cobalt standard by dissolving completely 1.246 g of potassium chloroplatinate (K_2PtCl_6) and 1.000 g of crystallized cobalt chloride ($CoCl_2 \cdot 6H_2O$) in a mixture of 100 mℓ concentrated HCl and about 200 mℓ distilled water contained in a 1 ℓ volumetric flask. Fill up to the 1 ℓ mark with distilled water. The color intensity of this solution is 500 units — sometimes called Hazen units after its originator.

The stock solution may be prepared also from metallic platinum, if potassium chloroplatinate is not readily available, by heating exactly 0.500 g of pure platinum in aqua regia until completely dissolved. Get rid of the nitric acid by repeatedly evaporating the solution to dryness after successive additions of a few mℓ of concentrated HCl. Dissolve the residue with 1000 g of crystallized cobalt chloride as described above and make up to volume with distilled water.

2. Procedure

Clean a series of 50 mℓ matched Nessler tubes with detergent and rinse with distilled water. Prepare a series of diluted standards to cover the expected sample range by making suitable dilutions of the platinum-cobalt stock solution. Table 4 gives a convenient dilution scheme covering a range of 1 to 70 color units. When it is not necessary to prepare standards lower than 5 color units, it would be more convenient to simply

Table 4
PREPARATION OF COLOR STANDARDS FROM STOCK SOLUTION

No. of Color Units	Volume of Stock Solution to be Measured (ml)	Final Volume with Distilled Water in Volumetric Flask (ml)
1	1.0	500
2	2.0	500
3	3.0	500
4	4.0	500
5	1.0	100
10	2.0	100
15	3.0	100
20	4.0	100
25	5.0	100
30	6.0	100
35	7.0	100
40	4.0	50
45	4.5	50
50	5.0	50
60	6.0	50
70	7.0	50

pipette 0.5, 1.0, 1.5, 2.0, 2.5, 3.0, 3.5, 4.0, 4.5, 5.0, 6.0, 7.0 ml of the stock standard directly into 50 ml Nessler tubes and make up the 50 ml mark with distilled water to produce standards of 5, 10, 15, 20, 25, 30, 35, 40, 45, 50, 60, and 70 color units, respectively.

Fill the Nessler tube up to the 50 ml mark with the sample and compare its color against the standards filled up to the 50 ml mark in a series of matched tubes. This is best done by looking vertically downward through the liquid column with the tubes placed on a white or reflective surface so that the light is reflected upward through the liquid. It is sometimes difficult to ascertain which of the color standards matches exactly the color of the sample under test, particularly at the lower end of the scale. However, reasonably good estimates are obtainable with proper care.

The number of color units of the standard matching the sample color is equivalent to that of the sample. Hazen,[3] in developing the platinum-cobalt standard, recommended that highly colored samples be either read by viewing through a shorter light-path or, more conveniently, after suitable dilution in order to minimize errors likely to occur with such samples. A color intensity of 70 units has generally been accepted as the upper limit for direct sample measurement. When the sample color exceeds this limit, the determination should be made on a sample aliquot that has been suitably diluted to 50 ml in the Nessler tube with distilled water.

3. Calculation

The color, C_x, of the diluted sample is calculated from equation

$$C_x = \frac{C_s \cdot 50}{V}$$

where C_s is the value of the color measured in the diluted sample, and V is the volume (ml) of sample diluted.

Report the results in accordance with the following criteria:

- for values between 1 and 50 report to the nearest unit.
- for values between 51 and 100 report to the nearest 5 units.
- for values between 101 and 250 report to the nearest 10 units.
- for values between 251 and 500 report to the nearest 20 units.

Measure and record the sample ph.

The use of instruments such as the Hellige® Aqua Tester or the Hellige® comparator permits a more convenient and faster method of color measurement than that described above. The Aqua Tester® method is a visual comparison method. The instrument uses pairs of matched Nessler tubes of 200 mm viewing depth housed in a metal cabinet. The instrument is constructed in such a way to permit color comparisons in the presence of extraneous light by bringing, by means of a prism and magnifying lens assembly, the viewed color fields in juxtaposition into one enlarged field. Color comparisons are made against removable color discs that are rotated to match the sample. The Water Quality Branch Analytical Methods Manual[4] reports a coefficient of variation of 0% at values of 10 and 80 color units, determined by making replicate measurements with the Aqua Tester®.

Standards Methods[2] describes a spectrophotometric method applicable to waters the colors of which, because of their hues, are difficult to measure by the visual comparison method. This procedure, although considered to be fairly accurate and precise, is somewhat complicated. A simplified spectrophotometric procedure, in which the intensity of the sample color is determined by measuring the attenuation of light of a wavelength that is most suitably attenuated, is normally satisfactory and is the preferred method for most practical purposes. The appropriate wavelength may be determined by means of a recording spectrophotometer or a filter photometer. In practice, measurements are made within the 430 to 460 nm range. Because of the good reproducibility of the mercury line, measurement at a wavelength of 436 nm is generally preferred for most natural waters. Measurements at 254 nm are sometimes made also as a first step in the qualitative determination of organic pollutants in water. When measuring waters with low color intensity it may be necessary to use cells having optical pathlengths of about 10 cm and sometimes even as high as 25 cm.

B. The Forel-Ule Scale/Secchi Disc Method

It is simpler and more convenient, for all practical purposes, to do color measurements *in situ* wherever feasible. This would preclude the possibilities of color changes due to pH and other changes likely to occur in the sample after it is collected. The Forel-Ule Scale/Secchi disc method is the one most often used for measuring color in the field and particularly in oceanographic and limnological studies. It is best suited for getting an approximate index of the apparent color or hue of natural waters, but the results bear no specific relationship with the platinum-cobalt color standards.

The Forel Scale comprises a series of 11 glass vials containing solutions of ammoniacal copper sulfate and neutral potassium chromate mixed in different proportions to give colors ranging from bright blue to pale greenish-yellow with an equivalent numerical scale of 1 to 11. The Forel-Ule Scale is an extension of the Forel Scale having a series of 22 solutions with colors ranging from bright blue to brown.

An increasing number of limnologists are relying on the use of another device, the Secchi disc, to get some indication of transparency/color readings of natural water bodies. The Secchi disc is basically a heavy circular plate made of metal, plastic, or wood and normally of approximately 30 cm in diameter. The upper surface is painted plain white or white and black with the bottom painted black. It is fastened to a rope through a ring located at the center of its upper surface. The rope is usually about 50 m long and is graduated in 1m lengths throughout. The disc is submerged vertically

FIGURE 1.   Secchi depth measurement aboard ship.

into the water being tested and is viewed from above to determine the depth at which it just disappears (Figure 1). It is then lowered past this point and pulled slowly back up to determine the point at which it is again just barely visible. The average of the two depth measurements observed, which should not be at too much variance, is recorded as the Secchi depth. This reading gives an indication of the water transparency.

The Forel-Ule Scale measurement of color uses the Secchi disc technique to compare the water color. The submerged Secchi disc is held at a point approximately 1 m below the water surface, and the color of the water column above the disc surface is compared with the colors in the vials of the Forel-Ule Scale. The color of stratified lakes generally increases markedly with depth and, under such conditions, it would be better to make the observation at a point about 1 or 2 m above the Secchi depth. It is important that the scale be shaded from open sunlight when determining the color, and the depth at which the observation was made should be recorded.

The Secchi disc measurement is susceptible to several drawbacks of which the following are the most prominent:

1. The disc diameter and the type of paint used will influence the visibility of the disc when lowered through the water column. It is best therefore, that the reflectance of the painted upper surface of the disc be maintained constant throughout the useful life of the disc.
2. The results may be affected by light reflection from and/or refraction at the water surface; angular movements of the disc due to currents or drifts; the position of the sun; shadows; weather conditions; the height of the observer above the water surface; the hue of the water; the spectral composition of the incident light, and the spectral sensitivity of the observer's eyes.

It is recommended, when making measurements from a boat, bridge, or platform, to do so from the shaded side in order to minimize the surface reflection effect. The observer should strive to keep the boat or platform stationary and to avoid unnecessary disturbance of the water being measured. It is essential also that the disc remain horizontal when making the measurements.

V. TURBIDITY

Turbidity is essentially an expression of the optical properties of the water which cause incident light to be scattered and/or absorbed instead of being transmitted straight through it. Although particle-free water itself absorbs and scatters light, this tendency is increased markedly in the presence of suspended material.

Water turbidity results from the presence of mineral and organic particulate matter, usually in a finely divided state (e.g., clay, silt, microscopic animals, and other organic and mineral material in the water) and held typically in suspension by turbulent flow and by Brownian movement. The type of substance contributing to turbidity depends to some extent on the water source. For example, rock and soil suspension may appear as turbidity, particularly in spring water. The amount of suspended material in natural water may stem from natural erosion and runoff, biological activities such as algal blooms and, directly or indirectly, from man's activities.

It is now a well known fact that light plays a highly significant role as a major limiting factor for photosynthesis and hence, primary production, in a water body. The abundance of microscopic plants (autotrophic phytoplankton) in a given water body is dependent on, among other things, the amount of light penetrating through that water. This in turn controls the populations of herbivores which serve as food for aquatic and terrestrial animals higher up in the food chain. Turbidity decreases the penetration of solar radiation incident on the water surface by its absorption and scattering effect on the light as it passes through the water. The temperature of a turbid water is likely to increase more than that of a turbidity-free (clear) water subjected to the same amount of direct solar radiation due to the greater absorbing capacity of the turbid water.

Turbidity may cause several serious problems in matters relating to water use and water quality in general. Water required for drinking as well as for industrial use such as in the manufacturing of beverage and food, must meet certain criteria established for turbidity. The maximum acceptable limit recommended by The Department of National Health and Welfare, Canada,[5] for Canadian drinking water in 5 units. High turbidity, apart from seriously detracting from the aesthetic characteristics of water, may render the water unsuitable for domestic, industrial, agricultural, and recreational uses. For example, high turbidity waters are often abrasive to pumps, pipes, and turbine blades and may cause inordinate plugging of water-sprinkler nozzles. It may also cause plugging of filters in water-intake pipes as well as placing an excessive load on water treatment facilities by interfering with the disinfection process and causing the generation of extra sludge. High turbidity reduces photosynthesis and hence biological activity (primary productivity) including fish population. Turbidity often poses a number of problems to practically every analyst engaged in water analysis and tends to make data interpretation more difficult. Recent studies have revealed that, under the right conditions, a considerable fraction of the total heavy metals and trace organics may be bound to suspended sediments which thus act as a "sink" or carrier for these substances. Acidification of turbid alkaline natural waters could cause the release of metals bound to clay suspensions present. Accordingly, a certain amount of toxicity may also be associated with turbidity depending on the conditions.

The turbidity value of a solution depends to some extent on the concentration of particulate material in suspension, but it is also strongly dependent on a number of other factors. The optical properties of the suspension depend on the size and number of particles as well as their distribution, shape, and the refractive indices of the particles and the liquid itself. The light-scattering process encompasses the phenomena of light diffraction, refraction, and reflection in varying degrees. The particle size and shape determine the extent of diffraction while refraction and reflection are basically governed by the composition of the medium, i.e., the refractive index.

The Rayleigh theory of the scattering dipole, although not rigorously applicable to liquids, formed the basis for the interpretation of the light-scattering phenomenon in an aqueous medium. When a particle, the size of which is small compared to the wavelength λ, is subjected to a homogeneous electrical field, E say, it behaves like a dipole with an induced moment F given in terms of the electrostatic equation, $F = pE$ where p is the polarizability of the particle. It is assumed then that the scattering effect is a result of the oscillation of the dipole at the frequency induced by the exciting radiation The basic equation linking the intensity I of a monochromatic unpolarized incident light with the intensity $I\theta$ of light scattered at an angle θ to the incident beam is given as

$$I\theta = \frac{I}{D^2 \lambda^4} \cdot 8\pi^4 \, p^2 \, (1 + \cos^2 \theta)$$

or

$$\frac{I\theta}{I} = \frac{8\pi^4 p^2}{D^2 \lambda^4} (1 + \cos^2 \theta)$$

where D is the distance between the point of observation and the particle and p, the polarizability of the particle.

It is evident from the above equation that the intensity of the scattered light varies with the reciprocal of λ^4.

Kullenberg[6] calls the ratio $I\theta/I$ the efficiency factor Q, given by the equation

$$Q = \frac{1}{\alpha^2} \int_0^\pi (i_1 + i_2) \sin \theta \, d\theta$$

where α relates to the particle diameter (d) and the wavelength (λ) expressed as $\alpha = \pi d/\lambda$ assuming a spherical particle; and i_1 and i_2 are the intensities scattered at an angle θ with the electric vectors perpendicular and parallel to the plane of observation, respectively. Different Q values, and hence different scattering zones, will result depending on the particle size distribution and the refractive index m given by the equation $m = 1 + \varrho/2\alpha$ where ϱ gives the phaseshift of a beam passing through the center of a given particle. Kullenberg's[6] table of limiting scattering cases is reproduced in Table 5 to demonstrate how the efficiency factor Q varies with particle size and refractive in-

Table 5
LIMITING SCATTERING CASES

Rayleigh and Fluc- tuation Theory	Rayleigh-Gans Theory	Anomalous Diffrac- tion
α, ϱ small	α large	α, ϱ large
$\lvert m\chi\,\alpha\rvert \ll 1$	ϱ small	ϱ fixed (intermediate)
$\beta \propto \lambda^{-4}$	α fixed	
$Q \propto (m\text{-}1)^2 X\alpha^4$	$m \to 1$	$m \to 1$
	$\lvert m\text{-}1\rvert \ll 1$	$\lvert m\text{-}1\rvert \ll 1$
	$\varrho \ll 1$	$Q \curvearrowright 2$
	$Q \propto (m\text{-}1)^2 X\alpha^2$	
	$Q \ll 1$ for $\alpha \gg 1$	

← in general: Mie theory→

Reproduced from Kullenberg, G., *Optical Aspects of Oceanography*, Academic Press, London, 1974. With permission.

dices. Table 5 shows that under the condition satisfying the applicability of the Rayleigh's theory, i.e., particle size is significantly smaller than the wavelength of light, the intensity of the scattered light decreases rapidly with decreasing particle size. The efficiency factor approaches a constant value of approximately 2 for large particle sizes, i.e., as α goes to infinity.

It follows, therefore, because of the variability of the scattering efficiency factor which is dependent on the particle size distribution and the refractive index of the medium, that two different suspensions giving the same turbidity value when measured under similar conditions may not necessarily contain the same concentration of particulate matter. The fact that, in practice, samples having diverse characteristics and containing essentially variegated suspended particulate matter are compared against turbidity standards of uniform particle size distribution and possibly different refractive index makes the meaning of turbidity readings somewhat nebulous. Furthermore, the inherent optical properties or the light absorption and scattering characteristics of a water sample, being a function of the kinds of pigments that make up the sample color, will depend as well on the planktonic species (e.g., blue-green algae) present. All of these factors make the extent of interpretation of turbidity values, from a limnological viewpoint at least, very limited indeed. Turbidity, nevertheless, serves as a good "indicator" parameter especially when correlating data for other related parameters for a given sample. Turbidity measurements are useful as a check in determining the efficiency of purification of untreated water. It is a good index in determining dosage rate of clarifying chemicals such as alum coagulant in water treatment. Turbidity data are also valuable for the planning and designing of water treatment plants. It is a useful parameter for monitoring the state of steam boilers and to determine the rate of corrosion occurring in boiler systems.

Because of the possibility of changes occurring in samples during and after sampling, measurements should preferably be made *in situ*. This, of course, is not always feasible and when in vitro measurements are to be made, it should be done as soon as possible after sample collection.

In general, turbidity is measured by comparing the optical interferences caused by the suspended material against the transmission of light in water by means of a precalibrated instrument. Measurements may be based on the Tyndall effect (scattering of incident light by the suspended particles) or opacimetry which relates to absorption of incident light by suspended particulate matter. Instruments based on the opacimetry principle are generally not sensitive to low turbidities and are susceptible to diffraction interferences. On the other hand, instruments based on the Tyndall effect are more sensitive and accurate and also cover a wider turbidity range.

Turbidity may be measured by a number of methods of which the following are best known; the Visual Method, the Spectrophotometric turbidimeter method, the Secchi disc method, the transmissometer (hydrophotometer) method, the scatterance meter (submarine photometer) method. The latter three methods enable only an indirect measurement of turbidity to be made, and the results produced through them are expressed in units other than the standard turbidity unit. However, each technique permits in its own way a means of making *in situ* measurements of turbidity which, in some cases, would be preferred to in vitro measurements.

Technological developments in remote sensing now make it possible to make turbidity measurements over large bodies of surface waters by means of spectro-optical techniques. Investigators have been able to demonstrate correlations between remotely-sensed digital satellite data and the suspended solids (turbidity) concentrations of some surface waters. This method of measuring turbidity, however, requires highly specialized and sophisticated equipment, making the cost prohibitive for routine application of the technique.

The transmissometer and scatterance meter methods are basically electronic instruments used to measure the relative amounts of light transmitted or scattered through a fluid. This gives a measure of the inherent optical properties of water masses and, in so doing, gives an indirect measure of the suspended and dissolved constituents. The transmissometer makes measurements of the attenuation of light emitted from a collimated light source based on its distribution through the medium. The percent transmission of the incident light through a known path length in the water is determined by means of a collimated detector on which the transmitted light falls. The exclusion of scattered light is a critical factor when making transmissometer measurements. The scatterance meter is a device used to determine the ratio of the radiant flux scattered from a beam, to the incident flux, and basically comprises a collimated light source and a detector in the form of a radiance meter. They may be free-angle, fixed-angle or integrating-angle meters. The free angle meter measures all values of the volume scattering function at a given point. The fixed-angle meter measures volume scattering function at a fixed angle, while the integrating scattering meter integrates directly the volume scattering function of a given optical medium over all angles and thus records the total scattering coefficient of the medium. Instruments of the latter type are not subject to particle size limitations and are considered suitable for determining what is sometimes referred to as the absolute turbidity of the medium. A wide variety of transmissometers and scatterance meters are commercially available, or they can be designed and constructed to satisfy specific needs.

The Secchi disc method is described in the section dealing with color. It simply gives an approximate index of water transparency and hence an indirect qualitative measure of turbidity. The Secchi disc reading serves primarily as an indicator parameter and, among other things, serves as an indicator of the trophic states of natural surface waters.

There are three basic visual methods available for turbidity measurements. They are the Jackson Candle turbidimetric method, the bottle standards comparison method, and the Baylis turbidimetric method.

A. The Jackson Turbidimeter Method

The Jackson turbidimeter, which in the earlier years was the standard instrument used for determining turbidity, is shown in Figure 2. It consists of a tripod support fitted with a receptacle for a standard candle and a brass sleeve which holds a glass tube graduated to read turbidity directly in Jackson turbidity units (JTU) or in centi-

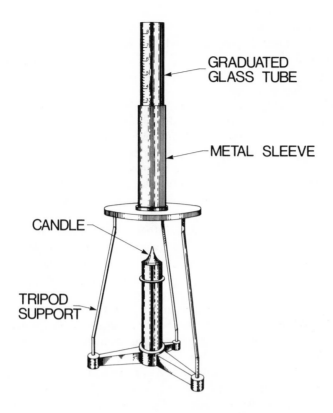

FIGURE 2.   The Jackson Candle turbidimeter.

meters. The metal parts are generally finished in baked enamel with a black oxidized brass finish inside the shield. The bottom of the glass tube is normally made of polished optical glass. The tubes come in standard lengths of 25 cm and 75 cm to measure turbidities in the ranges of 100 to 5,000 and 25 to 5,000 JTU, respectively.

Turbidity measurements are made by comparing maximum depth of visibility of a standard candle such that the depth of sample in the tube just at the point of extinction corresponds to the sample turbidity. The candle is inserted in its holder so that it is directly underneath and in line with the vertical axis of the turbidity tube. It is maintained at a fixed distance from the bottom of the tube by means of a spring-loaded mechanism on which it is mounted. It is thus possible to make empirical measurements of turbidity based on the principle of measuring the light path through the suspension that causes the image of the candle flame underneath to become just indistinguishable against the general background illumination when the flame is viewed from above through the suspension.

Erroneous results may be obtained if the samples contain debris or coarse particles that settle rapidly. Air bubbles, sample color, and vibrational effects causing visibility disturbances may also lead to spurious results. Low values may result in the presence of very fine or black particles.

An operable range of 25 to 5,000 JTU is claimed by some manufacturers. However, because of the extremely short light path involved at the upper levels, it is recommended that measurements be made on a diluted sample when its turbidity exceeds 1000 JTU.

1. Equipment and Reagents

A Jackson candle turbidimeter equipped with Jackson glass tubes or equivalent, and standard candle is used.

Turbidity-free water: distilled water kept in an air-tight container should be satisfactory for most practical purposes.

Stock Turbidity Suspension Standard: about 5 g of Kaolin is added to 1 ℓ distilled water. The mixture is shaken vigorously for about a minute and allowed to stand for about 24 hr. Then the supernatant is withdrawn without disturbing the bottom particles. The turbidity of the suspension is then measured using the candle turbidimeter. The suspension is preserved by adding 1 g of mercuric chloride to every liter of suspension. After standardization of the turbidity suspension, other less-turbid suspensions can be prepared by appropriate dilution of the freshly checked stock suspension with turbidity-free water.

In dealing with natural waters, it is best to use water from the same source as the samples to prepare the stock turbidity suspension standard. This, however, will not be practical in cases where samples from several different sources are to be measured.

2. Procedure

Set up the turbidimeter in an area where the effects of drafts and extraneous light are minimal if not non-existent. It is also better to set up the instrument in a darkened room. Check to ensure that the candle is well positioned, making sure that it is at the right distance from the turbidity tube. Trim off charred portions of the wick in order to keep the size of the flame as constant as possible. Light the candle and proceed immediately to the next step.

Shake the sample to ensure it is thoroughly mixed and slowly pour it into the turbidity tube until the image of the flame just disappears from view. The sample should be poured very slowly as the point of image disappearance is approached, and the field of view should be uniformly illuminated with no visible bright spots. Use a pipette to gradually remove just enough of the sample to make the flame image just visible again. Then, gradually add small amounts of the sample contained in the pipette to get a more refined reading. Remove the tube and read the turbidity of the sample.

3. Limitations

The candle turbidimeter is applicable to samples with turbidities in excess of 25 JTU, making it unsuitable for most natural and treated waters. The measurements are subject to outside interference such as drafts and diffuse light, and the results are generally not easily reproducible. The inevitable accumulation of soot or moisture on the bottom of the tube will, undoubtedly, affect precision and accuracy. Another major flaw with the candle turbidimeter is the fact that the candle light, being in the yellow-red or longer wavelength range of the visible spectrum, is not scattered by very fine particles including the fine siliceous particles. It must be noted, however, that the opinion of most investigators is that light scatterance by suspended particles is effected predominantly by the relatively larger particles, i.e., above 1 to 2 μm in size. In addition, the instrument is incapable of measuring turbidity resulting from black particles such as charcoal dust.

In spite of the foregoing, reasonable results are attainable with due care. In particular, the following points should be noted. The turbidity tube, the bottom specifically, should be spotlessly clean both inside and out and free from scratches and smears. The tube should be enclosed throughout most of its length to minimize the effect of extraneous light. Therefore, when using the longer 75 cm tube, it should be in conjunction with the extended sleeve (extension tube) provided for this purpose. The flame should not be allowed to burn for periods in excess of a few minutes at a time, to prevent it from increasing too much in size. Each time before relighting the flame, charred portions of the wick that are easily removable should be broken off with the fingers. Furthermore, frequent trimming of the wick is essential for satisfactory results.

B. The Bottle Standards Comparison Method

Samples with turbidity in the 5 to 100 JTU range could be estimated by the following procedure.

Prepare a series of turbidity standards to cover the expected sample range by appropriately diluting concentrated standard suspensions with known amounts of turbidity-free water. Shake the sample and standards vigorously and pour equal amounts into clear glass bottles being of the same size, shape, and type and fitted with stoppers. The side walls of the bottles should preferably be plain and the bottles should not be filled to the top in order to facilitate shaking of the sample and standards before each reading. Arrange so that any artificial lighting is directly above or below the bottles so that there is no direct light reaching the eye from these sources.

Shake each bottle vigorously and immediately compare the sample turbidity against the standards by looking horizontally through the bottles at some fixed object and noting the distinctness with which the object outline can be seen. The turbidity value of the standard producing the visual effects most closely matching those of the sample is taken as the turbidity reading of the sample.

C. The Baylis Turbidimetric Method

Baylis[7], in 1925, described a visual comparison turbidimeter designed for measuring turbidity in liquids contaning very small amounts of suspended matter. A pure blue light source is positioned underneath a clean turbidity optical glass tube containing the liquid being tested. When viewed through a turbidity-free solution, only a blue light will be observed. However, suspended matter would reflect white light and thus cut off some of the blue light transmitted through the solution. The extent to which the blue light intensity is reduced is proportional to the sample turbidity, which is determined by comparing with standards of known turbidity.

The Baylis turbidimeter is suited for measuring turbidities below approximately 2 JTU and has a detection limit of about 0.1 JTU.

Results determined by the visual methods may be reported in accordance with the following guidelines:

- Turbidity values up to 1.0 JTU, report to the nearest 0.1 JTU.
- Turbidity values between 1 and 10, report to the nearest 1 JTU.
- Turbidity values between 10 and 100, report to the nearest 5 JTU.
- Turbidity values between 100 and 400, report to the nearest 10 JTU.
- Turbidity values between 400 and 700, report to the nearest 50 JTU.
- Turbidity values above 700 JTU, report to the nearest 100 JTU.

D. The Spectrophotometric (Photoelectric) Turbidimeter Methods

As previously pointed out, the candle turbidimeter method suffers from a number of limitations. The inapplicability to most natural and treated waters and the questionable precision are, undoubtedly, the two most significant deficiencies. This necessitates the use of other more suitable methods for turbidity measurements.

Several spectrophotometric methods, which are essentially adaptations of the visual methods, are available for measuring turbidity. The nephelometer and the absorptometer are the two basic types of photoelectric turbidimeters in existence. Absorptometers are based on the opacimetry principle — absorption of light by suspended particles — and, due to certain inherent features, are generally found to be less sensitive and less accurate than the nephelometers. Accordingly, the nephelometric methods are the ones most commonly used.

There are generally two basic problems associated with these so-called secondary methods; viz., (1) they do not give results that coincide exactly with those obtained with the Jackson candle turbidimeter for all samples and (2) the results obtained from different turbidimeters having different optical systems do not always agree with one another, even if they were all individually precalibrated against the Jackson candle turbidimeter. Moreover, the practice of using different standard suspensions, containing different types of particulate matter, to calibrate the instruments further complicates matters.

The majority of the nephelometric turbidity meters available are based on the Tyndall effect and give a relatively good indication of the light intensity scattered in a particular direction, usually at a 90° angle or at the 90° and 270° angles to the incident beam. These instruments generally give results that are sufficiently comparable with one another and as such, are considered to be the standard for measuring low and moderately high turbidities. They are not suited for very high turbidity measurements, because the samples become increasingly more opaque with increasing turbidity, and the amount of light penetration through the sample is progressively reduced. The response of the nephelometer is zero at zero turbidity and is generally linear over the instrument operating ranges. Some of the better instruments available are capable of making measurements over full scale ranges of as low as 0 to 0.2 FTU and have relatively high sensitivity. Thus the nephelometric method is more sensitive, precise, and less time-consuming than the visual methods. In addition, it is applicable to measurements over a wide enough range to make it the method of choice. A variety of the flow-through type of nephelometric turbidimeters now available facilitate on-line continuous monitoring of turbidity.

The Hellige® and Hach® turbidimeters are among the typical brands of nephelometric turbidimeters commercially available. The Hellige® turbidimeter is essentially a visual method covering a range of about 0.5 to 150 mg/ℓ SiO_2 on the silica turbidity scale. It consists of a black light-tight metal housing containing an opal glass bulb, a vitreous enamelled reflector, a precision slit regulated by a graduated dial, a frame with two glass light filters, and a platform to support the turbidity tube. An eyepiece, complete with lens and diaphragm, mounted at the top of the housing, permits the tester to look vertically down through the sample, contained in a special cell, at a central field of view. The instrument compares the intensity of two light beams from the same opal glass lamp. Light from one beam passes through the sample and is scattered while the other beam passes through filters and controlled slits and serves as a reference field of view. The instrument is the calibrated-slit type of turbidimeter, which makes measurements based on a fixed ratio of scattered light to the transmitted light intensity. The amount of light transmitted from a single light source is adjusted by means of a calibrated slit to the point where the operator matches the final intensity of the light for the different paths.

The turbidity of the sample is measured by slowly rotating the graduated dial from its lowest to highest readings until a black spot in the center of the field of view just disappears and merges with the surrounding field. The reading on the dial scale is noted, and the turbidity value is determined from a special calibration graph provided. This method is simple and much more rapid then the other visual methods and, by virtue of certain instrumental features, yields more reproducible results.

Direct-reading instruments such as of the Hach® or Monitek® turbidimeter types are preferred for making turbidity measurements in water. Our experience with the

FIGURE 3.   The Hach model 2100A turbidimeter.

Hach® 2100 and 2100A models (Figure 3), used both in our main and mobile field laboratories for well over a decade, has been generally quite good. The following procedure, which may be considered as the standard for turbidity measurements, is being given with the Hach® Model 2100A turbidimeter in mind. It should not be construed, however, that this particular brand of instrument is the only suitable one that is now commercially available.

The nephelometric method is applicable to the measurement of turbidity in all water types over a practical working range of about 0.05 to 1000 FTU.

A strong light beam is transmitted upward through a turbidity optical tube containing the sample (Figure 4). The amount of light reflected at 90° angle to the incident light beam by the suspended particulate matter is directly proportional to the turbidity present and is received by a photomultiplier tube. This diffracted light is transformed to an equivalent electrical signal, which is measured on the instrument meter graduated in turbidity units.

A literature search will reveal that several different turbidity standards, each with different characteristics, may be used for calibration purposes. These include gum arabic, infusorial earth (kieselguhr), kaolin, fuller's earth, natural soil, finely divided titanium dioxide mixed in partially polymerized polystyrene, formazin, and solid acrylic rods.

In general, standards generated from kaolin and fuller's earth may be better suited

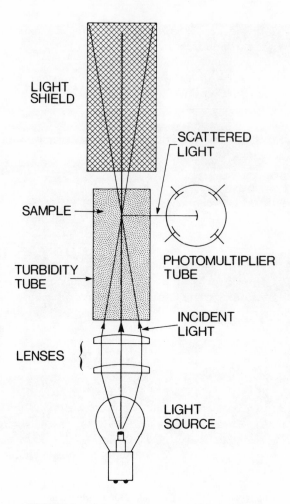

FIGURE 4.   Light path diagram for Hach turbidimeter.

in cases where sample turbidity stems predominantly from non-sediment colloidal clay particles. The siliceous infusorial earth standard is suited for samples containing predominantly sedimenting materials such as fine sand and calcareous particles, whereas the gum arabic standard is better suited in the case of relatively low turbidity caused principally by non-sedimenting colloidal material. The units in which turbidity is expressed, e.g., mg/ℓ silica, drops of gum arabic, formazin turbidity units, will obviously depend on the type of turbidity standard used for calibration purposes. It is worth noting, however, that the unit of turbidity is an arbitrary unit and should not be confused with concentration.

The formazin suspension is the preferred standard and is the one recommended. It is relatively simple to prepare and, due to its consistency (particles are uniform in number, size, and shape), is more reproducible in its light-scattering properties than the other liquid standards. On the other hand, the recently announced AMCO AEPA-1 ® styrene divinylbenzene polymer suspension is reputed to have a number of advantages over the formazin suspension. The units of turbidity measured against the formazin standard is the FTU (Formazin turbidity unit), and these approximate the units derived from the Jackson turbidimeter. The use of the term nephelometric turbidity unit to express turbidity results is becoming increasingly popular. Because other standards could possibly be used with the nephelometer, the use of FTU is preferred since it specifically denotes that measurements were made on instruments that have been calibrated against the formazin standard.

As is the case for other methods, turbidity readings are affected by debris and coarse particles that settle rapidly. Dirty turbidity cells, air bubbles, and sample color also cause erroneous results. The color interference effect is eliminated in some instruments using the signal ratioing approach, whereby the scattered and transmitted light passing through the sample are both measured and a ratio taken of the two. Since color will affect both beams equally, its effect is thus cancelled. Instruments in which the stray light effect is minimal should be capable of producing good results.

1. Equipment and Reagents

The Hach®-model 2100A turbidimeter or equivalent may be used.

Turbidity-free water — Distilled water kept in an air-tight container should be satisfactory for most practical purposes. If necessary, the water is filtered through 0.45 μm membrane filters before storing in a well-stoppered container.

Stock Turbidity Standard — (a) analytical reagent grade hydrazine sulfate, $(NH_2)_2 \cdot H_2SO_4$, 1.000 g, is dissolved in about 50 mℓ distilled water in a 100 mℓ volumetric flask, and the solution is made up to the mark with more distilled water. (b) Hexamethylenetetramine, $(CH_2)_6N_4$, 10.00 g, is dissolved in about 50 mℓ distilled water in a 100 mℓ volumetric flask, and the solution is made up to the mark with distilled water. A formazin turbidity suspension of 4,000 FTU results when equal volumes of solutions (a) and (b) are mixed and allowed to stand for about 24 hr. Solutions (a) and (b) may be kept for up to about 4 weeks, after which fresh solutions should be prepared.

Intermediate Turbidity Standard — exactly 5.0 mℓ each of solutions (a) and (b) is pipetted into a 100 mℓ volumetric flask. After mixing thoroughly the resulting solution is allowed to stand for about 24 hr at 25°C±3°C, during which time a white suspension forms. It is then made up to the mark with turbidity-free water and mixed thoroughly. This suspension has a turbidity of 400 FTU and is stable for about two to four weeks.

Working Turbidity Standards — working turbidity standard suspensions are prepared as required by diluting homogeneous aliquots of the intermediate turbidity standard. The dilution scheme given in Table 6 is provided as a guide. Other working solutions, specifically those lower than 1 FTU, should be prepared from working standards of higher concentrations. Working standard suspensions should be kept for no longer than a week.

2. Procedure

Because the inherent optical properties of aqueous solutions are dependent on the wavelength of the incident light, it is recommended that a monochromatic light source be used. Furthermore it is recommended that measurements be made at a wavelength of approximately 800 nm, notwithstanding the somewhat lower diffused radiation flux intensity at 800 nm compared to lower wavelengths, in order to minimize, if not eliminate, the possibility of color interference. This can be achieved by the use of tungsten lamps or luminescence diodes.

The manufacturer's instructions should be followed for instrument calibration and sample measurement. The use of the solid turbidity suspension rod supplied by some manufacturers is discouraged because of its high susceptibility to scratches and because it is not always feasible to standardize the instrument with a rod in which turbidity value is close to the sample turbidity. Use the prepared liquid standard turbidity suspension for instrument calibration. The use of two standards, bracketing the expected sample values, is recommended for calibration within each of the instrument operating

Table 6
PREPARATION OF WORKING TURBIDITY STANDARDS FROM THE 400
FTU INTERMEDIATE STANDARD

Working Turbidity Standards (FTU)	Preparation[a]
100	Pipette 25.0 ml intermediate standard into 100 ml volumetric flask and make up to mark with turbidity-free water.
50	Dilute 25.0 ml intermediate standard by making up to mark in a 200 ml volumetric flask.
40	10.0 ml intermediate standard made up to mark in a 100 ml volumetric flask.
20	5.0 ml intermediate standard made up to mark in a 100 ml volumetric flask.
10	5.0 ml intermediate standard made up to mark in a 200 ml volumetric flask.
5	25.0 ml intermediate standard made up to mark in a 2.0 ℓ volumetric flask.
4	10.0 ml intermediate standard made up to mark in a 1.0 ℓ volumetric flask.
2	5.0 ml intermediate standard made up to mark in a 1.0 ℓ volumetric flask.
1	5.0 ml intermediate standard made up to mark in a 2.0 ℓ volumetric flask.

[a]   The intermediate turbidity standard should be shaken well to ensure homogeneity at the point of pipetting out aliquots for dilution. Use turbidity-free water for all dilutions.

ranges in which measurements are to be made. The standards should be shaken thoroughly each time before being poured into the turbidity tube.

a. Calibration of Turbidimeter

Turn the instrument on and allow a few minutes for it to warm up and stabilize. Shake the turbidity standard thoroughly and pour a homogeneous aliquot into a clean turbidity tube to fill it to about 1 or 2 cm from the top. Remove any liquid that may be on the outside of the tube with a soft absorbent tissue. Place the tube into the instrument and cover with the light shield provided. If in using the Hach® instrument, measurements are being made in the 0 to 100 or 0 to 1000 FTU ranges, then the special cell riser provided should be inserted before insertion of the turbidity tube. The light path is thus decreased resulting in increased linearity at these higher levels. Select the correct range by turning the range change turret accordingly. After ensuring complete absence of air bubbles in the suspension, adjust the standardization control so that the meter registers the exact turbidity value of the standard. Check the calibration by using another standard suspension within the same range.

b. Making Turbidity Measurements

There is the likelihood of experiencing appreciable differences in high turbidity values if measured directly. Accordingly, samples with turbidities greater than 40 FTU should be suitably diluted before making measurements. The reading is then multiplied by the dilution factor to determine the sample turbidity. If feasible, use a portion of the sample, filtered through 0.45 μm membrane filter, to make dilutions. It should be borne in mind though, that measurement errors are likely to increase with dilution although potential interferences such as may be attributed to color may be reduced. Because of the potential dilution error, it would be better to make direct measurements on samples having higher turbidities if the instrument measuring range permits.

Mix the sample thoroughly by inverting the stoppered container repeatedly for about 30 sec taking care not to generate an excessive amount of, if any, air bubbles in the process. Fill a clean turbidity tube to about 1 to 2 cm from its top with a homogeneous sample aliquot, insert it into the instrument, and place the light shield over it. Read the instrument meter after ensuring that there are no bubbles trapped in the sample.

In a strict sense, the turbidity effect caused by the scattering properties of even optically pure water itself, i.e., water free from any suspended particles, should be subtracted from the instrument reading. This value is, however, negligible from a practical viewpoint.

Report results in accordance with the following scheme:

- Turbidity values up to 1.0 FTU, report to the nearest 0.05 FTU
- Turbidity values between 1 and 10 FTU, report to the nearest 0.1 FTU
- Turbidity values between 10 and 40 FTU, report to the nearest 1 FTU
- Turbidity values between 40 and 100 FTU, report to the nearest 5 FTU
- Turbidity values between 100 and 400 FTU, report to the nearest 10 FTU
- Turbidity values between 400 and 1000 FTU, report to the nearest 50 FTU
- Turbidity values above 1000 FTU, report to the nearest 100 FTU

3. Precision and Accuracy

Reasonably accurate data are attainable with the better turbidimeters having negligible stray light effect. The stray light effect should be corrected for, particularly at the lower turbidity ranges where it is most significant.

The mean standard deviation for three testers using a Hach® model 2100A turbidimeter for measurements aboard a limnological research vessel was ±0.01 FTU at 1.0 FTU. Relative standard deviation at levels below 1 FTU was generally in the order of 10 to 20%. The accuracy and precision of the method depends to some large extent on the condition of the turbidity tube and the measuring technique used. The following points are noteworthy.

The inner and outer surfaces of the turbidity tube should be scrupulously clean and free of scratches or cracks. The cell should be washed inside and out before use with a good detergent followed by rinsing thoroughly with distilled water.

The instrument should be standardized frequently with suspension standards that have not exceeded the age limit. Instrument lenses including light entrance and exit windows should be free from scratches, dirt, grease, and condensation. This is particularly crucial when measuring low turbidities. Some meter readings tend to fluctuate immediately after switching from one range to another. It may be necessary, therefore, to wait a few seconds for the instrument to stabilize before reading the meter, after switching from one range to another.

The sample and standard turbidity suspension should always be thoroughly mixed and freed of air bubbles when making measurements. It may be necessary to wait several minutes to allow bubbles in the sample under test to rise past the line of measurement before measurements are made. However, there may be appreciable settling of the suspended particles occurring during that time. Alternatively, trapped bubbles may be rapidly expelled by dipping the end of the tube containing the sample into an ultrasonic cleaning bath.

VI. TEMPERATURE

The temperatures at which a sample is collected and at which physiochemical measurements are made are important for data correlation and interpretation purposes. For example, temperature readings are essential when performing tests such as color, pH, and specific conductance, as well as for calculating parameters such as saturation and stability indices, salinity, and forms of alkalinity. The measurement of ambient temperature in surface waters is of vital importance in the field of limnological and oceanographic studies. Temperature affects the odor/taste characteristics of water, the stability of salts, and, in particular, gases; it influences the bioproductivity in the aquatic environment, determines the degree of dissociation of dissolved salts in a given water system, and controls to some extent the rate of oxidation of organic matter. Temperature and pH are two important factors, in addition to certain microbial activities within an aquatic medium, that govern the methylation of elements such as lead and mercury in the aquatic environment.

Water temperature is, in general, a critical parameter for aquatic life in any water body. Significant changes in the thermal regime of water systems, such as are likely to occur through waste heat discharges, result in major implications to all forms of aquatic life existing in that water.

Recent increasing concerns about the possible side effects of constructing and operating electric power generating plants with the attendant thermal pollution that may result, require that accurate and frequent temperature measurements be made in major water bodies. Temperature measurements are necessary for studying water circulation and mixing behavior, the identification of source water supply such as deep wells, identifying water stratification and thermal pollution sources, and for determining the effects of entrainment and waste heat discharges on the aquatic environment. The parameter is also useful for industrial plant process control and heat exchange calculations.

A. Mercury Thermometer Immersion Method

In order for temperature data intended for ecological and other studies to be meaningful, it is imperative that the measurements be made on-site immediately after sample collection and, preferably, *in situ*.

There are several methods available for measuring the temperature of water. The simplest and most straightforward procedure is with the immersion type mercury thermometer. On the other hand, advances made in remote sensing technology now make it possible to measure the surface temperatures of large bodies of surface water by means of airborne devices such as infrared radiometers. There are several makes of mercury thermometers available on the market, some of which would not be best suited for this measurement. Care must therefore be exercised in selecting a suitable instrument. The thermometer should preferably be graduated in °C with a scale marked at every 0.1°C or less and covering the expected range of sample temperatures. The readings of the thermometer should be certified accurate to within ±0.1°C. It is also recommended that it be further checked against a guaranteed precision thermometer such as are available from the National Bureau of Standards. Some thermometers are equipped with a magnifying glass to facilitate readings of the scale.

1. Procedure

Immerse the stem of the thermometer at least three inches below the surface of the water being measured. Hold vertically in place until there is no further movement of the mercury column. The time required for this equilibration may vary from about 1 to 10 min, depending on the thermal capacity of the instrument. With the thermometer still immersed vertically in the sample, read the temperature to the nearest 0.1°C.

2. Procedure for Special Cases

This procedure will obviously permit *in situ* measurements on surface waters only. When *in situ* measurements are impracticable, as is the case for depth samples, withdraw about 5 or 10 ℓ of the water in a suitable container and measure the temperature immediately. Air temperature and wind speed are two of the factors likely to influence the results. In special cases, the use of special maximum-minimum or sheathed immersion thermometers may help to minimize the effects of temperature variation during the measurements.

Unless the point chosen for the measurement is truly representative of the sampling location, it is advisable that measurents be made at several points a few meters apart at the specified sampling site. Some studies may also require a concurrent air temperature measurement. The sling psychrometer is the instrument recommended for this measurement. If a sling psychrometer is not available, the air temperature may be measured with an ordinary thermometer mounted with a fixation ring. The thermometer is rotated for about 2 min opposite to the wind direction. The average of two readings measured to the nearest 0.1°C is taken as the air temperature at the time of measurement. It is important that when making this measurement, all necessary precaution be taken to minimize, if not totally eliminate, extraneous sources of heat such as the body heat of the operator or direct exposure of the measuring device to the sun rays.

The immersion mercury thermometer procedure for water temperature measurements is indeed simple and economical. Its main disadvantage, however, is its inapplicability to *in situ* measurements of depth (below-surface) temperature, a feature that is required for most limnological and oceanographic studies. There are, however, several other measuring devices, albeit more costly and complex, now available on the market. The reversing thermometer; the mechanical, electronic, or Sippican® Expendable Bathythermograph; the sea-bucket; the thermistor; and the infrared radiation thermometer as the ones most commonly used. Each of these has its own specific features which govern when and how they should be used.

B. The Reversing Thermometer

Reversing thermometers first became available just over a century ago. The reversing thermometers in current use are precision instruments and have remained basically unchanged from that which was first developed. The instrument is comprised of a main (reversing) and an ordinary (auxiliary) mercury thermometer through which very accurate and precise water temperature measurements could be made at predetermined depths. There are two types of reversing thermometers available, viz., the "protected" and the "unprotected" reversing thermometer (Figure 5). In the protected thermometer, a main and an auxiliary thermometer are encased in a heavy glass jacket sealed at both ends, and from within which the air is partially evacuated. The space around the reservoir of the main thermometer is filled with mercury which, through its high thermal conductivity, makes the thermometer more sensitive to temperature changes. The large reservoir of mercury at the lower end is connected to the small bulb at the upper end by means of a fine capillary which is constricted and branched at a point just above the reservoir. The point at which the capillary is branched is called the appendix dead arm, by means of which the mercury in the thermometer stem is separated from that contained in the reservoir. The thermometer is bent through a 360° loop called the pigtail at a point slightly above the appendix dead arm. When the thermometer is reversed, the mercury column breaks off at the appendix and flows down into the bulb to fill it up to some point in the stem to denote the temperature at the point of reversal. The mercury level remains at this temperature reading until the thermometer is restored back to the upright position. The difference between the *in situ* water temperature and the air temperature causes an error in the reading of the main thermometer for which a correction can be made from the reading of the auxiliary thermometer, which is basically an ordinary mercury thermometer mounted beside the main thermometer.

As shown in Figure 5, the unprotected thermometer has one open end in the heavy glass jacket covering the main and auxiliary thermometers.

Readings from the protected thermometer, unlike those from the unprotected instrument, are not influenced by the hydrostatic pressure encountered at different depths. Accordingly, the protected reversing thermometer gives readings of the true water temperature at the specific depth of measurement. On the other hand, if the relationship

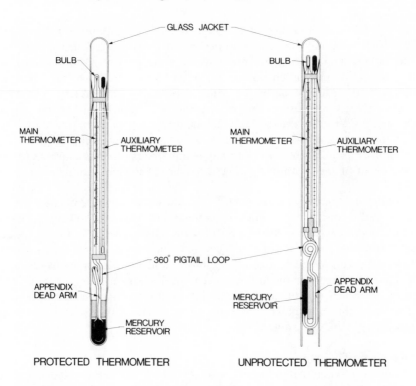

FIGURE 5. Protected and unprotected reversing thermometer in the upright or lowering position.

between temperature, pressure, and depth is established for a given water body, the unprotected thermometer serves essentially as a pressure gauge and, when used in conjunction with the protected instrument, permits accurate measurements of the depths at which the temperature has been measured. This is of particular advantage where accurate depth measurements are otherwise difficult to make, as in cases where the cable onto which sampling bottles are attached cannot be maintained vertical during the sampling process.

In practice, the reversing thermometers are fitted onto a suitably designed sampling bottle such as the Knudsen or Nansen bottle by means of a detachable frame as shown in Figure 6. When the sample bottle is triggered to collect a sample at a specific depth, the top and bottom ends of the bottle are reversed, and in so doing, the thermometer frame is inverted so that the reversing thermometer becomes reversed at that point to read the water temperature at the actual sampling point.

When the sampling bottle/thermometer assembly is returned to the surface, the auxiliary thermometer is read first to the nearest 0.1°C, followed by the main thermometer which is read to the nearest 0.01°C. It is advisable to equip each sampling bottle with frames capable of accommodating two protected and one unprotected reversing thermometers, but it is not essential to install an unprotected thermometer in every frame. The readings of any two corresponding main thermometers on the same bottle must be within 0.10°C of each other. If not, they should be checked against a third instrument. Although instruments are normally calibrated by the manufacturer before being shipped, it is recommended that they be further checked and recalibrated periodically. The operator, in reading the instruments, should exercise care in locating and reading the position of the mercury meniscus correctly. This is facilitated by use of a magnifying glass. Variations of ±0.03°C or less between readings taken by different operators are acceptable.

FIGURE 6.   Sampler equipped with reversing thermometer.

Reversing thermometers are very delicate instruments and must be handled with extreme care. It is important that the manufacturer's operational and maintenance instruction be fully complied with in order to ensure satisfactory instrument performance. In particular, the instruments should always be carried or stored in an upright position in the special carrying cases normally provided.

C. The Mechanical Bathythermograph, MBT

The bathythermograph (BT) is a device used for recording graphically the *in situ* water temperature relative to the depth at which the measurements are made. The BT is supposedly capable of measuring temperature with an accuracy of ±0.05°C. It is particularly useful for limnological and oceanographic work and can be towed from a

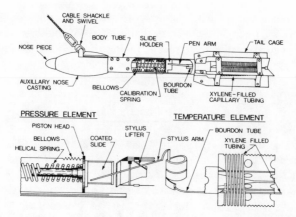

FIGURE 7.   The Bathythermograph.

ship while travelling at speeds of up to about 18 knots, or it may be used to obtain a temperature-depth profile at a specific point. Thus, the instrument can be used for measuring temperature and temperature gradients, locating thermoclines, identifying thermal regimes of different water masses, as well as providing supplemental data for water current studies.

The BT (Figure 7) comprises three basic parts; viz., (1) a heavy cone-shaped nose-piece to which a cable shackle and a swivel are attached, (2) a body tube housing for the temperature, pressure, and mechanical recording components, and (3) a tail cage assembly containing the temperature element capillary tubing. The instrument is towed through or lowered into the water by means of a cable attached to the cable shackle. A weighted replaceable auxiliary nose casting is used to ensure that the instrument is maintained at the correct depth below the water surface, when being towed.

The temperature element corresponds to the mercury column in a mercury thermometer and consists of about 15 m of fine copper tubing filled with xylene and protected by being wound around a hexagonal support inside the tail cage while at the same time well exposed to the water. Changing water temperature causes the xylene to expand or contract, which in turn causes the pressure in the tube to increase or decrease accordingly. Pressure changes occurring in the xylene-filled tube result in the actuation of a Bourdon® tube, which is a hollow brass coil spring carrying a pen arm and stylus assembly at its free end. The stylus records temperature changes on a coated, usually gold-plated, glass slide upon actuation of the Bourdon® tube.

The pressure or depth element assembly consists of an evacuated copper bellows, fixed at one end, which compresses lengthwise with increasing water pressure. An accurately calibrated spring coil enclosed within the bellows counteracts any tendency for the bellows to compress. The slide is held rigidly in place by means of a slide holder attached to the free end of the bellows.

When subjected to increasing water pressure, as when lowered in the water column, the bellows compress proportionately with the depth to which the instrument is lowered. This compressing action causes the slide to move towards the nose of the BT at right angles to the direction in which the stylus moves as it concurrently records temperature changes. The resulting trace on the slide is a combined recording of water temperature and depth.

1. Procedure

First take an approximate measurement of the depth at the location at which the

BT is to be taken. Select the BT with the appropriate depth range. Slide the BT sleeve forward towards the BT nose, and check to make sure that the slide has been correctly inserted with the bevelled corner in first and facing the nose of the instrument. Push the slide all the way in. Check the grooves of the slide holder to ensure that they are free of dirt and glass chips and make sure that the slide is firmly in place. Fasten the instrument to the cable by means of the shackle or other suitable means.

Cover the stylus assembly and slide holder by sliding the BT sleeve back towards the BT tail until it reaches its stop. From this moment on, the stylus assembly is in contact with the glass slide and is ready for recordings. The instrument must now be handled very gently and carefully to avoid disturbance of the recording mechanism.

Use a suitable winch to lower the BT to just below the water surface, and allow it to remain in that position for approximately 30 sec to allow the instrument to equilibrate with the surface temperature. During this period, measure the water surface reference temperature by the bucket surface temperature method. This is done by collecting a surface water sample in a specially designed wood-bottomed rubber cylinder, inside of which is a rubber sheath containing a precalibrated mercury immersion thermometer encased in a brass shield that can slide up or down over it. Stir the water and read the thermometer while still immersed. The average of several readings is recorded as the surface reference temperature.

Set the counter on the winch to zero and carefully lower the BT at a speed of about 1 to 2 m/sec to the required depth. It is recommended to lower the BT more slowly when going through the thermocline because of the more rapid temperature changes occurring in that region. Haul in the BT at an approximate speed of 1 to 2 m/sec. As soon as the BT has been retrieved, slack off the cable, set the winch brake, and remove the glass slide carefully from its holder for immediate inspection of the trace.

Hold the slide by its edges between the thumb and forefinger, being careful not to smear it or obscure the trace with smudges or fingerprints, and examine the trace. Any noticeable difference, i.e., hysteresis, between the up and down trace should be negligible. If not, repeat the measurement.

Using a sharp implement or pencil, record all necessary identification data, including the serial number of the BT used, on the slide. Rinse the labelled slide with a little fresh water and determine the temperature/depth readings by placing it in a BT gridmount viewer and reading against the BT grid specified for the particular BT used. The BT surface temperature reading should compare satisfactorily with the bucket surface reference temperature. If not, the BT calibration should be checked.

Record all information on the appropriate field log sheet.

2. Notes

BTs are manufactured to operate at certain maximum depths, and proper selection of instrument is important. Being a sensitive instrument, it should be treated appropriately and should not be subjected to unnecessary shock or excessive heat. It is important that the manufacturer's operational and maintenance instruction manual be followed for the proper use and care of the BT and slides. Ensure that the instrument is securely fastened to the cable before it is lowered in order to minimize the chances of loss.

D. The Electronic Bathythermograph, EBT

The EBT enables direct instantaneous readings and graphical recordings of temperature/depth measurements to be made at a specified location. It comprises two basic components: the transducer or sensor element, and a remote electronic display equipment. The transducer (Figure 8) is connected to the electronic display by an electrical cable and, being relatively light, can be lowered manually to depths of up to about

FIGURE 8.   The EBT transducer.

230 m. The display unit may be a portable panel meter indicator and/or an X-Y graphic chart recorder located at some suitable place on the surface away from the water.

The transducer is essentially a thermistor temperature probe comprising a semiconductor resistor system, the resistance of which is a function of temperature, and a depth sensor. The depth sensor works on the principle that water pressure increases uniformly with depth and is made up of a Bourdon® type tube connected to a variable resistor, through which pressure changes are converted to an equivalent change in electrical resistance. The resistance change is measured as a corresponding voltage change in the display unit. The temperature and depth are read on the abscissa and ordinate, respectively, of the recorder chart.

1. Procedure

Before attempting to use the instrument for measurements, it should be calibrated in accordance with instructions provided by the manufacturer. The instrument calibration should immediately precede each EBT measurement. Switch the calibrated instrument to "run" and place the transducer in the water such that the sensor end is touching the water. After about 5 sec, lower the recorder pen onto the chart paper and note the surface temperature. This reading may be cross-checked by concurrently taking a bucket surface reference temperature. Gradually lower the transducer to the required depth and then haul it up at about the same rate. Look for and note any difference between the up and down trace recorded on the chart. Switch off the instrument and chart recorder after use and store the transducer in a bucket or other suitable container filled with water.

2. The EBT/Rosette Sampler Combination

Figure 9 shows a more sophisticated electronic display arrangement on board one of the limnological survey vessels used at the Canada Centre for Inland Waters (CCIW) for Great Lakes studies. An EBT/Rosette sampler assembly also in current use at CCIW is shown in Figure 10. The latter, when used with the unit shown in Figure 9, permits automatic EBT recordings and triggering of specific sample bottles when certain predetermined depths or water temperatures are reached. Such an arrangement enables one to "view" the thermal structure of the water column being sampled, and is particularly useful when sampling in the thermocline region in which relatively large temperature changes may occur for small depth changes.

E. The Sippican® Expendable Bathythermograph, XBT

The Sippican® expendable BT permits direct readings and/or graphical recordings of temperature/depth measurements to be made from some stationary platform or while in motion, as from a moving vessel or helicopter.

The XBT comprises three basic components, viz., (1) the expendable probe or thermal element (2) the launcher and (3) the recording/display unit. The probe (Figure 11) is a ballistically shaped device which contains a calibrated thermistor in its nose. This is connected to a pair of fine two-conductor wires wound on two spools, one in the probe itself and the other within the upper portion of an outer housing called the deck or shipboard canister. The resistances of the pair of spooled wires are made equal by the addition of suitable resistors in the deck canister. The nose of the probe housing is weighted, and the housing itself is hydrodynamically designed and has fins strategically placed to stabilize its motion and to ensure good reproducibility of its rate of fall when launched. The probe is held in place within the canister by means of a probe launch pin which when pulled, causes the probe to be launched leaving the canister in

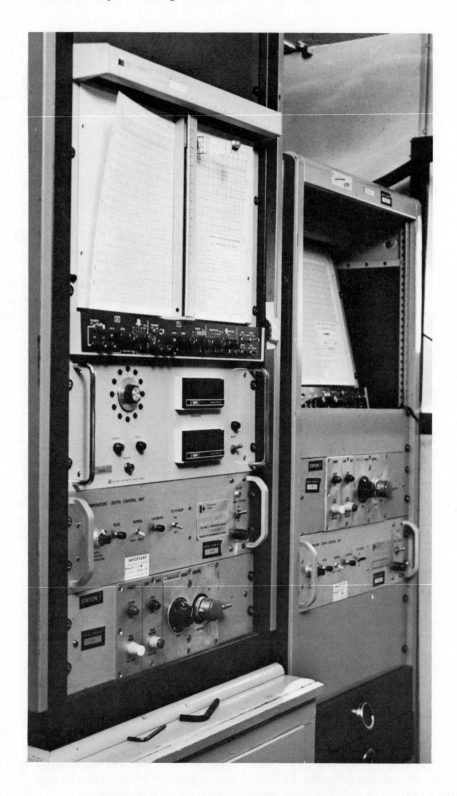

FIGURE 9.   Electronic remote control/readout device for EBT.

FIGURE 10.   EBT/Rosette sampler assembly.

the launcher. The launcher is a spring-loaded breech mechanism which includes a discharge tube, the breech, the stanchion, and a cable connecting the launcher and recorder through which the probe thermistor is electrically connected to the recording device via the spooled wire. When the probe is launched, the wire spools unwind freely enabling the probe to free-fall through the water at some fixed rate. When all of the

FIGURE 11.   The Sippican expendable BT.

spooled wire is completely unwound, the wire breaks and the probe continues its descent to the bottom.

As the probe sinks, the wires are unwound from both ends, and even as the vessel steams away from the point of launching, the wire in the water remains stationary. Changes in the water temperature are denoted by corresponding changes in the electrical resistance of the thermistor with a resulting voltage change recorded on the recording/display unit. Thus the recording device senses and converts temperature changes into equivalent electrical voltage changes. As in the case of the EBT, temperature/depth profiles may be recorded on the appropriate axis of the recorder chart. Present-day technology enables ready transformation of the analog output of the XBT to the more convenient digital readout form. The device is capable of measurements to within $\pm0.2°C$ at depths of up to 1000 m with a depth measuring accuracy of within ±5 meters or about 2%, whichever is greater.

F. The Thermistor

The EBT transducer described above uses a thermistor probe to detect temperature changes in a water body. The thermistor temperature probe consists of a semiconductor resistor system, the resistance of which varies with temperature. When connected to a suitable electrical recording device, changes in electrical resistance resulting from temperature changes are monitored. The electrical signals are then transformed and recorded as a temperature reading on an accurately calibrated scale. The nickel or platinum resistance wire thermometer is one of the most sensitive and reliable temperature sensors available. Its sensitivity is potentially of the order of about $0.0001°C$ and accurate absolute temperature measurements can be made by virtue of the high stability of the instrument. A variety of electric thermometers with up-to-date features such as direct digital readout and/or continuous recording devices are now commercially available at moderate cost. These cover variable ranges with resolution ranging from about $0.1°C$ to $1.0°C$. Readings with accuracy and repeatability of the order of about $\pm0.2°C$ are obtainable with the low-range digital thermistor thermometers. Thermistors have relatively large negative temperature coefficients of resistance, and although they normally require frequent calibration, they are relatively cheap, durable, and accurate enough for most practical purposes.

G. The Infrared Radiation Thermometer

This instrument is used to measure the surface water temperature by measuring the infrared radiation emitted at the surface. The thermometer consists of three basic parts: (1) an optical system, (2) an electronic system and (3) interconnecting cables and power cables. Through the optical system, the amount of energy emitted by the water is compared with that emitted by an internal, controlled reference source. The differential energy is converted into an equivalent voltage by means of the electronic component. The instrument recording device records the equivalent temperature on an accurately calibrated scale. The instrument responds rapidly to the naturally emitted 1-R radiation from water surfaces, clouds, and other emitting sources within its field of view. Care must be taken, therefore, to minimize, if not totally eliminate, the effect of extraneous radiation sources during measurements. The manufacturer's operational procedures as given in the instruction manual should be followed when setting up and using the instrument.

Infrared radiometry has been used in recent years for measuring the temperature of surface waters by means of airborne or spaceborne remote sensing techniques. The present state of the art enables temperature readings to within about .5°C under optimum conditions.

VII. pH

The pH of a substance is defined as the logarithm to the base 10 of the reciprocal of the hydrogen ion activity (i.e., the negative of the $\log_{10}$ of the hydrogen ion activity) at a specific temperature. Application and practical interpretation of this definition has certain limitations from an operational standpoint. This has led investigators to accepting what is now referred to as the operational definition of pH expressed by the equation

$$pH_{(x)} = pH_{(s)} + \frac{(E_x - E_s)F}{RT1_n 10}$$

where pH_x is the pH of the unknown X, pH_s is the assigned pH of a reference standard S, E_x and E_s are the values of the e.m.f. of a pH cell with electrodes immersed in X and S, respectively, F is the Faraday's constant, R the gas constant, and T the absolute temperature in degrees Kelvin (Bates[8]).

pH is an important parameter in assessing water quality. It is used for calculating carbonate, bicarbonate and CO_2 concentrations, and stability index. It may be used also in calculating the hydrogen ion concentration (H^+) and affords an indirect measure of the acidity or alkalinity of certain waters. It governs the solvent properties of water and determines, among other things, the extent and type of physical, biological, and chemical reactions likely to occur within a water system or between the water and surrounding rocks and soils. For example, low-pH water reacts with calcareous materials (e.g. limestone) in surrounding rocks and soils. The resulting decrease in the bicarbonate reserves in such systems invariably leads to water bodies having relatively low buffering capacities. pH has significance in controlling corrosion and scaling tendency of water, and its measurement is useful in monitoring for compliance with set standards (drinking water for example) and to ensure that industrial effluents meet certain specified criteria. For example, Ryznar[9] expressed the stability index of a given water as twice the saturation pH (pH_s) with respect to calcium carbonate minus the actual pH of the sample (i.e., $2pH_s - pH$). This gives a measure of the degree to which that water is likely to cause corrosion or incrustation under certain specific conditions. The saturation or corrosion index of a water type is given by the empirical expression $pH - pH_s$ and gives a measure of the degree of corrosiveness of that water. pH greatly affects the toxicity and/or synergistic effects of certain water constituents and specifically ammonia and cyanide; it affects the metal complexing capacity of some natural waters and may determine, to some considerable extent, how much of a metal is in the labile or dissolved state. It is not uncommon, for example, for a variety of heavy metals to be leached from particulate matter in contact with water of low pH. Thus, the final equilibrium in the dissolved concentration of a given ion depends in part on the final pH of the solution. Biological activities in most natural waters are influenced by their pH. Recent studies have shown lakes with excessively low pH exert a negative impact on sensitive fish species, such as the salmonids or trout, with resulting diminishing population. Productivity and community composition of aquatic species are liable to be adversely affected by a decrease of about 0.5 to 1.0 pH units in natural surface waters with pH normally within the range of 6.0 to 8.0 units. The effect becomes progressively more pronounced with further decreases to the point where, at below a pH of about 4.5, fish and other aquatic life become virtually nonexistent. The Department of National Health and Welfare, Canada,[5] has set the recommended maximum acceptable limit for Canadian drinking water between 6.5 to 8.5 pH units.

The basicity of water increases with increasing concentrations of carbonates, bicarbonates, and hydroxyl ions; while the presence of free mineral acids and carbonic acids tends to increase water acidity. The pH of most natural waters falls approximately within the range of 4 to 9 depending on the concentrations of the carbonate, bicarbonate, and hydroxyl ions present. Alkaline waters are generally more common than acid waters. However, there may be instances where the effect of acid rain on small lakes causes a lowering of their pH to values below 4. Similarly the influx of acidic or alkaline industrial wastes may cause considerable changes in the normal pH of a water system. Effective disinfection of drinking water is best carried out at a pH no greater than about 8.3. Thus it is evident that the monitoring of pH in the aquatic environment is valuable for water pollution control and process control purposes, particularly with respect to water and waste treatment.

In general, there are two basic methods, colorimetric and electrometric, available for pH measurement. Colorimetric methods are based on the principle that certain dyes in solution undergo specific color changes relative to pH. The colorimetric methods suffer from interferences of which color, turbidity, salinity, colloidal material, and several oxidants and reductants are the most prominent sources. These methods are generally more time-consuming and less precise and accurate than the electrometric methods and are being now commonly replaced by the latter. The colorimetric methods should be used only in cases where application of the preferred electrometric method is unfeasible.

A. Colorimetric Method

The simplest colorimetric method available for measuring the pH of an aqueous solution involves the use of pH test papers which have been impregnated with certain pH-sensitive dyes. Measurements are made by simply immersing the paper into the sample and comparing its color with an accompanying chart. This method, supposedly, has an accuracy of ±0.3 pH units and better if papers of overlapping ranges are used.

Another procedure uses what is called a universal indicator color chart. The chart has a number of translucent color standards, each of which represents a specific pH value, permanently mounted adjacent to openings of the same size. A sample aliquot is mixed with a measured volume of the universal indicator solution and the resulting color is matched against the chart.

Other noteworthy colorimetric methods involve the addition of a prepared indicator to the sample and comparing the resultant color with that of a set of color standards of known pH. Different indicators are required for different pH ranges. A number of the more common indicators and their pH ranges are given in Table 7. Color comparison may be made visually with a color comparator such as the Hellige® comparator, the Taylor® pH slide comparator, or other suitable equipment including the spectrophotometer.

Although, with proper care, a precision and accuracy of about 0.05 to 0.1 pH unit is attainable, the colorimetric method has several disadvantages, notably the following.

1.   It is fairly involved and generally requires preparation and standardization of

Table 7
COLOR INDICATORS AND pH RANGE COVERED

Indicator (%)	Approximate pH Range	Visual Color Range	Method of Preparation
Cresol Red (0.02)	0.2—1.8; 7.2—8.8	Red to yellow; yellow to red	Dissolve 0.05 g in 26.2 mℓ of 0.01 M NaOH and make up to 250 mℓ with distilled water.
Methyl Violet (0.04)	0.2—1.8; 2.0—3.2	Yellow to blue; blue to violet	Dissolve 0.1 g in 250 mℓ distilled water.
Thymol Blue (0.04)	1.2—2.8; 8.0—9.6	Red to yellow; yellow to blue	Dissolve 0.1 g in 21.5 mℓ of 0.01 M NaOH and make up to 250 mℓ distilled water.
Metacresol Purple (0.04)	1.2—2.8; 7.6—9.2	Red to yellow; yellow to purple	Dissolve 0.1 g in 26.2 mℓ of 0.01 M NaOH and make up to 250 mℓ with distilled water.
Benzopurpurine 4B(0.1)	2.2—4.2	Violet to yellow	Dissolve 0.1 g in 100 mℓ distilled water.
Bromophenol Blue (0.04)	3.0—4.6	Yellow to blue	Dissolve 0.1 g in 14.9 mℓ of 0.01 M NaOH and make up to 250 mℓ with distilled water.
Bromocresol Green (0.04)	3.8—5.4	Yellow to blue	Dissolve 0.1 g in 14.3 mℓ of 0.01 M NaOH and make up to 250 mℓ with distilled water.
Methyl Red (0.02)	4.2—6.2	Pink to yellow	Dissolve 0.02 g in 60 mℓ ethanol and make up to 100 mℓ with distilled water.
Chlorophenol Red (0.04)	5.2—6.8	Yellow to red	Dissolve 0.1 g in 23.6 mℓ of 0.01 M NaOH and make up to 250 mℓ with distilled water.
Bromocresol Purple (0.04)	5.2—6.8	Yellow to purple	Dissolve 0.1 g in 18.5 mℓ of 0.01 M NaOH and make up to 250 mℓ with distilled water.
Bromothymol Blue (0.4)	6.0—7.6	Yellow to blue	Dissolve 0.1 g in 16 mℓ of 0.01 M NaOH and make up to 250 mℓ with distilled water.
Phenol Red (0.02)	6.8—8.2	Yellow to red	Dissolve 0.05 g in 28.2 mℓ of 0.01 M NaOH and make up to 250 mℓ with distilled water.
Thymolphthalein (0.04)	9.4—10.6	Colorless to blue	Dissolve 0.04 g in 50 mℓ ethanol and add 50 mℓ of distilled water.
Alizarin Yellow R(0.01)	10.1—12.0	Yellow to red	Dissolve 0.05 g in 500 mℓ distilled water.
1,3,5-Trinitrobenzene (0.04)	12.0—14.0	Colorless to orange	Dissolve 0.04 g in 100 mℓ ethanol.

Note: A more comprehensive list of indicators may be found in the *CRC Handbook of Chemistry and Physics*,[10] or as given by Bates.[8] Ready-made color indicator solutions of comparable concentrations are normally available commercially.

several solutions. However, standardized pH indicator solutions are now commercially available.

2.   The indicators, and sometimes the color standards, change with time.
3.   Color, turbidity, salinity, chlorinated compounds, colloidal material, and several oxidants and reductants interfere.
4.   The sharpness of color changes is contingent on the type of indicator used.
5.   The approximate pH of the sample must be known to enable the selection of the right indicator.
6.   The indicators, being themselves weak acids and bases, must have their own pH individually adjusted to values close to that of the sample to avoid changing the true pH of the sample and particularly so for weakly buffered samples. This could be a lengthy and tedious process.
7.   The absorbance is dependent on the wavelength of the light used for color measurement.
8.   The method is susceptible to variations in the ionic strength and solvent composition of solutions and requires correcting the results for salt effect.
9.   Reaction between sample constituents, including metal ions, proteins and colloidal substances, may result in erroneous results.

Because of the foregoing, and because there are now several relatively inexpensive but accurate pH meters commercially available, the electrometric method is the one of choice. This method reqires a comparatively high initial capital outlay, but in the long run, the net benefits derived from the overall manpower savings coupled with the superior quality of data produced, will undoubtedly make such an investment worthwhile.

B. Electrometric Method

This method is applicable to the measurement of pH of solutions over the full pH scale. It requires the use of either a potentiometric/null-detector type or a direct-reading type pH meter equipped with a glass and a suitable reference electrode. This combination affords a means of measuring the difference between the pH values of two solutions, a standard reference (buffer) and the unknown, maintained at the same temperature. Measurements are based on the principle of the Nernst's equation, relating the electrode potential difference between two electrodes in the same solution as a function of the hydrogen ion activity at a given temperature. In other words, the measured potential difference or e.m.f. (E) is given as

$$E = E_o + 2.3026 \frac{RT}{nF} \log A_H$$

where E_o = a constant equivalent to the standard potential of the glass-saturated calomel pH cell, R = the gas constant, T = absolute temperature in degrees Kelvin, n = charge on the ion, F = Faraday's constant, and A_H = hydrogen ion activity.

This equation can be rearranged to incorporate the definition given above for pH; and taking n = 1, we have:

$$E = E_o + 2.3026 \frac{RT}{F} \cdot pH$$

$$\text{whence } pH = (E - E_o) \frac{F}{2.3026 \, RT} = pH_s + \frac{EF}{2.3026 \, RT}$$

where pH_s represents the known pH of a reference buffer solution and E is the electromotive force (e.m.f.) between the glass and reference electrodes when immersed in the sample. And so we have pH = K_1 + $K_2(E/T)$ where K_1 and K_2 are constants. The factor 2.3026 RT/F is called the Nernst or slope factor and has a specific value for a given temperature T. It is actually the ratio of millivolts per pH unit determined by the number of millivolts difference in the pH cell e.m.f. that corresponds to each pH unit difference observed at a given temperature. For example, at 25°C and 60°C a difference of 1 pH unit corresponds to a cell e.m.f. difference of approximately 59 mV and 66 mV, respectively.

The approach to be taken in calibrating the pH meter and in measuring sample pH will depend largely on the fundamental question of whether the sample pH is to be measured *in situ,* or on a grab sample at some specific temperature.

Where an *in situ* measurement is impracticable and it is desirable to determine pH at the ambient temperature of the sample, the best approach is to collect the sample in wide-mouth Thermos® flasks of about 250 mℓ volume and to do the measurement on-site as soon as possible thereafter. Fill the Thermos® flasks to overflowing, while taking care not to aerate the sample, and then stopper immediately. Immerse the electrode/thermometer assembly directly into the flask. The practice of heating and maintaining the sample at a specific temperature (usually 20°C or 25°C) while measuring the pH is not recommended due to the changes, such as evolution of dissolved gases, that are liable to occur in the sample. This approach, however, may be the only reasonable one to take in some cases where, in order to facilitate data interpretation, it is necessary to make measurement at a specified constant sample temperature. Alternatively, the measured pH may be converted to its equivalent value at a particular temperature by applying predetermined conversion factors following the procedure described by Langelier.[11] The procedure is somewhat involved, however, and its application is valid only if certain specified conditions are met.

The use of the glass electrode is preferred to the universally accepted primary standard — the hydrogen electrode, because the former is less susceptible to interferences. The only two significant sources of error are (1) interference of sodium ions at pH above 10 and (2) temperature effect. Glass electrodes may also be susceptible to "acid error" at pH below 1.0, as well as possible electrode "poisoning" by some substances.

Most suppliers account for the sodium effect by providing special "low sodium error" electrodes. If such electrodes are not available, then the sodium ion error should be compensated for by applying a correction factor based on the manufacturer's specifications.

It is obvious from the Nernst equation that the electrode potential, and hence the measured pH, is temperature dependent. Basically, temperature has two specific effects on the measurement of pH of a solution. The potential change per pH unit (pH potential) as well as the ionization in the sample vary with temperature. The more expensive pH meters are equipped with manual and/or automatic temperature compensators to compensate for the temperature/pH potential effect. Otherwise, the temperature effect may be minimized by standardizing and maintaining the electrode assembly at a temperature as close as possible to the sample temperature. The ionization effect is intrinsic to the solution and as such, it is necessary to determine and record the solution temperature at which the pH has been measured.

1. Equipment and Reagents

An electronic pH meter with a glass electrode and preferably equipped with temperature compensator and clamps for the electrodes and a mercury immersion thermometer is required. The use of a calomel-saturated KCl or silver-silver chloride reference electrode is recommended. Combination electrodes are also available and may be used to advantage, especially when working with a limited volume of sample. Other needed equipment includes a thermostatted water bath, a mercury immersion glass thermometer, a magnetic stirrer and Teflon®-coated magnetic stirring bars or equivalent, and wide-mouth Thermos® flasks.

Standard buffer solutions having specific pH values may be prepared by dissolving prepackaged or capsulated certified powders or tablets in the prescribed volume of CO_2-free distilled water. Alternatively, the required solutions could be prepared as noted in Table 8 by weighing the required amount of pure certified reagent grade salts and dissolving in a volumetric flask with CO_2-free distilled water. The solution, maintained at 20°C or 25°C, is made up to mark to yield the buffer of the required pH value. Store solutions in clean capped polyethylene bottles and avoid unnecessary exposure to the air.

Buffer solutions, the borate and carbonate buffers especially, tend to deteriorate with age, and in cases where precise calibration is required, it is recommended that they not be used if stored for periods in excess of about 4 weeks. The useful life of the tartrate solution may be extended by the addition of a few crystals of thymol.

Prepackaged buffers are supplied with a table giving the solution pH values at different temperatures whereby a pH/temperature graph could be drawn. For the benefit of analysts wishing to prepare their own buffer solutions, Table 9, a reproduction from Standard Methods,[2] is provided for reference.

2. Procedure

a. pH Meter Calibration

Since the quality of the results obtained is governed primarily by the calibration of the pH meter, particular attention must be paid in ensuring proper instrument calibration. This is best accomplished by the use of at least two buffer solutions of pH values close to and bracketting the expected sample pH. The wide variety of instruments now available on the market precludes the provision of any precise instructions here. Although the following general guidelines may apply to most meters, it is recommended that the instructions contained in the manufacturer's operation manual be followed.

Adjust and maintain the temperature of a water bath close to that expected for the samples to be measured. Place the buffer solutions, contained in well-stoppered bottles, in the water bath for temperature equilibration. Attempt to maintain the electrodes at the same temperature as that of the buffer and samples.

Assuming that the instrument and electrodes are already set up in accordance with the manufacturer's instructions, wash the electrode with distilled water. Blot off any excess water from the electrodes with a soft absorbent tissue. It is usually convenient to attach a mercury thermometer to the electrodes to facilitate temperature measurements.

Select two buffer solutions having pH values close to and bracketting the expected pH value of the sample. It is best to select buffers of pH values different by about 2 to 4 pH units if that would adequately cover the expected sample range. Pour enough of one of the buffer solutions in a small beaker to completely cover the bulb of the glass electrode and the tip of the reference electrode when immersed. Immerse the electrode/thermometer assembly into the solution and ensure that the electrodes are not touching the sides and/or bottom of the beaker. Read the temperature of the buffer solution while stirring it gently and, if required, set the manual temperature compensator control to that temperature. Check Table 3, or the list of tables or graph provided with the buffer, to determine the actual pH value of the solution at the above-noted temperature. Unlock and adjust the calibration control to make the meter read the exact pH value computed in the preceding step and relock the calibration control. There may be slight variations to this step for instruments with "expanded scale"

Table 8
STANDARD pH VALUES ASSIGNED BY THE NATIONAL BUREAU OF STANDARDS[a]

Temp. °C	Primary Standards							Secondary Standards	
	Tartrate (Saturated)	Citrate (0.05 m)	Phthalate (0.05 m)	Phosphate (1:1)	Phosphate (1:3.5)	Borax (0.01 m)	Carbonate (0.025 m)	Tetroxalate (0.05 m)	Calcium Hydroxide (Saturated)
0		3.863	4.003	6.984	7.534	9.464	10.317	1.666	13.423
5		3.840	3.999	6.951	7.500	9.395	10.245	1.668	13.207
10		3.820	3.998	6.923	7.472	9.332	10.179	1.670	13.003
15		3.802	3.999	6.900	7.448	9.276	10.118	1.672	12.810
20		3.788	4.002	6.881	7.429	9.225	10.062	1.675	12.627
25	3.557	3.776	4.008	6.865	7.413	9.180	10.012	1.679	12.454
30	3.552	3.766	4.015	6.853	7.400	9.139	9.966	1.683	12.289
35	3.549	3.759	4.024	6.844	7.389	9.102	9.925	1.688	12.133
38	3.548		4.030	6.840	7.384	9.081		1.691	12.043
40	3.547	3.753	4.035	6.838	7.380	9.068	9.889	1.694	11.984
45	3.547	3.750	4.047	6.834	7.373	9.038	9.856	1.700	11.841
50	3.549	3.749	4.060	6.833	7.367	9.011	9.828	1.707	11.705
55	3.554		4.075	6.834		8.985		1.715	11.574
60	3.560		4.091	6.836		8.962		1.723	11.449
70	3.580		4.126	6.845		8.921		1.743	
80	3.609		4.164	6.859		8.885		1.766	
90	3.650		4.205	6.877		8.850		1.792	
95	3.674		4.227	6.886		8.833		1.806	

[a] Reproduced from Standard Methods for the Examination of Water and Wastewater, 14th ed., APHA-AWWA-WPCF, American Public Health Association, Washington, D.C., 20036, 1975, 123. With permission.

Table 9
PREPARATION OF pH STANDARD BUFFER SOLUTION

Buffer pH at 25°C	Weight of Salt(s) per Liter Solution Needed
1.679	12.61 g potassium tetroxalate dihydrate, i.e., 0.05 $MKH_3C_4O_82H_2O$ solution.
3.557	Prepare a saturated solution containing approximately 6.4 g/l of potassium hydrogen tartrate ($KHC_4H_4O_6$) by vigorously shaking about 8 g of finely crystalline salt in about 250 ml distilled water at 25°C and making up the decanted clear solution to 1 l. This solution is not stable for long.
3.776	11.41 g potassium dihydrogen citrate, i.e., 0.05 $MKH_2C_6H_5O_7$ solution.
4.008	10.12 g potassium hydrogen phthalate, i.e., 0.05 $MKHC_8H_4O_4$ solution.
6.865	3.388 g potassium dihydrogen phosphate (0.025 MKH_2PO_4) and 3.533 g disodium hydrogen phosphate (0.025 MNa_2HPO_4).
7.413	1.179 g potassium dihydrogen phosphate (0.008695 MKH_2PO_4) and 4.302 g disodium hydrogen phosphate (0.03043 MNa_2HPO_4).
9.180	3.814 g sodium borate decahydrate, i.e., 0.01 M $Na_2B_4O_710H_2O$ solution.
10.012	2.092 g sodium bicarbonate (0.025 M $NaHCO_3$) and 2.640 g sodium carbonate (0.025 M Na_2CO_3).
12.454	Prepare a saturated calcium hydroxide solution containing approximately 1.5 g Ca(OH)$_2$/l by following the procedure given in Standards Methods[2] or as described by Bates.[8]

Note: (1) The potassium dihydrogen phosphate and disodium hydrogen phosphate should be dried by heating at about 110°C for roughly 2 hr and then cooled in a desiccator before weighing.

(2) Ready-made certified buffer solutions with pH accuracy of ± 0.02 units are also commercially available.

reading. Remove the electrode/thermometer assembly from the buffer solution. Wash thoroughly with distilled water and blot off excess water.

Repeat the above procedure using the other buffer solution. The meter should read to within 0.02 unit of the pH value computed for the second solution. If not, adjust the temperature compensator control until the meter reads the exact pH value expected. Compare the new setting of the temperature compensation control with the actual measured temperature of the buffer. There shouldn't be an appreciable difference, if any, between the two. If there is, it may be indicative of defective electrodes and/or contaminated buffer solutions or an improperly adjusted or defective temperature compensator. These should be checked and replaced if found defective.

The pH meter should be calibrated regularly at about 2 hr intervals when being used continuously. In cases of sporadic use, less frequent calibration would be required, but the instrument should, at the very least, be calibrated each time immediately before and after use.

b. Sample pH Measurement

After the pH meter has been satisfactorily calibrated, remove the electrode/thermometer assembly from the buffer and wash thoroughly with distilled water. Blot off excess water and rinse the electrodes with a little of the sample to be measured. Immerse the electrode/thermometer assembly into the sample following the precautions noted above. Record the sample temperature to 0.1°C when the thermometer reading becomes steady, and set the temperature compensator accordingly. Read and record the sample pH to the appropriate number of decimal places. Check the sample temperature to determine if it has changed. If so, take the average of the two temperature readings as the sample temperature.

3. Precision and Accuracy

The better makes of electronic pH meters now available are capable of producing results with a precision of about ±0.005 pH unit and, under the right conditions, the significance of the differences between any two measurements would be limited only by the analyst's skill and the precision of his or her measurements. It should be noted, nevertheless, that due to uncertainties inherent in actual pH determinations, the fundamental meaning of these so-called accurate and precise pH readings is considerably less certain than the optimum precision attainable. A repeatability of about ±0.02 unit is generally adequate for most routine measurements in well-buffered solutions; while about ±0.1 unit is considered acceptable for weakly buffered solution. Mean standard deviation of 0.01 pH unit at mean pH of 8.75 has been determined for three different analysts (Philbert and Traversy[12]). Results obtained from several interlaboratory comparison studies indicate that a between-lab relative standard deviation of about ±2.5% is reasonable.

4. Notes

It may be difficult in some instances to obtain an accurate and precise pH reading due to meter fluctuation or drift. The electrode characteristics may be a contributing factor, and some electrodes may require more time than others for stabilization. This is likely to occur from the effect of the residual liquid junction potential of the reference electrode when the electrodes are immersed successively in solutions containing significantly different types of ions and having much different ionic strengths. The liquid junction potential effect can be minimized or cancelled, especially when attempting to measure small differences in pH between solutions of similar composition, by using the same pH meter with the same reference electrode and buffer solution throughout and by standardizing the instrument and electrodes at a pH close to that of the sample. Nevertheless, significant meter drift has been observed when measuring pH of some types of poorly buffered natural waters due to changes occurring upon exposure of the sample to the air. In both cases, it is desirable that the actual sample measurement be done as rapidly as possible or, preferably, under special conditions that preclude sample-air contact. Where the observed meter fluctuations are not extreme, the average of the maximum and minimum readings noted may be recorded as the sample pH. Sample aliquots used for pH measurements should not be used afterwards for specific conductance or mineral analyses.

The accuracy and precision of the results obtained depend largely on the condition of the pH electrodes, which must be handled and maintained with due care. The instructions provided by the manufacturers should be followed. The bulbs of glass electrodes are very sensitive and should not be handled unnecessarily. They should not be rubbed vigorously when attempting to remove excess water adhering to them as required above. The level of the electrolytic solution within the calomel electrode must be maintained at the required height and care should be taken to guard against any contamination or possible plugging up of the electrode in order to ensure free flow of the electrolyte at the liquid junction. It is recommended, wherever feasible, that the millivolt scale of a new pH meter be calibrated against a known source of e.m.f. before it is put into use.

C. Automated Procedure for pH Measurement

Technicon[13] introduced, about 3 years ago, an automated ion-selective electrode (ISE) AutoAnalyzer® II method for the measurement of pH in water. The method uses the basic AutoAnalyzer® components comprising a sampler, proportioning pump, ISE analytical test cartridge and pH combination electrode, ISE detector module, recorder, and an optical digital printer/ readout.

According to Philbert et al.[14] the method works satisfactorily and is capable of yielding results that are comparable to that obtained by the manual electrometric procedure.

Mean relative standard deviation of ±0.16% and ±0.18% were determined at pH values of 7.90 and 8.24, respectively.

VIII. SPECIFIC CONDUCTANCE

The ability of a solution to conduct an electrical current is governed by the migration of solute ions and is dependent on the nature and number of the ionic species in that solution. This property is called the electrical conductivity which, when expressed appropriately, is known as the specific conductance of the substance. The amount of current carried by, and hence the conductivity of, a solution varies inversely with its resistance. It depends also on the number, mobility, valency, and types of ions present as well as the temperature of the solution. Thus specific conductance is a characteristic property of an electrolytic aqueous solution and is directly related to the thermo-chemical state of a given water.

The electrical conductivity of a solution is defined as the reciprocal of the electrical resistance (resistivity) measured across two parallel non-polarized electrodes each measuring 1 cm × 1 cm and immersed 1 cm apart in the solution. The unit of electrical conductivity, therefore, is the inverse ohms/cm or mhos/cm and, for practical purposes is expressed as micromhos/cm (μmhos/cm) or millimhos/cm (mmhos/cm). However, new international units Siemens (S), milli-Siemens (mS), and micro-Siemens (μS) are now used instead. Thus 1 S corresponds to a resistance of 1 ohm, 1 mS corresponds to 1,000 ohms and 1 μS corresponds to 1,000,000 ohms.

The specific conductance of a sample correlates with the concentration of dissolved minerals or with what is more commonly known as the total dissolved solids (TDS) of that sample. An empirical relationship exists between the specific conductance and TDS of most natural waters. Thus TDS may be computed from the much more accurate, precise, and more rapid specific conductance measurement if the appropriate conversion factor is known for that particular water. This factor may vary from about 0.55 to 0.9 but has been found to fall somewhere between 0.55 and 0.7 for most natural waters depending on the soluble constituents and the temperature at which the measurement has been made. For example, the factor for water containing relatively high concentrations of free acids or hydroxide alkalinity may be appreciably lower than 0.55, while it may be much higher than 0.7 for saline and boiler waters. It is therefore advisable that the right conversion factor be established experimentally for each water type being examined. A conversion factor of approximately 6.5 is generally applicable to most natural waters. Furthermore, the specific conductance in μS/cm multiplied by 0.01 gives an approximation of either the cationic or anionic concentration in meq/ℓ.

Conductivity measurements serve as a useful indicator of the degree of mineralization in a sample. Variations in the dissolved mineral concentration and movements of water regimes can be readily determined. It provides a means of checking the purity of distilled water and serves as a guide in making appropriate sample dilutions for other chemical tests and checking analytical results with which it correlates. Freshly distilled water has a specific conductance of between 0.5 to 2 μS/cm, which may increase to about 2 to 4 μS/cm after standing for a few weeks, due mainly to the absorption of CO_2 and, to a lesser extent, ammonia from the atmosphere. Specific conductance of most natural surface waters generally range from about 50 to 1500 μS/cm, and some industrial wastewaters may have specific conductance of up to 10,000 μS/cm and over.

A. Measurement of Specific Conductance

The principle of the conductivity meter is that of an ohmeter, and measurements are based on the Wheatstone bridge principle. In terms of Ohm's law the electrical resistance of the sample is measured by means of a conductance cell and a Wheatstone bridge, or alternatively, the conductance may be measured as the ratio of the electrical current flowing through the cell to a fixed voltage applied across opposite faces of a centimeter-cube of the test solution. A galvanometer or a cathode ray oscilloscope may be used as a null point indicator. Measurements could be made by using the conventional audio-frequency alternating current bridge method, the direct current method, or by the radio-frequency method. Most modern meters provide for the specific conductance of the sample to be measured at some specific temperature by reading directly from an indicating meter the conductivity of the conductance cell immersed in the sample, and then correcting that reading by multiplying by the cell constant predetermined for that particular cell.

The electrical conductivity of a solution, logically, increases with temperature. It is therefore essential to make accurate measurements of the sample temperature when measuring its conductivity. Specific conductance results are generally reported at a standard temperature of 25.0°C but other standard temperatures, 20.0°C and 18.0°C specifically, are also used. Because the temperature/conductivity relationship or temperature coefficient of conductivity varies for different waters, and because it is not always practical to measure conductivity at the standard temperature, it may be necessary to develop a set of temperature correction factors for each water type, such as has been done for the Laurentian Great Lakes (Rodgers).[15]

While there are no known interferences in this test, the effect of electrode polarization could lead to erroneous results. This effect may be obviated in most conductivity meters by

1. Using an AC voltage for the measurement
2. Using the proper frequency
3. Providing for a suitable current density
4. Ensuring that the platinum electrodes used are coated with platinum black

The above-noted requirements become increasingly more important with higher conductivity readings. It is recommended, however, that the specific operational instructions provided by each manufacturer be carefully followed to ensure optimum instrument performance.

1. Equipment and Reagents

The use of a conductivity meter preferably with error not exceeding ±1% or 1 $\mu S/$ cm, whichever is the greater, is recommended. A conductivity cell equipped with platinum electrodes and having suitable cell constant is required. Immersion and flow-through type cells are commercially available. The choice of cell will depend on the expected range of conductivity and resistance range of the conductivity meter. The following general guideline may aid in the selection of a proper cell. Use a conductivity cell with a cell constant of approximately 0.1, 1, and 10 for readings in the ranges of 0 to 100 $\mu S/cm$, 100 $\mu S/cm$ to about 100 mS/cm, and above 100 mS/cm, respectively.

Conductivity water — Freshly boiled and cooled deionized distilled water having specific conductance less than 1 $\mu S/cm$ is desirable.

Standard potassium chloride, 0.010 M — Heat some anhydrous potassium chloride crystals of certified reagent grade at about 105°C for approximately ½ hr. Allow to cool in a desiccator. Dissolve 0.7456 g in conductivity water in a liter volumetric flask and make up to 1-liter mark with more water at 25°C. Store in a glass-stoppered Pyrex® bottle. This standard reference solution has a specific conductance of 1413 $\mu S/cm$ at 25°C and is usable with cells having constants between one and two. Table 10 gives the corresponding values of this solution at different temperatures. Stronger

Table 10
VARIATION OF SPECIFIC CONDUCTANCE WITH
TEMPERATURE FOR STANDARD 0.010 *M* KCl SOLUTION

Temp. (°C)	Spec. Conduct. (Microsiemens)	Temp. (°C)	Spec. Conduct. (Microsiemens)	Temp. (°C)	Spec. Conduct. (Microsiemens)
5	897	22.3	1340	26.2	1445
6	922	22.4	1343	26.3	1448
7	947	22.5	1346	26.4	1451
8	972	22.6	1348	26.5	1453
9	997	22.7	1351	26.6	1456
10	1022	22.8	1354	26.7	1459
11	1047	22.9	1356	26.8	1462
12	1072	23.0	1359	26.9	1464
13	1097	23.1	1362	27.0	1467
14	1122	23.2	1364	27.1	1470
15	1147	23.3	1367	26.2	1472
16	1173	23.4	1370	27.3	1475
17	1199	23.5	1373	27.4	1478
18	1225	23.6	1375	27.5	1481
19	1251	23.7	1378	27.6	1483
20	1278	23.8	1381	27.7	1486
20.0	1278	23.9	1383	27.8	1489
20.1	1281	24.0	1386	27.9	1492
20.2	1283	24.1	1389	28.0	1494
20.3	1286	24.2	1391	28.1	1497
20.4	1289	24.3	1394	28.2	1500
20.5	1292	24.4	1397	28.3	1502
20.6	1294	24.5	1400	28.4	1505
20.7	1297	24.6	1402	28.5	1508
20.8	1300	24.7	1405	28.6	1511
20.9	1302	24.8	1408	28.7	1513
21.0	1305	24.9	1410	28.8	1516
21.1	1308	25.0	1413	28.9	1519
21.2	1310	25.1	1416	29.0	1522
21.3	1313	25.2	1418	29.1	1524
21.4	1316	25.3	1421	29.2	1527
21.5	1319	25.4	1424	29.3	1530
21.6	1321	25.5	1427	29.4	1532
21.7	1324	25.6	1429	29.5	1535
21.8	1327	25.7	1432	29.6	1538
21.9	1329	25.8	1435	29.7	1541
22.0	1332	25.9	1437	29.8	1543
22.1	1335	26.0	1440	29.9	1546
22.2	1337	26.1	1443	30.0	1549

or weaker KCl reference solutions (Table 11) may be used, depending on the cell constant of the conducting cell being used and/or the sample range.

2. Procedure

Some of the modern conductivity meters are equipped wth special features such as a cell constant adjustment control, manual and/or automatic temperature compensator, and compensating selector switch, whereby specific conductance at 25.0°C can be read directly. When using such instruments, it is best to follow the calibration and measuring instructions provided by the manufacturer. The following procedure is applicable to the typical less-sophisticated instrument equipped with the immersion type conductivity cell.

Connect the cell to the meter and turn the power on, allowing meter to warm up

Table 11
SPECIFIC CONDUCTANCE OF STANDARD
KCl SOLUTION OF VARYING STRENGTHS
AT 25.0°C AND 20.0°C

Concentration (moles/l) of KCl	Equivalent Conductance[a]	Specific Conductance (μS/cm)	
		at 25.0°C	at 20.0°C[b]
0.0	149.94		
0.0001	148.98	14.90	13.45
0.0005	147.83	73.92	66.86
0.001	146.97	147.0	133.0
0.005	143.58	717.9	649.7
0.010	141.26	1413.0	1279.0
0.020	138.30	2766.0	2503.0
0.050	133.39	6670.0	6036.0
0.10	128.92	12892.0	11678.0
0.20	124.08	24820.0	22520.0
0.50	117.27	58640.0	53310.0
1.0	111.87	111900.0	102200.0

[a] All values except the last three were calculated from the revised Fuoss-Onsager equation proposed by Fuoss (Hills[16]). Other data are taken from Robinson and Stokes.[17]

[b] Computed from values at 25.0°C based on factors derived from data presented by Hamer and Wood.[18]

for about 5 min. Clamp the thermometer with the conductivity cell, and rinse them several times with deionized distilled water. Finally, rinse with a little of the sample to be measured. Immerse the cell and thermometer into the sample contained in a beaker, making sure that the platinum electrodes are fully submerged and no air bubbles are trapped within. The latter is best achieved by repeatedly lowering and raising the cell in the sample before attempting to make measurements. Select the appropriate range for measurement. Set the "measure-calibrate" switch to "calibrate" and adjust the calibration switch as required. Read the sample temperature to the nearest 0.1°C while gently stirring the sample. Set the "measure-calibrate" switch to "measure" and read the meter, making sure to read the scale corresponding to the range at which the range switch is set.

Measure sample temperature again immediately after the conductivity measurement. There should be no significant change; but if there is, take the average of the two readings as the sample temperature. Large changes in sample temperature may be avoided by keeping samples thermostatted in a water bath while making measurements. Remove the cell and thermometer from sample and wash thoroughly with deionized distilled water, and repeat the procedure if other samples are to be measured. The cell and thermometer should be rinsed with each new sample or standard to be measured just before they are immersed.

3. Calculations

The conductivity of the sample, as measured above, must be corrected for (1) the cell constant of the conductivity cell used and (2) temperature.

Thus, for a measured conductivity value of C μS/cm at t°C, the corrected conductivity value C' is given as C' = (Cf$_1$f$_2$)μS/cm at 25°C, where f$_1$ = correction factor for the conductivity cell, and f$_2$ = temperature factor for computing the conductivity measured at t°C to its value at 25°C.

a. Calculation of Cell Constant

Conductivity cells are normally supplied with their cell constant marked on them. However, the value of the cell constant for a particular conductivity cell is likely to vary depending on the concentrations of dissolved salts present. It is recommended, therefore, that the cell constant be checked regularly and preferably at specific conductance values close to those expected for the samples. The 0.1 M, 0.005 M, and 0.001 M KCl solutions (Table 11) are the ones most likely to be used. Moreover, it would be convenient to present the cell constant/conductivity variations in a tabular or graphic form for future reference.

The procedure for determining the cell constant is identical to that prescribed above, except that the standard reference KCl solution is measured instead of the sample. Assuming that a conductivity reading of 1310 μS/cm at 21.6°C was observed, the theoretical value at 21.6°C is deduced from Table 10 to be 1321. Hence the cell constant is 1321/1310 = 1.008.

In order to ensure accurate determination of the cell constant, it is advisable to make about five or more conductivity measurements on the standard KCl solution, preferably at different temperatures. In a real case situation, a relative standard deviation of 0.26% was obtained for a mean cell constant value of 0.9779 determined by successive daily determinations over a 7-day period.

b. Correction for Sample Temperature

Where it is unfeasible to measure the conductivity directly at 25.0°C or to develop the temperature correction table for a particular sample, it is reasonable to make the assumption that the conductivity of the sample increases with temperature at a rate of approximately 2% per °C. Thus to deduce the specific conductance of the sample at 25.0°C, add 2% of the value determined at the test temperature for each °C that temperature is below 25.0°C, or conversely, subtract 2% of the observed value for each °C that the test temperature exceeds 25.0°C. A temperature coefficient of 2% per °C at 25.0°C has been assumed to produce Table 12 which will facilitate temperature correction for routine measurements.

The accuracy of specific conductance measurements may be improved considerably by employing a sample dilution technique. The objective of this is to bring the sample conductivity to within a reasonably narrow range (about 90 to 120 μS/cm) and to allow for, as much as possible, the contribution of each individual ion to the total measured conductivity. Application of this technique is recommended particularly where samples contain TDS in excess of about 700 mg/ℓ.

First, determine roughly the specific conductance of the undiluted sample. Using conductivity water, dilute the sample precisely in a suitable known ratio to bring its conductivity within the 90 to 120 μS/cm range. This is best achieved by preparing successive dilutions, since specific conductance does not vary in an exact ratio with sample dilution. Measure the specific conductance (S_d) of the diluted sample at 25.0°C as described above.

Deduce the specific conductance (S) at 25°C of the sample, otherwise termed the "theoretical" or "dilution" conductance, from the equation:

$$S = D \cdot S_d - (D - 1) C_w$$

where D is the dilution ratio = sum of volumes of sample and dilution water used/ volume of sample used, and C_w is the specific conductance of the dilution water.

C. Precision and Accuracy

Conductance measurements are among the most precise of physical analytical chemistry. A precision of about 0.01% or better is attainable with proper equipment and

Table 12
TEMPERATURE CORRECTION FACTORS TO COMPUTE SPECIFIC CONDUCTANCE VALUES AT 25.0°C

Multiplication factor

°C	0	0.1	0.2	0.3	0.4	0.5	0.6	0.7	0.8	0.9
4	1.521	1.518	1.515	1.512	1.509	1.506	1.503	1.500	1.497	1.494
5	1.491	1.488	1.485	1.482	1.479	1.476	1.474	1.471	1.468	1.465
6	1.462	1.459	1.456	1.453	1.450	1.447	1.444	1.442	1.439	1.436
7	1.433	1.430	1.427	1.424	1.421	1.419	1.416	1.413	1.410	1.407
8	1.405	1.402	1.399	1.396	1.393	1.391	1.388	1.385	1.382	1.380
9	1.377	1.374	1.371	1.369	1.366	1.363	1.360	1.358	1.355	1.352
10	1.350	1.347	1.344	1.341	1.339	1.336	1.333	1.331	1.328	1.325
11	1.323	1.320	1.318	1.315	1.312	1.210	1.307	1.304	1.302	1.299
12	1.297	1.294	1.292	1.289	1.286	1.284	1.281	1.279	1.276	1.274
13	1.271	1.268	1.266	1.263	1.261	1.258	1.256	1.253	1.251	1.248
14	1.246	1.243	1.241	1.238	1.236	1.234	1.231	1.229	1.226	1.224
15	1.221	1.219	1.216	1.214	1.212	1.209	1.207	1.204	1.202	1.199
16	1.197	1.195	1.192	1.190	1.188	1.185	1.183	1.180	1.178	1.176
17	1.173	1.171	1.169	1.166	1.164	1.162	1.159	1.157	1.155	1.153
18	1.150	1.148	1.146	1.143	1.141	1.139	1.136	1.134	1.132	1.130
19	1.129	1.125	1.123	1.121	1.118	1.116	1.114	1.112	1.110	1.107
20	1.106	1.103	1.101	1.099	1.096	1.094	1.092	1.090	1.088	1.085
21	1.084	1.081	1.079	1.077	1.075	1.073	1.070	1.068	1.066	1.064
22	1.062	1.060	1.058	1.056	1.053	1.051	1.049	1.047	1.045	1.043
23	1.041	1.039	1.037	1.035	1.033	1.031	1.028	1.026	1.024	1.022
24	1.020	1.018	1.016	1.014	1.012	1.010	1.008	1.006	1.004	1.002
25	1.000	0.998	0.996	0.994	0.992	0.990	0.988	0.986	0.984	0.982
26	0.980	0.978	0.976	0.974	0.972	0.971	0.969	0.967	0.965	0.963
27	0.961	0.959	0.957	0.955	0.953	0.951	0.949	0.947	0.946	0.944
28	0.942	0.940	0.938	0.936	0.934	0.932	0.931	0.929	0.927	0.925
29	0.923	0.921	0.919	0.918	0.916	0.914	0.912	0.910	0.908	0.907
30	0.905	0.903	0.901	0.899	0.898	0.896	0.894	0.892	0.890	0.889

care. The precision and accuracy with which specific conductance could be determined depends mainly on the type and condition of the equipment used, the accuracy of control and measurement of the test solution temperature, and on how much care is exercised in instrument calibration and in taking the actual readings. It is not unreasonable to expect an accuracy of within about 1% or better and a within-laboratory relative standard deviation of about ±0.2% or less if proper equipment and techniques are employed. An accuracy of within about ±3% is considered acceptable for most routine work.

The mean relative standard deviation for three different analysts was determined to be ±0.33% at the 300 μS/cm level. The Water Quality Branch *Analytical Methods Manual* records a relative standard deviation of ±0.5% at the 520 μS/cm level for a single lab.

d. Measurement of Specific Conductance by Automated Methods

There are several automated procedures in use for measurement of specific conductance, either singly or simultaneously with other determinands. Sekerka and Lechner,[19] for example, describe a relatively simple arrangement for the simultaneous determination of water hardness, pH, and conductance. This procedure is advantageous where there are large numbers of samples to be analyzed at any one time.

IX. DISSOLVED AND PARTICULATE MATTER

The physical, chemical, and biological characteristics of water and wastewater are determined by the types and amounts of particulate matter and dissolved substances present, which include mineral constituents such as sodium, potassium, calcium, and magnesium salts. Standard Methods[2] uses the term residue to denote "solid matter suspended or dissolved in water or wastewater". The most convenient way to separate the two phases is by filtering through a filter of suitable pore size. The fraction remaining on the filter medium is called suspended solids or, more appropriately, nonfilterable residue. The filtrate contains the "dissolved" solid fraction otherwise called filterable residue or total dissolved solids (TDS). It is suggested by some that separation be done by centrifugation, but we do not recommend this approach except in cases where sample filtration is impracticable, such as in the presence of large amounts of colloidals. In such cases, when sufficent sample is available, the use of an ultracentrifuge would be preferred for separation of the particulate matter.

A number of factors, including the physicochemical properties of the suspended material, the pore size, area and thickness of the filter, and the concentration of suspended matter present, influence the separation of the filterable and nonfilterable residue of a water sample. In particular, the choice of filter pore size, which has been the subject of much discussion, is a very important criterion in defining the two parameters. Glass fiber filters are the ones generally used, but these have no definitive pore size and their average pore size is of such order that the very fine particles, colloidals for example, are liable to pass through. However, it is now possible to get glass fiber filters with pore sizes in the range of 0.7 to 0.9 μm. Whether or not the effective pore size of the filter remains constant throughout sample filtration is another question. The type and concentration of the suspended matter will determine also what volume of sample that could be filtered before blockage of the filter occurs.

Dissolved and particulate matter may enter surface water via atmospheric outfalls, municipal and industrial effluents, agricultural runoffs, soil erosion, or by other natural hydro-geological interactions. As already noted, suspended particulate matter bears some relationship with turbidity. Suspended solids may consist of living and dead organic matter as well as minerals, such as plants and animals, silt, and eroded soil. Thus particulate organic carbon and nitrogen data are being used extensively in studying physical, chemical, and biological processes in aquatic ecosystems. Sewage and industrial waste discharging into surface water systems are likely to add significant amounts of solids to these waters, which in turn is likely to destroy aquatic life by cutting off sunlight and by silting over benthic fauna and flora. High suspended solids render water aesthetically unsatisfactory for recreational uses and causes serious problems if used for certain domestic and industrial purposes. For example, they cause excessive loads on water treatment plants and cause plugging of water intake pipes or of the nozzles of water sprinklers. It has been shown too that suspended solids may act as a carrier of toxic chemicals including heavy metals and trace organics. Dissolved solids on the other hand, relate to the concentration of dissolved salts in, and hence the specific conductance of, a given sample.

The TDS value gives a reasonably good indication of the suitability of a water type for a particular use. Water high in TDS is less palatable than water with moderate mineral content and may induce an unfavorable physiological reaction when consumed. For example, water with excessive amounts of sulfates may have a laxative effect on man and animal while high nitrate concentration may cause irritation of the mucous membrane of the stomach. The Department of National Health and Welfare, Canada,[5] recommends a maximum acceptable TDS concentration of 500 mg/ℓ for Canadian drinking water. Water intended for industrial use must also not exceed certain limits set for TDS. Boiler scale formation and accelerated corrosion are due in part to excessive TDS. High dissolved solids may also interfere with the clarity, color, and taste of certain manufactured products. Industrial users of water generally prescribe limits of 1000 mg/ℓ TDS, but the limits, naturally, would depend on each particular requirement and are highly variable. Limits for suspended solids or nonfilterable residue for most domestic, agricultural, and industrial applications vary from 0 to about 25 mg/ℓ depending on the specific purpose.

The TDS value determines in general terms the physico-chemical properties of water, which, as shown in Table 13, may be categorized in a general way according to TDS Concentraton. Because of the relative ease of making turbidity measurements, one is likely to succumb to the temptation to relate the nonfilterable residue, expressed in mg/ℓ, with the measured turbidity of a sample. While turbidity could be used as some sort of empirical measure of nonfilterable residue, any attempt to establish a correlation factor must be restricted to specific cases where the validity and limitations of such a factor have been established through data derived from several successive measurements of the two parameters for the same sample type. The fact that turbidity data, being dependent on a number of variables, is so nebulous, complicates matters even further.

Similarly, conversion of the more easily determined and more accurate specific conductance data to TDS data is becoming increasingly popular. In practice however, the TDS value obtained by evaporating and weighing the dried filterable residue is not likely to equal exactly that deduced from conductivity measurements, or by summation of the mineral salt content determined individually. This disparity is attributed to the inherent errors associated with each determination. For example, the salts may undergo changes during the evaporation and drying process. These and other factors are discussed in greater detail below. Conductivity results, on the other hand, are dependent on a number of factors including sample pH, the valency of the ions in solution and the degree of dissociation of the dissolved salts.

Direct determination of the concentration of dissolved and particulate matter which, henceforth, will be referred to as filterable and non-filterable residue, are generally accomplished gravimetrically. A measured volume of the well-shaken homogeneous sample is filtered through a 0.45 μm membrane filter or glass fiber filter, evaporated, and then dried and weighed repeatedly until the weight remains constant.

The choice of the 0.45 μm membrane filter is preferred to the glass fiber for the following reasons: more of the finer suspended particles are likely to be separated, the pore size of the membrane filter is more consistent, and it brings the filterable residue data more in line with the true concentration of the dissolved constituents of the sample. Its main drawback is that it plugs much more quickly than the glass fiber filter, and it may interfere with the fixed nonfilterable residue determination, which necessitates application of a correction to the weights determined. Residue data may relate specifically to six distinct parameters, depending on what fraction of sample has been measured and the procedure followed. These are as follows:

1. Filterable residue. This gives a measure of the amount of dissolved salts present after a measured volume of the filtered sample is evaporated, dried, and weighed.
2. Filterable residue, fixed. This is the residue remaining after subjecting the filterable residue fraction, as determined above, to a temperature of about 550°C to 650°C for approximately 1 hr before being weighed.
3. Nonfilterable residue. This is the residue remaining on the filter medium after it is dried to constant weight.
4. Nonfilterable residue, fixed. This is the residue remaining after the nonfilterable

Table 13
DISTINCTION BETWEEN WATER
TYPES ACCORDING TO TDS
CONCENTRATION

mg/l TDS	Water type
0—1000	Fresh, non-saline
1000—3000	Slightly saline (brackish)
3000—10,000	Moderately saline
10,000—100,000	Saline
100,000 and above	Brine

fraction, as determined above, is subjected to a temperature of about 550°C to 650°C for approximately 1 hr before being weighed.

5.   Total residue. This is equivalent to the sum of the filterable and nonfilterable residue fractions, determined either separately as above or by evaporating a measured volume of a homogeneous unfiltered sample to dryness and weighing to constant weight.

6.   Total residue, fixed. This is equivalent to the sum of the fixed filterable and nonfilterable residue fractions determined either separately as above or by evaporating a measured volume of a homogeneous unfiltered sample to dryness, then subjecting it to a temperature of about 550°C to 650°C for approximately 1hr before being weighed.

Differences between the fixed residue and residue data give a measure of the "loss on ignition" which serves as an indication of the amount of organic matter present. This at best is no more than an approximation and is sometimes evaluated on the basis of the color changes taking place in the evaporation and ignition products. For example, a colorless residue after evaporation may, upon ignition, remain colorless or change to a pale brown or intense black color depending on the relative concentration of organic matter present.

As already noted, a number of factors, including filter pore size, as well as the chemical and physical nature and concentration of the dissolved and particulate matter present, govern interpretation of the final results. The sample evaporation process, and specifically the drying temperatures, is very important also. The residue remaining after evaporation is likely to differ appreciably in its composition from the filterable and/or nonfilterable material actually present in the first case. The extent of change will depend on how much of the volatile compounds is lost during evaporation, as well as the temperature and duration of drying. Drying temperatures in the range of 103 to 105°C and 178 to 182°C are common, each having its own merits. Residues dried at the lower temperature tend to retain water of crystallization as well as some mechanically occluded or insterstitial water. Bicarbonates may dissociate and convert to carbonates by loss of CO_2. Ammonium carbonate, if present, may decompose at temperatures below 105°C. Removal of occluded water is only slight at 103 to 105°C, making the attainment of constant weight very slow. On the other hand, loss of volatile organic matter is drastically reduced, if not eliminated, at the lower drying temperatures.

Residues dried at the higher temperature tend to retain some water of crystallization, but most, if not all, of the mechanically occluded water is expelled. Volatile organics are partly lost and bicarbonates are converted to carbonates with even the potential of partial decomposition of carbonates, ammonium carbonate for example, to oxides or basic salts. Chloride and nitrate salts may be lost also. Magnesium chloride, for instance, is likely to hydrolyze and be converted to the hydroxide. Generally speaking, samples dried at 180°C yield results that agree better with those obtained by summation of the mineral salt content determined individually than the values derived by drying the residue at the lower temperature.

The fixed residue results are subject to considerable error due to the possible loss of CO_2, the decomposition and/or volatilization of mineral salts during ignition, and the hygroscopicity of the residue due to the presence of calcium oxide in the ash. Accordingly, the loss on ignition results give only an approximation of the amount of organic material present. A better approximation of this parameter may be made by determining the total organic carbon content, which itself, can be measured much more quickly.

In general, the following statements apply to residue measurements. Some evaporation and ignition products, particularly from highly mineralized samples high in calcium, magnesium, chloride, or sulfate content, are likely to be hygrosopic and should be weighed as quickly as possible. Despite the drying process, some residues may contain liquids such as glycerol and sulfuric acid, which may dissolve some of the salts present. Carbonates, bicarbonates, organic matter, expulsion of interstitial water and/or water of hydration, nitrites, nitrates, chlorides, and sulfates are among the list of possible causes of interferences. Dissolved substances are likely to precipitate or vice versa. For instance, dissolved iron or manganese may precipitate on sample exposure to the air. Also particulate matter may be lost by adsorption onto the container walls during sample storage.

Because of the foregoing and other practical constraints, residue data are less likely to meet all of the precision and accuracy criteria that usually apply to other analytical methods. Whatever the situation, however, it is imperative that the analyst record and report the critical features (filter medium used, drying and ashing temperatures, etc.) of the procedure followed; and the data user must take all of these factors into consideration when interpreting the results.

A. Filterable Residue

The method being described is applicable to potable, surface, ground, and waste waters over a practical range of about 5 or 10 mg/ℓ to about 20,000 mg/ℓ residue. The method does not include water constituents that are volatile at the boiling temperature or at the drying temperature employed.

A suitable volume of a well-mixed sample is filtered through 0.45 μm membrane filter or glass fiber filter. The filtrate is evaporated and dried to constant weight at 103°C to 105°C or 178°C to 182°C. The dried residue may be hygroscopic and must be weighed as quickly as possible. Depending on the sample composition, one or more of the factors discussed above may affect the accuracy and precision of the results. If too large a sample volume is filtered, it may result in too much residue remaining in the evaporating dish. This may cause formation of a crust trapping water underneath and thus prolonging the process of drying to constant weight. It is best to take a sample aliquot such that the residue is around 200 mg.

1. Equipment

The equipment normally required includes the following: membrane or glass fiber filter discs of about 4.7 or 4.25 cm diameter; filter holder and membrane filter funnels; pressure or vacuum filtration apparatus; evaporating dishes of approximately 100 mℓ to 200 mℓ capacity and made of porcelain, Vycor®* or platinum; a temperature-controlled water bath or electric hot plate; a temperature-controlled drying oven; desiccators; a muffle furnace; and an analytical balance of about 200 g weighing capacity and capable of weighing to 0.1 mg.

---

* produced by Corning Glass works, Corning, N.Y.

2. Procedure

If a glass fiber filter is used, prepare it as follows: place the disc on the membrane filtration apparatus. Apply sufficient vacuum or pressure and wash the disc three times with about 20 ml of distilled water each time. After completion of the final wash allow all excess water to drain out by continuing to apply the vacuum or pressure for a few minutes longer, and then discard the washing. Dry for about 1 hr in an oven set at about 104°C or 180°C depending on the temperature at which the residue is to be dried, and then store in a desiccator. If the membrane filter is used, wash two or three times with distilled water before filtering the sample.

Prepare the evaporating dishes by heating at 600 ± 25°C in a muffle furnace for about 1 hr. Allow to cool partially in the air and then put into a desiccator and keep stored therein until ready for use. The dishes must be weighed just before use.

Assemble the filtering apparatus and turn on the vacuum/pressure apparatus. Shake the sample thoroughly to dislodge any particles clinging onto the side walls of the container and to obtain a homogeneous sample aliquot. Using a graduated measuring cylinder, immediately measure a suitable volume (100 to 500 ml depending on the filterable residue content) of the well-mixed sample to be filtered. Filter all of the measured sample through the membrane or glass fiber filter and continue to apply the suction or pressure for about 2 or 3 min after filtration is complete to remove as much water as possible.

Place the weighed evaporating dish on the water bath or hot plate and transfer successive suitable portions of the filtrate into it. Evaporate and keep adding the remaining sample until completely transferred. Rinse the vessel (usually the graduated measuring cylinder) that contained the sample with turbidity-free deionized distilled water and pour washings into the dish. Evaporate contents to near dryness while ensuring that none of the dish contents are lost due to splattering.

Place the dish containing the residue into an oven preset at either 104 ± 1°C or at 180 ± 2°C and dry residue for at least 1 hr. Cool dish and contents in a desiccator and weigh. Repeat the drying and weighing cycle until the weight is constant or the weight loss is less than 0.5 mg. Replace the dish residue into the desiccator and save for determination of filterable fixed residue.

3. Calculation

Calculate the filterable residue in mg/ℓ from the formula;

$$\text{filterable residue} = \frac{(W_2 - W_1)\,1000}{V}\ \text{mg/ℓ}$$

where W_1 = weight of the empty dish, W_2 = weight of dish and dried residue, and V = volume (ml) of filtered sample evaporated. The residue weight given by the difference between W_2 and W_1 is to be expressed in mg when applying the formula. Record and report the temperature at which the residue has been dried and the type of filter used.

4. Precision and Accuracy

For a variety of reasons, the precision and accuracy of this method is at best only fair. In general, however, observation of the following points should improve the precision and accuracy of the results. (1) It is important to keep the samples under close scrutiny during evaporation and particularly towards the end in order to guard against losses by splattering. (2) Because, in practice, several samples are likely to be evaporated concurrently, care should be exercised to ensure that the remaining portions of the samples are poured into the correct dish while being evaporated. (3) During evaporation, it is best to heat the dish just sufficiently to effect evaporation without boiling the sample. (4) Care should be taken to guard against the possibility of iron being deposited on the walls of the vessel containing the filtrate when samples containing iron compounds in sufficient amounts are being filtered or transferred into the evaporating dish.

Data furnished by Standard Methods[2] indicate that a between-lab relative standard deviation of 9.7% was determined for 18 laboratories measuring a synthetic sample containing 134 mg/ℓ filterable residue.

B. Fixed Filterable Residue

The filterable residue is heated to approximately 600°C whereby substances that are volatile at this temperature are expelled. The remaining residue is weighed to determine the fixed residue content and/or loss on ignition. The ignition loss gives a rough approximation of the amount of organic matter present in the filtered sample. This test would logically be done after the filterable residue of the sample has been determined. It is subject to the potential interferences noted under the filterable residue method and more, as discussed above.

1. Equipment

The equipment is similar to that used for filterable residue.

2. Procedure

Turn on the muffle furnace and maintain at 600 ± 25°C. Use a pair of platinum-tipped tongs to place the dish containing the dried filterable residue, after it has been weighed to constant weight, into the furnace and leave for about 30 min. Allow dish and contents to cool partially in the air until most of the heat has been dissipated and then transfer into a desiccator. Weigh the cooled dish.

3. Calculation

Calculate the fixed filterable residue from the formula:

$$\text{fixed filterable residue} = \frac{(W_3 - W_1)\ 1000\ \text{mg}/\ell}{V}$$

where W_1 = weight of empty dish, W_3 = weight of dish and residue after ignition, and V = mℓ of sample used.

The weight loss on ignition gives a measure of the filterable volatile residue, which may be calculated from the formula:

$$\text{filterable volatile residue} = \frac{(W_2 - W_3)\ 1000}{V}\ \text{mg}/\ell$$

where W_2 is the weight of the dish and residue before ignition.

4. Precision and Accuracy

A coefficient of variation of about 7 to 10% is considered reasonable for this test.

C. Nonfilterable Residue

The following method is applicable to potable, surface, ground and wastewaters over a practical range of about 5 or 10 mg/ℓ to about 20,000 mg/ℓ residue. The method does not include nonfilterable residue that is volatile at the drying temperature.

The sample is filtered through a membrane or glass fiber filter; the particulate matter remaining on the filter is dried at either 140 ± 1°C or 180 ± 2°C and then weighed.

Buoyant floating particles, such as leaves and sticks or large particulate agglomerates that cannot be dispersed throughout the sample by vigorous shaking, are considered to be extraneous material and should be excluded. The volume of sample to be filtered should be chosen judiciously since too much residue on the filter disc may entrap water and prolong the drying process. Airborne particulate matter may interfere during the filtration step specifically. Some of the particulate matter may adhere to the walls of the filtration cup and must be washed down with filtered distilled water. Because of the impracticability of preserving the sample, undue delays in measurements may cause errors. For example, the results are liable to be biased high if samples with high nutrient contents are left standing for lengthy periods due to algal growth or mold formation.

1. Equipment

The equipment is similar to that used for filterable residue, except that a filter disc support is required also. Platinum, porcelain, or Vycor® evaporating dishes, or aluminum or stainless steel planchets are convenient disc supports.

2. Procedure

The filter disc used to filter the sample for the filterable residue may be used for this test. Otherwise, proceed as follows.

If the glass fiber filter is used, prepare it as described above. After the excess water has been drained from the filter, use a pair of clean forceps to remove it from the filter holder and place it into the evaporating dish or on the planchet. Place the dish or planchet with the filter into a muffle furnace and heat for about 1 hr at $600 \pm 25°C$. Allow to cool in a desiccator and keep stored therein until ready for use. Weigh the filter disc just before use.

If the membrane filter is used, follow the procedure described above, but dry in an oven either at $104 \pm 1°C$ or $180 \pm 2°C$ depending on the temperature at which the residue is to be dried. Care should be taken when preheating the membrane filters not to exceed the manufacturer's specifications. Cellulose nitrate membrane filters are thermally resistant up to about 130°C while the cellulose acetate filters are resistant up to about 180°C. If the filters show any signs of deterioration when heated to the higher temperature, then preheat at a lower safe temperature.

Set up the vacuum or pressure filtration assembly. Use a pair of clean forceps to transfer a pretreated, weighed filter disc onto the membrane filter holder. Shake the sample vigorously to ensure dislodgment of any particulate matter clinging onto the container walls and the transference of a well-mixed aliquot into the graduated measuring cylinder. Filter a suitable measured volume of the sample depending on the nonfilterable residue content. Wash the measuring cylinder and the side of the filter funnel well with filtered distilled water while continuing filtration in order to ensure that all of the particulate matter contained in the sample aliquot is completely transferred onto the filter disc.

Using a pair of forceps, carefully pick up the filter disc by its edge and transfer it from the filter holder into the evaporating dish or onto the planchet. When the membrane filter is used, it is best to weigh it directly in a porcelain, Vycor® or platinum dish in order to facilitate determination of fixed nonfilterable residue later. Place the supported filter disc into a drying oven preset at $104 \pm 1°C$ or $180 \pm 2°C$ and dry for about 1 hr. Cool in a desiccator and then weigh. Repeat the drying and weighing cycle until the weight is constant or the weight loss is less than 0.5 mg. Store the filter disc containing the residue in a desiccator for the determination of fixed nonfilterable residue if it is not feasible to proceed immediately.

3. Calculation

Calculate mg/ℓ nonfilterable residue from the equation:

$$\text{nonfilterable residue (mg/\ell)} = \frac{(W_2 - W_1)\,1000}{V}\ \text{mg/\ell}$$

where W_1 = initial weight of filter, W_2 = weight of filter and residue, and V = mℓ of sample filtered. Record and report the temperature at which the residue has been dried and the type of filter used.

4. Precision and Accuracy

The precision of the method is particularly bad at the lower end of the range. According to Standard Methods[2] coefficient of variation at around the 15 mg/ℓ residue level was determined to be about 33%. Normally, however, the precision of the test should improve considerably at higher residue values. Apart from the potential sources of errors already discussed above, one of the main factors affecting the precision and accuracy of the results is the problem of getting homogeneous aliquots that are truly representative of the sample. Accumulation of airborne particulate matter on the filter during or after sample filtration may also cause errors.

D. Fixed Nonfilterable Residue

The nonfilterable residue remaining after its determination is heated to approximately 600°C, whereby substances that are volatile at this temperature are expelled. The remaining residue is weighed to determine the fixed residue content and/or loss on ignition. The ignition loss gives a rough approximation of the amount of organic matter present in the nonfilterable fraction of the sample. The measurement is made after the determination of nonfilterable residue and is subject to the potential interferences noted for that procedure and others as discussed above.

1. Equipment

The equipment required is the same as for the nonfilterable residue procedure.

2. Procedure

Turn on the muffle furnace and maintain it at 600 ± 25°C. Place the supported filter with the weighed residue into the furnace and leave for about 30 min. Allow filter and residue to cool partially in the air until most of the heat has been dissipated, transfer into a desiccator and weigh when cool.

3. Calculation

Calculate mg/ℓ fixed nonfilterable residue from the equation:

$$\text{fixed nonfilterable residue} = \frac{(W_3 - W_1)\,1000}{V}\ \text{mg/\ell}$$

where W_1 = initial weight of filter, W_3 = weight of filter and residue, and V = mℓ of sample filtered.

Determine the loss on ignition or the nonfilterable volatile residue from the equation:

$$\text{nonfilterable volatile residue} = \frac{(W_2 - W_3)\,1000}{V}\ \text{mg/\ell}$$

where W_2 = weight of filter and residue, W_3 = weight of filter and residue after ignition, and V = mℓ of the sample filtered.

When the membrane filters are used for this test, they are completely ashed at 800°C, and a weight correction should be applied. However, the relatively large weight of the dish containing the filter and ashed residue must be taken into account, and the results, being derived by weight difference, are subject to greater errors than would otherwise be the case. Since some manufacturers provide specification data on filter weight and ash weight per unit cm² of filter, the correction can be made by using the data provided.

Assume for example that a filter of area 1 cm² weighs w mg and has an ash content of 0.005 mg per cm;² then total ash weight of paper = 0.005a mg. Let the equivalent loss in weight per filter disc = (w − 0.005a) mg = w′ mg.

Now let the initial weight of dish + filter = W_1 mg; the weight of dish + filter and residue before ignition = W_2 mg; and W_3 = weight of dish + filter and residue after ignition.

$$\text{then fixed nonfilterable residue} = \frac{(W_3 + w' - W_1)\,1000}{V} \text{ mg/}\ell$$

$$\text{and nonfilterable volatile residue} = \frac{(W_2 - W_3 - w')\,1000}{V} \text{ mg/}\ell$$

The value of w′ will remain reasonably constant for a particular brand of membrane filters of specific dimension and need not be recalculated for every filter disc used.

4. Precision and Accuracy

The statements made for the nonfilterable residue determination apply here also. According to data furnished by Standard Methods,[2] coefficient of variation of about 6.5% is attainable at the 170 mg/ℓ volatile residue level.

E. Determination of Total Residue, Total Fixed Residue, and Total Volatile Residue

These parameters may be deduced by summation of the respective residue values determined for the filterable and nonfilterable fractions as described above. The parameters may be measured directly as follows.

1. Determination of Total Residue

The method is applicable to potable, surface, ground, and wastwaters over a practical range of about 5 or 10 mg/ℓ to about 20,000 mg/ℓ residue. The method does not include water constituents that are volatile at the boiling temperature or at the drying temperature employed. A measured volume of the well-mixed sample is evaporated in a preweighed evaporating dish, and the dried residue measured by weight difference.

Buoyant floating particles such as leaves and sticks or large particulate agglomerates that cannot be dispersed throughout the sample by vigorous shaking are considered to be extraneous material and should be excluded from the sample aliquot taken. Floating oil and grease, if present, should be first dispersed with a blender before extracting aliquots for evaporation. Too large a volume of sample should not be evaporated lest the formation of a crust over entrapped water necessitates prolonged residue drying. Airborne particles may affect the results if not prevented from entering the dish during sample evaporation. It is possible to avert this by use of a dust shield. If samples are stored for lengthy periods, those high in nutrient concentration are particularly likely to show signs of molds or algal growth, which will affect the results.

a. Equipment

The equipment required is the same as for the filterable residue determination excepting the filtration equipment.

b. Procedure

Heat a number of clean evaporating dishes at 600°C ± 25°C in a muffle furnace for about 1 hr. Allow to cool partially in the air to dissipate most of the heat, then put into a desiccator. Weigh the preheated evaporating dish just before use.

Shake the sample container vigorously for several minutes and immediately measure a suitable volume into the graduated measuring cylinder. It is best to select a volume, estimated from the sample specific conductance and turbidity reading, that will give residue in the range of about 25 to 250 mg. Place the weighed evaporating dish on the water bath or hot plate and fill to within about 6 mm from the top. Apply heat to evaporate the sample, but do not boil. Keep adding the remaining sample into the evaporating dish as the volume reduces until the total volume measured has been transferred. Rinse the graduated cylinder two or three times with turbidity-free deionized distilled water, and pour washings into the evaporating dish. Allow contents of dish to evaporate to near dryness ensuring that none of it is lost due to splattering. This is particularly likely to occur towards the end.

Transfer the dish and residue into an oven preset at either 104°C ± 1°C or at 180°C ± 2°C and allow to dry for at least 1 hr.

Cool dish and contents in a desiccator and weigh. Repeat the drying and weighing cycle until the weight is constant or until the loss of weight noted is no greater than 0.5 mg or 4% of the previous weight, whichever is less. Replace into the desiccator and save for total fixed residue determination.

c. Calculation

Calculate the total residue in mg/ℓ from the equation:

$$\text{total residue} = \frac{(W_2 - W_1)\,1000}{V}\ \text{mg/ℓ}$$

where W_1 = weight of dish empty, W_2 = weight of dish plus dried residue, and V = mℓ of sample evaporated. Record and report the temperature at which the residue has been dried.

d. Precision and Accuracy

A coefficient of variation of about 5 to 10% is attainable.

2. Determination of Total Fixed Residue and Total Volatile Residue
a. Procedure

Place the dish and residue remaining from the total residue determination into a muffle furnace preset at 600°C ± 25°C, and leave for about 30 min. Allow dish and contents to cool partially in the air until most of the heat has been dissipated, transfer into a desiccator, and weigh when cool.

b. Calculation

Calculate the total fixed residue from the equation:

$$\text{total fixed residue} = \frac{(W_3 - W_1)\,1000}{V}\ \text{mg/ℓ}$$

where W_1 = weight of dish empty, W_3 = weight of dish plus residue after ignition, and V = mℓ of sample evaporated.

Calculate total volatile residue or total loss on ignition from the equation:

$$\text{total volatile residue} = \frac{(W_2 - W_3)\,1000}{V}\ \text{mg/ℓ}$$

where W_2 = weight of dish plus residue before ignition.

3. Determination of Residue by the Ion-Exchange Method

Filterable residue data may be obtained and/or checked by means of ion-exchange techniques. The method is useful as a check on the chemical analytical data produced from dissolved minerals determination. It is also applicable for routine measurements of the dissolved mineral content of samples having more or less similar known characteristics. The method entails the passage of the sample over a strongly acidic cation-exchange resin whereby the cations in the sample are replaced by an equivalent amount of hydrogen ions. The resulting acid formed is titrated against standard sodium hydroxide. If the total alkalinity of the sample was previously determined, the filterable residue and fixed filterable residue may be calculated from the appropriate formulae. The following description is based on that given in Standard Methods.[2]

a. Equipment and Reagents

A variable speed magnetic or mechanical stirrer is needed. The titration requires a 10 ml burette with 0.05 ml graduations. Standard solutions of 0.01 M sulfuric or 0.02 M hydrochloric acid and 0.02 M sodium hydroxide are required. Analytical grade strongly acidic cation exchange resin such as Dowex® 50 W-X 8 or Amerlite® IR-120 is used.

b. Procedure

Pipette a sample aliquot containing about 0.1 to 0.2 meq cations into a 250 ml Erlenmeyer flask. Add enough distilled ater to make volume up to 100 ml Add 2.0 g of the resin, and stir for about 15 min at a moderate speed. Filter through a plug of glass wool placed in the neck of a glass funnel, and wash twice with about 15 ml of distilled water each time. Alternatively, the above steps may be replaced by setting up a resin column similar to that shown in Figure 12.

While stirring the combination of filtrate and washings, titrate against the standard NaOH to pH 4.5, using a pH meter or a few drops of methyl orange indicator to denote the end point. Pipette a suitable volume of sample into another Erlenmeyer flask. Determine the sample alkalinity by titrating it against the standard acid solution using a pH meter or a few drops of methyl orange indicator to denote the end point at pH 4.5. Keep the sample stirred continuously while being titrated. Calculate the total alkalinity T of the sample in meq/l.

c. Calculation

Calculate the sample cationic concentration C in meq/l from the expression:

$$C = \frac{V_{NaOH} \times M \times 1000}{V_s} + T$$

where V_{NaOH} = ml of standard NaOH used to titrate the sample after ion exchange, M = molarity of the standard NaOH, and V_s = volume (ml) of sample used for ion exchange.

The filterable residue content is given by the equation:

$$\text{filterable residue} = (C - T) F + 50 T$$

where F is a factor varying between 70 and 90 depending on the water type and the resin used for ion-exchange. When sample contains significantly greater amounts of chlorides than sulfates, F has a value of about 70, and when sulfates and/or bicarbonates predominate, or the organic content is unusually high, F has a value of approximately 90. The use of an average F value of 80 may suffcice for most natural waters however.

Conversely, a more accurate F value may be derived for water with the same general characteristics from direct measurements of the filterable residue (R) dried at 104 ± 1°C and the fixed filterable residue (R_x).

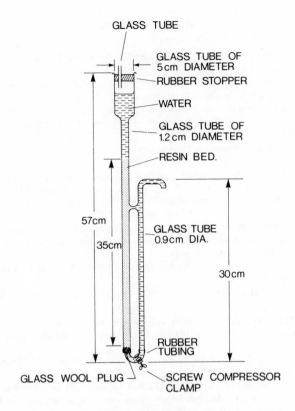

GLASS TUBE

GLASS TUBE OF
5cm DIAMETER

RUBBER STOPPER

WATER

GLASS TUBE OF
1.2 cm DIAMETER

RESIN BED.

57cm

35cm

GLASS TUBE
0.9cm DIA.

30cm

RUBBER
TUBING

GLASS WOOL PLUG — SCREW COMPRESSOR
CLAMP

ION−EXCHANGE COLUMN

NOTE:− THE ABOVE GIVEN DIMENSIONS ARE NOT CRITICAL AND
SERVE AS GUIDELINES ONLY. SMALLER COLUMNS
MAY BE USED IF WARRANTED BY THE SAMPLE SIZE.

FIGURE 12. Ion-exchange column.

Thus we have:

$$F_1 = \frac{R - 50\,T}{C - T}$$

and

$$F_2 = \frac{R_x - 50T}{C - T}$$

With F_1 and F_2 established for given types of water, the filterable residue and fixed filterable residue values may be calculated more accurately from the ion-exchange method using the following equations:

$$\text{Filterable residue (mg/}\ell\text{)} = (C - T)\,F_1 + 50\,T$$

$$\text{Fixed filterable residue (mg/}\ell\text{)} = (C - T)\,F_2 + 50\,T$$

REFERENCES

1. **Leithe, W.,** *The Analysis of Organic Pollutants in Water and Waste Water,* Ann Arbor Science Publishers, Mich., 1973, Chap. 4.
2. Standard Methods for the Examination of Water and Wastewater, 14th ed., APHA-AWWA-WPCF, American Public Health Association, Washington, D.C., 20036, 1975, 123.
3. **Hazen, A.,** The measurement of the colors of natural waters, *J. Am. Chem. Soc.,* 18, 264, 1896.
4. *Analytical Methods Manual,* Water Quality Branch, Inland Waters Directorate, Environment Canada, Ottawa, 1979.
5. *Guidelines for Canadian Drinking Water Quality,* 1978, Department of National Health and Welfare, Ottawa, 1979.
6. **Kullenberg, G.,** Observed and computed scattering functions, in *Optical Aspects of Oceanography,* Jerlov, N. G. and Steemann Nielsen, E., Eds., Academic Press, London, 1974, Chap. 2.
7. **Baylis, J. R.,** Turbidimeter for accurate measurement of low turbidities, *J. Indust. and Eng. Chem.,* 18, 311, 1926.
8. **Bates, R. G.,** *Determination of pH, Theory and Practice,* John Wiley & Sons, N.Y., 1965.
9. **Ryznar, J. W.,** A new index for determining amount of calcium carbonate scale formed by a water, *J. Am. Water Works Assoc.,* 36, 472, 1944.
10. *CRC Handbook of Chemistry and Physics,* 59th ed., Weast, R. C. and Astle, M. J., Eds., CRC Press, Boca Raton, Fla., 1978, D-188.
11. **Langelier, W. F.,** Effect of temperature on the pH of natural waters, *J. Am. Water Works Assoc.,* 38, 179, 1946.
12. **Philbert, F. J. and Traversy, W. J.,** Methods of sample treatment and analysis of Great Lakes water and precipitation samples, in *Proc. 16th Conf. Great Lakes Res.,* Int. Assoc. Great Lakes Res., 1973, 294.
13. Technicon Corp., Industrial Systems, Analytical methodology, No. 378-75WE, Technicon Instruments Corp., Tarrytown, N.Y. 10591.
14. **Philbert, F. J., Smith, M. N., and Elkei, O.,** Evaluation of an automated ion-selective electrode system for fluoride, ammonia, and pH determinations in water, in *Advances in Automated Analysis,* Technicon International Congress, 1976, Industrial Symposia, 2, Barton, E. C., Conetta, A., Du Cros, M. J. F., Kabot, F. J., Love, R. K., Maddix, C., Murdock, J. F., Perlman, L., Eds., Mediad Inc., Tarrytown, N.Y., 1977, 43.
15. **Rodgers, G. K.,** Temperature adjustment of conductivity measurements in waters of the Great Lakes, Physical Limnology Report PL-31, Great Lakes Institute, University of Toronto, Canada, 1962.
16. **Hills, G. J.,** *Electrochemistry, Specialist Periodical Reports,* 1, The Chemical Society, Burlington House, London WIV OBN., 1970, 79.
17. **Robinson, R. A. and Stokes, R. H.,** *Electrolyte Solutions,* 2nd ed., revised, Butterworths, London, 1965, 466.
18. **Hamer, W. J. and Wood, R. E.,** Electrolytic conductivity and electrode processes, in *Handbook of Physics,* 2nd ed., Condon, E. U. and Odishaw, H., Eds., McGraw-Hill, N.Y., 1967, 4-150.
19. **Sekerka, I. and Lechner, J. F.,** Automated simultaneous determination of water hardness, specific conductance, and pH, *Anal. Lett.,* 7 (6), 339, 1974.

Chapter 2

MAJOR IONS

F. J. Philbert

TABLE OF CONTENTS

I. INTRODUCTION*

Practically all of the elements or substances making up the earth's crust may be found in virtually all natural waters in varying amounts depending primarily on the source and other factors such as the physical, chemical, geological and biological interactions taking place. In the majority of cases, the major constituents, and hence the major ion concentration, are predominantly constituted of the dissolved mineral salts present in a given water.

As a general rule, elements occurring at concentration levels of about 1 mg/ℓ or higher are categorized as major constituents. Hydrogen, calcium, magnesium, sodium, potassium, carbon, nitrogen, oxygen, silicon, phosphorus, sulfur, chlorine, and possibly iron, are the main elements of the periodic table of which the major constituents of natural water may be comprised. More specifically, calcium, magnesium, sodium, potassium, carbonates, bicarbonates, chlorides, sulfates and sometimes nitrates and silicates are classified as the major constituents of most surface and ground waters. Practically every water quality monitoring program involves analysis of one or more of these parameters which are fundamentally essential for the general assessment of water quality irrespective of prevailing local conditions. The amounts of these substances present in water are governed largely by geo-chemical and other natural factors, such as the properties of drainage basins, the type of soil in the drainage area, and climatic conditions; but inputs from municipal, industrial, agricultural and atmospheric discharges could be major sources also.

The chemical composition of natural waters is governed primarily by the natural environmental factors to which the water is exposed in the hydrologic cycle. Specifically, the composition of natural water is influenced by the type and amount of soluble products of rock weathering and decomposition and by the terrain traversed by that

---

* The use of specific names of equipment and material is not to be construed as an endorsement or promotion of that product by the author or Environment Canada.

water. The chemical composition of water may be subjected to significant changes due to, for instance, chemical interaction and/or precipitation, adsorption or ion exchange, admixture of other waters, natural biological process in aquatic plants and animals, and the direct or indirect results of man's activities. Major ions analyses have been a prerequisite for hydrological investigations conducted over the years. Studies on the occurrence and movements of water from various sources depend primarily on major ions analyses. The data may be used for, among other things, the classification and assessment of water types, and provides ancillary geological information such as may be required to determine the origin and buffering capacity of water bodies. The major ions concentration gives a measure of dissolved solid contents. Monitoring/surveillance data on major ions provide information on trends and for identifying areas of non-compliance with standards and water quality objectives.

The above-noted ions fall into two groups, viz., cations which are the postively charged ions (e.g., Ca^{++}) and anions, the negatively charged ions (e.g., Cl^-). The total concentration of the dissolved major ions relates directly to the total mineral content and hence bears a strong direct correlation with the filterable residue and specific conductance of a given water.

Summation of the individually determined major ions is expected to result in a more accurate and consistent measure of the total mineralization or the filterable residue concentration of a given water than is obtainable by the gravimetric residue method. The ionic balance test, one of the most important and frequently used quality control checks in the chemical analysis of water, can be done only after all of the major ions have been determined. The test involves the summation of the cationic and anionic concentrations, expressed in milliequivalents per liter, and calculation of the percent difference between the sums. All in all, determination of major ions in a sample forms the basis for deducing several other related parameters including the sodium absorption ratio, percent sodium, sum of ionic constituents, and total hardness. It provides data that may be used by the analyst in a variety of ways for in-house quality control checks on related physical and/or chemical parameters and for determining correlations among them.

While the specific impact of the major ions on water quality is best considered separately for each component, the major ion content is particularly important when considering water required for industrial applications where the effects of corrosion and/or incrustation could be of major importance. Dissolved calcium and magnesium salts are particularly critical in the case of water intended for use in high pressure boilers. Furthermore, water with excessively high major ion concentrations are likely to be corrosive and may pose serious operational and economic problems when used in some municipal, industrial, agricultural, and manufacturing applcations. For instance, all of the major ions have been known to cause problems in the plating baths used in electrochemical and metal finishing plants. Calcium and magnesium are particularly troublesome because of their tendency to precipitate and form scales on heated surfaces or to form a sludge in the water itself.

The parameters that best relate to the major constituents of a water are calcium, magnesium, sodium, potassium, total alkalinity, acidity, chloride, sulfate, and total hardness. Each of these parameters will be considered separately below.

It is convenient to collect a single sample of about one or two liters in volume for the major ion measurements. Whether the sample is filtered or not will depend on the objectives of the sampling program. However, when a filtered sample is required, it is recommended that the sample be filtered through 0.45 μm membrane filters on site immediately after sample collection or as soon as possible thereafter if on-site filtration is not practical. Plastic or glass sample bottles have been found suitable. However, it should be remembered that water has a considerable amount of solvent action on glass and the dissolved mineral concentration is likely to increase if the sample, alkaline waters especially, is stored in nonresistant glass containers. For instance, the use of borosilicate or soda-lime glass containers to store samples intended for reactive silicates and sodium analyses may cause significant increases in their concentrations and should be avoided in such cases. Acidity and total alkalinity should be done on site immediately after sample collection. When this is not feasible, they should be determined as soon as possible after receipt in the laboratory. The samples should be transported and stored refrigerated at approximately 4°C in the interim period between sampling and analysis. Care should be taken not to freeze the samples as the results are likely to be affected (Philbert[1]). Alternatively, a separate aliquot of the sample may be preserved for acidity measurement by addition of phenolphthalein and a known amount of a standard sodium hydroxide solution in excess. The acidity may then be determined in the lab by back titration. In general, refrigeration of the complete major ion sample should be a satisfactory means of sample preservation for periods of up to about a week or so. The cations of the major ions group may be determined on separate sample fraction such as that prepared for trace metal analysis. In this case, stabilization could be satisfactorily effected for periods of up to six months or more by acidifying the sample with metal-free concentrated nitric acid to bring its pH to below two. The ionic balance calculation is, however, likely to be hampered somewhat when the cations and anions are determined on two separate sample fractions treated differently. Nevertheless, it is a good practice to do the cation determination on both the acidified and non-acidified sample fractions as a cross check to ensure that there has been no removal of cations from solution such as is likely to be the case through precipitation of dissolved components. In the case of calcium, organisms in the water may induce the precipitation of calcium as $CaCO_3$ compounds if conditions are right. The acidified sample should therefore yield more authentic results relative to the unacidified sample.

II. DETERMINATION OF CALCIUM

Calcium is one of the most abundant metals in natural surface and ground waters and exists mainly as the bicarbonate and, to a lesser degree, in the form of the sulfate and chloride. Calcium may dissolve readily from rocks or be leached from soils. In the presence of CO_2, calcium carbonate in water is dissolved, in which case the resulting buffered system is likely to maintain the pH of most natural waters somewhere between six and eight. Other sources include primarily industrial and municipal discharges. Water in contact with the earth's crust and, specifically, flowing through deposits of limestone, dolomite, gypsum, and silicate minerals such as the pyroxenes and amphiboles, may dissolve calcium ions and its salts in varying quantities, depending on its source and prevailing hydrogeologic conditions. Significant amounts of calcium may also enter water systems through run-offs of calcium salts used for road deicing during the winter season. Typically, calcium concentrations in natural fresh waters fall below 10 mg/ℓ Ca, although waters in the proximity of carbonate rocks and limestones may contain calcium ranging from about 30 to 100 mg/ℓ Ca. Waters in contact with gypsiferous shales are likely to contain several hundred mg/ℓ dissolved calcium, while seawater concentration is of the order of about 400 mg/ℓ Ca. The presence of calcium carbonate polymorphs and hydrates plays an important role in the well-known $CaCO_3$ - H_2O - CO_2 equilibrium system of natural waters. Water having an under-saturation of $CaCO_3$ is likely to promote reactions enhancing iron pick-up with consequential development of what has been referred to as "red water." An over-balance of $CaCO_3$, on the other hand, poses other problems for water users.

Calcium is one of the two principal contributors to water hardness. Large amounts of dissolved calcium create problems for domestic uses, including washing, bathing, laundering and cooking, as well as industrial problems. Although small amounts of calcium salts (calcium carbonate particularly) may produce an anticorrosion effect on metallic pipes by forming a layer of protective coating over the surface, they may also cause serious scaling in boilers, plating baths, or service pipes. The eventual formation of an incrustation on the surfaces of cooking utensils where water containing substantive amounts of calcium is used, is a very common experience. The criteria set for calcium concentration in water intended for industrial uses is of the order of 50 to 75 mg/ℓ Ca. Calcium is an essential nutritional element for animal life and it aids in maintaining the structure of plant cells and soils. A high concentration of the metal is relatively harmless to organisms, and it is believed that the toxicity effect of certain chemicals to fish may be reduced in the presence of calcium salts. On the other hand, while low calcium intake may have adverse physiological effects on animal and plant life, excessive consumption may predispose the formation of concretions in the body and particularly in organs such as the kidney and bladder (kidney and bladder stones are common examples) or it may lead to irritation of the urinary passages which attendant discomfort to one when urinating. Based on current available information, however, maximum limits for drinking water have been considered unnecessary (Department of National Health and Welfare, Canada[2]).

The choice of analytical methods for calcium determination is relatively wide. These methods fall into the general categories of (1) gravimetric, (2) colorimetric, (3) titrimetric, (4) atomic absorption spectrometry, (5) atomic emission spectrometry, (6) ion-selective electrodes, and (7) ion chromatography.

There are other analytical techniques, including neutron activation analysis, mass spectrometry, and atomic fluorescence, available for calcium determination. While these may have the desirable sensitivity and specificity, they are not convenient and cost-effective enough for routine use and are better suited for specialized applications.

The gravimetric method is based on the principle of precipitating calcium in the form of the oxalate in an acetic acid medium after removal of phosphates, silica, aluminum, iron, manganese, strontium, or suspended matter as necessary. The washed precipitate is then ignited and weighed. The method is not as precise or sensitive as the others, and the procedure is a lengthy one. It is, therefore, not favored.

The colorimetric method entails the precipitation of calcium by a known volume of a 0.1% chloranilic acid ($C_6H_2Cl_2O_4$) at a pH of between four and five. The mixture is allowed to stand for about 3 hr with frequent stirring, after which it is made up to a specific volume and then filtered. The excess chloranilic acid is determined by measuring the optical density of the solution at 535 nm and comparing the readings against a calibration curve developed by treating a series of standard calcium solutions in a similar manner. This method is also not favored because of the lengthy procedure. Brittain[3] describes an improved spectrophotometric method in which he used a 5.85×10^{-5} M solution of 1-(2-Naphtholazo-3,6-disulfonic acid)-2-naphthol-4-sulfonic acid, otherwise known as hydroxynaphthol blue (HNB), in an ammonia buffer solution of pH 10. The method, which has a detection limit of about 80 μg/ℓ Ca, is comparatively fast and is amenable to automation. It is reported, however, to be susceptible to interferences from iron (ferrous iron particularly) copper, nickel, and aluminum, which must first be removed when analyzing natural waters.

There are two basic titrimetric methods, viz., permanganate titration and EDTA complexometric titration. The atomic absorption spectrophotometric or the plasma atomic emission spectroscopic methods are preferred for general use, but when the required instrumentation is not available, the EDTA titration method, being relatively simple and fast and capable of producing satisfactory results for control and routine applications, is considered to be the next best choice. The permanganate method entails first removing potential interferences in the form of strontium, phosphates, silica, aluminum, iron, manganese, and suspended matter followed by quantitative precipitation of calcium as calcium oxalate under controlled pH conditions. The precipitate is redissolved in dilute sulfuric acid and titrated immediately against standard potassium permanganate solution. The calcium concentration is then calculated knowing the amount of titrant used to oxidize the precipitated oxalate and the calcium equivalence of each ml of the standard $KMnO_4$. Like the gravimetric method, the permanganate titration method is much too involved and lengthy and is not recommended for practical reasons. Szekeres[4] describes a variation of the EDTA complexometric titration method involving back-titration after addition of a known excess amount of standard EDTA solution. The end point detection is not interfered with by the precipitated magnesium and while the method overcomes the problems of possible interferences from magnesium, phosphate, or iron, it could hardly be considered for routine measurements because of the lengthy procedure involved.

A. EDTA Complexometric Titration Method

The EDTA complexometric method is normally applicable to the determination of calcium in most waters including natural surface and ground waters, as well as domestic and industrial waters. It covers a practical working range of about 0.5 mg/l Ca to about 100 mg/l Ca, although the lower or upper limits may be extended by use of micro titration apparatus or by appropriate dilution of the sample, respectively.

If ethylenediamine tetraacetic acid (EDTA) or its salts is added to a sample containing calcium and/or magnesium, it can be made to combine selectively with either calcium alone or with the calcium and magnesium present, depending on the pH. At a sufficiently high pH of between 12 and 13, EDTA complexation takes place with the calcium ions only, since at this pH the magnesium is precipitated as the hydroxide. When the pH is adjusted to 10, both the calcium and magnesium present are complexed sequencially in that order. The use of a suitable end-point indicator enables quantification of the calcium and magnesium concentration. A variety of indicators such as ammonium purpurate (murexide), Eriochrome® Blue Black R, solochrome dark blue (calcon) or calcon carbonic acid, is available for the titration of calcium. The calcon indicators yield more distinctive end-points than murexide and their use is recommended, particularly in cases where difficulties are experienced in accurately detecting end-points with the murexide indicator. It is possible, however, for some of these indicators be absorbed by the magnesium hydroxide precipitate formed with the resulting loss in end-point sharpness. There is also the possibility for the precipitated magnesium hydroxide to entrain some of the calcium present, in which case the results may be biased low. Eriochrome® Black T is used to indicate end-point in determining total hardness (Ca + Mg) by EDTA complexometric titration.

The method is susceptible to interferences when applied to waters of very high salt contents such as seawater and brines, or highly colored waters. Metal ions, including iron, manganese, copper, zinc, lead, cobalt, nickel, strontium, barium, and aluminum, that react with EDTA may interfere by yielding high results and/or obscuring the end-point if they are present in sufficient quantities. Orthophosphate in excess of about 1 mg/l will interfere by precipitating calcium at the pH of the test. Alkalinity in excess of about 30 mg/l may reduce the distinctiveness of the end-point. Sulfate in excess of 10,000 mg/l interferes. Also, calcium may be removed from solution through precipitation of calcium carbonate if the titration is performed too slowly or in the presence of too much calcium. It is recommended that the sample aliquot taken for titration contain not more than about 25 mg $CaCO_3$. Detection of the end-point may be obscured, if not made impossible, in highly colored samples, and the atomic absorption spectrophotometric method should be used in such cases.

The effect of heavy metal interference may be minimized by addition of hydroxyl-amine and cycanide. The presence of up to 5 mg/ℓ or more of iron and about 10 mg/ℓ of manganese, copper, lead, and Zn can then be considered to be ineffective. Manganese interference may be eliminated by addition of two or three drops of a 5% ascorbic acid solution or a few crystals of potassium ferrocyanide.

1. Equipment and Reagents

A titration assembly, consisting of a 10 mℓ microburette or of other suitable size graduated in 0.05 mℓ and equipped with a shaded incandescent lamp, is required. The use of a magnetic stirrer with white upper surface is recommended.

2 M sodium hydroxide: this solution is prepared by dissolving 80 g NaOH in about 800 mℓ deionized distilled water, cooling the solution and diluting to 1000 mℓ. The solution is stored in a polyethylene or equivalent bottle.

0.010 M standard EDTA solution: analytical reagent grade disodium EDTA dihydrate ($Na_2H_2C_{10}H_{12}O_8N_2 \cdot 2H_2O$) is dried over concentrated H_2SO_4 in a desiccator, after which 3.723 g of the salt is dissolved in deionized distilled water in a 1 ℓ volumetric flask and made up to the 1 ℓ mark. The molarity of the solution is checked by titrating 25.0 mℓ of it against standard calcium solution as described below for the sample analysis. One mℓ of this solution is equivalent to 0.4008 mg Ca, 0.243 mg Mg, or 1.0 mg $CaCO_3$. The solution, which should be stored in a polyethylene or Pyrex® glass container, is stable for several weeks. It should be checked for possible deterioration regularly and compensated for by applying a suitable factor when calculating the calcium concentration of the sample. In standardizing the EDTA, titrate against 15.0 mℓ of Ca standard pipetted into an Erlenmeyer flask and diluted to 50 mℓ with deionized distilled water.

Calculate the molarity *(M)* of the standard from the formula:

$$M = \frac{0.01 \, V_1}{V_2}$$

where V_1 = volume of standard calcium solution taken for titration, and V_2 = volume of EDTA titrant used.

400 mg/ℓ Ca stock standard calcium solution: stock standard solutions are commercially available on the market. These, however, should always be checked for their accuracy before use. To prepare a stock standard solution, a few grams of anhydrous analytical reagent grade calcium carbonate powder that preferably contains negligible amounts or no magnesium and/or heavy metals is dried by heating overnight at approximately 180°C. It is cooled in a desiccator, after which exactly 1.000 g is weighed and transferred into a 500 mℓ Erlenmeyer flask. A funnel is placed in the mouth of the flask and 1 + 1 HCl added gradually until all of the $CaCO_3$ is dissolved. About 200 mℓ of water is added and the solution boiled gently for a few minutes to expel CO_2. The solution is cooled, a few drops of methyl red is added, and the pH adjusted to an intermediate orange color by adding 3 M NH_4OH or 1 + 1 HCl as necessary. The solution is transferred quantitatively into a 1000 mℓ volumetric flask and made up to the mark with deionized distilled water. A well-stoppered polyethylene bottle is used for solution storage. Each 1.0 mℓ of the standard is equivalent to 0.4008 mg Ca or 1 mg of $CaCO_3$.

1000 mg/ℓ Mg standard magnesium solution: exactly 1.000 g of pure magnesium metal turnings is weighed out and transferred quantitatively into a 500 mℓ Erlenmeyer flask. A funnel is placed in the neck of the flask, and about 150 mℓ deionized distilled water is added. 5.0 mℓ 1 + 1 H_2SO_4 or HCl is added to the flask at the rate of about 1 mℓ at a time. After each addition of acid, the solution is mixed well allowing enough time for the reaction to subside before adding more acid. After addition of the acid is completed and the reaction has subsided, the contents of the flask is boiled for about 10 min to ensure all of the metal is completely dissolved. The solution is cooled and transferred quantitatively to a 1 l volumetric flask. The solution is made up to the 1000 ml mark with deionized distilled water. This solution contains 1000 mg/l Mg or an equivalent of 4116 mg/l $CaCO_3$.

Mixed calcium/magnesium intermediate standard solution: a mixed standard containing calcium and magnesium in a ratio comparable to the samples to be analyzed may be prepared by diluting appropriate volumes of each of the above stock standard solutions. For example, by pipetting 25.0 ml of the stock Mg into a 500 ml volumetric flask containing exactly 250.0 ml of the Ca stock standard and making up to mark with deionized distilled water, a standard containing 200 mg/l Ca and 50 mg/l Mg is obtained. Each ml of this solution now contains 0.2004 mg Ca and 0.05 mg Mg or an equivalent of 0.7058 mg $CaCO_3$. When this solution is used to standardize the EDTA solution, it should be remembered that the solution, now being only half as strong, would require only half the volume of EDTA that would otherwise be needed. Thus the molarity *(M)* of the EDTA standard is given by the formula:

$$M = \frac{0.01 \, V_1}{2V_2}$$

Assuming therefore that the EDTA is exactly 0.010 *M*, and V_2 ml were required to titrate the 15.0 ml Ca standard used,

$$V_2 \text{ m}l \text{ of EDTA} = 15.0 \times 0.2004 \text{ mg Ca}$$

$$= 3.006 \text{ mg Ca}$$

$$\text{so each m}l \text{ of EDTA} = \frac{3.006}{V_2} \text{ mg Ca}$$

$$\text{and the calculation factor f} = \frac{3.006}{V_2}$$

Ammonium purpurate (murexide) indicator: this is prepared by thoroughly mixing 1.0 g ammonium purpurate ($C_8H_8N_6O_6$) with 200 g sucrose or alternatively, using a mortar and pestle, 0.200 g is ground with 100 g powdered sodium chloride crystals. About 0.2 g of the mixture is used for each titration. During the titration, the color changes from pink to purple at the end-point. The indicator is unstable in alkaline solutions, and titration should be done immediately after its addition.

Eriochrome® blue black R indicator: Eriochrome® blue black R, otherwise called solochrome dark blue (color index No. 15705) or calcon, ($C_{20}H_{13}N_2NaO_5S$), 0.200 g, is ground with 100 g powdered NaCl and stored in a well-stoppered, wide-mouth bottle. About 0.2 g of the mixture is used for each titration. During titration, the color changes sequentially from red to purple to a bluish purple and finally to pure blue at the end-point.

Calcon carbonic acid indicator: calcon carbonic acid ($C_{21}H_{14}N_2O_7S \cdot 3H_2O$), 0.2 g, is ground with 100 g NaCl, and the mixture is stored in a dark-colored, wide-mouth bottle. About 0.2 g of the mixture is used for each titration. The end-point is reached when the solution changes to a distinctive blue color.

Suitably prepared solutions of the above indicators could also be used, but they are stable for only a few days, whereas the dry powdered mixtures are stable for several months if stored properly.

Hydroxylamine hydrochloride solution: hydroxylamine hydrochloride ($NH_2OH \cdot HCl$), 30 g, is dissolved in deionized distilled water and diluted to 1000 mℓ.

Potassium ferrocyanide: analytical reagent grade potassium ferrocyanide [$K_4Fe(CN)_6 \cdot 3H_2O$] crystals is required.

Sodium cyanide solution: sodium cyanide (NaCN), 25 g, is dissolved in deionized distilled water and diluted to 100 mℓ. This solution is poisonous and should be used with due caution. It should not be brought in contact with any acid or acidified solutions and should be used in a well-vented fume hood. Alternatively, the NaCN in the powder form (250 mg) may be added to the solution to be titrated.

2. Procedure

Pipette 50.0 mℓ of sample into a 250 mℓ Erlenmeyer flask. Use a smaller volume and dilute to 50 mℓ if the sample contains high concentration of Ca or if the total alkalinity is greater than about 500 mg/ℓ $CaCO_3$. Select a sample volume containing somewhere between 5 to 10 mg calcium for titration.

Place the flask on the magnetic stirrer and stir while adding 1 mℓ of hydroxylamine hydrochloride, followed by 1 mℓ of NaOH solution and 1 mℓ of the sodium cyanide solution, in that order.

Add about 0.2 g of the calcon carbonic acid, Eriochrome® Blue Black R, or the murexide indicator mixture and titrate immediately with standard EDTA solution.

Record the volume of titrant used to reach the end-point and determine a reagent blank by repeating the above procedure with 50.0 mℓ deionized distilled water in place of the sample. It is best to do at least one repeat titration on each sample to confirm results. If the readings are within ±0.1 mℓ, take the average volume of titrant used to calculate the calcium concentration or else do a third titration to obtain two readings that agree.

3. Calculation

Calculate the concentration of calcium in the sample using the formula:

$$Mg/\ell\ Ca = \frac{1000\ Vf}{m\ell\ sample\ titrated}$$

where V is the average volume (mℓ) of EDTA titrant used minus the volume (mℓ) used for the blank titration, and f is a calculation factor based on the EDTA standardization and denoting the amount (mg) of Ca that is equivalent to each 1.0 mℓ of standard EDTA solution.

Assume, for example, that the molarity of the EDTA was found to be 0.0095 M instead of 0.010 M, we know that 1.0 mℓ of 0.010 M solution is equivalent to 0.4008 mg Ca, then 1.0 mℓ of 0.0095 M solution will be equivalent to

$$\frac{0.4008 \times 0.0095\ mg\ Ca}{0.01} = 0.381\ mg\ Ca$$

hence f = 0.381

The results would be biased high in the presence of significant amounts of strontium. If the strontium concentration is known, correct the calcium results by subtracting a value derived by multiplying the mg/ℓ Sr value by 0.46.

4. Precision and Accuracy

Within-lab coefficients of variation of ± 2.9% and 0.5% have been reported at levels of 13 mg/ℓ and 45 mg/ℓ Ca, respectively.[5] Standard Methods[6] reports a between-lab relative standard deviation of ± 9.2% and a relative error of + 1.9% for a synthetic sample containing 108 mg/ℓ Ca.

5. Notes

In standardizing the EDTA solution, at least three titrations should be made, and the average reading used for calculation purposes after being corrected for reagent blank.

It is best to remove turbidity in highly turbid samples by centrifugation or filtration before proceeding with the measurement. This should be denoted when reporting the results.

It is essential that the titration be conducted immediately after addition of the NaOH solution. Although about 1 mℓ is generally enough to adjust the sample pH to between 12 and 13, this should be ascertained by checking with a pH meter periodically, especially when dealing with a sample of peculiar compositions. The titration step should not take longer than about 5 min for completion.

It is advisable to take a smaller sample aliquot and repeat the measurements if it is found that more than about 15 mℓ of EDTA is required for titration.

B. Atomic Absorption Spectrophotometric Method

During the past few years, the use of atomic absorption spectrophotometric techniques for water analysis has increased rapidly, to the extent where the availability and use of at least one such instrument is now commonplace in practically all water chemistry analytical laboratories. All in all, the technique is simple, fast, versatile, and relatively free from interferences. In most cases, all that is required is to aspirate an acidified sample directly into the burner/nebulizer. However, pretreatment by the addition of suitable reagents to concentrate the sample and/or obviate interference may be necessary before sample aspiration. A more detailed description of the technique of atomic absorption spectroscopy is given in Chapter 5.

Calcium concentration in water can be determined accurately, precisely, and fairly rapidly by means of atomic absorption spectrometry. The method, when properly applied, may be used to measure calcium in most types of waters including surface and ground water, precipitation, wastewaters, as well as brackish water and seawater. The method has a practical working range of about 0.0005 mg/ℓ to about 50 mg/ℓ Ca, although linearity and resolution tend to fall off at about 20 mg/ℓ Ca and above. The upper range can be extended by aspirating a suitably diluted sample. Colored samples can be analyzed without any major problems, granted that sample turbidity is not excessive.

The sample is treated with lanthanum chloride to suppress interference and then aspirated into the air-acetylene flame of an atomic absorption spectrophotometer equipped with a calcium hollow cathode lamp. The sample is atomized in the flame, and calcium atoms in their ground state absorb light (energy quanta) from the hollow cathode lamp source, emitting resonance energy characteristic of the calcium absorption wavelength. The light beam, after passing through the flame, goes to a monochromator, where the light of the selected wavelength is isolated and its intensity is then measured by a photomultiplier. The amount of light absorbed in the flame is equal to the loss of light intensity and is proportional to the number of calcium atoms present, which in turn is proportional to the sample concentration. In the double-beam instruments, the energy of the light beam passing through the flame is compared with that of a reference beam emitted from the same source, but which bypasses the flame.

Ionization, chemical, and nonspecific background interferences are the three most common types of interferences to which atomic absorption spectrophotometric methods are most susceptible. The method is relatively free from atomic spectral interferences. In the determination of calcium, chemical interferences from phosphate, aluminum, and sulfate are encountered. The presence of magnesium in excess of about 1000 mg/ℓ may cause low calcium results. Addition of lanthanum eliminates chemical interferences from up to about 600 mg/ℓ and 100 mg/ℓ phosphate and aluminum, respectively. Chemical interferences could be minimized by use of the hotter nitrous oxide-acetylene flame which, on the other hand, causes significant ionization interference. If the nitrous oxide-acetylene flame is used, an alkali metal should be added to suppress any potential interferences that may result from ionization. Samples containing significant amounts of turbidity should be first filtered or centrifuged to avoid burner and nebulizer clogging, which will result in erratic results. The analysis of brackish water, salt water, and brines is subjected to non-specific background interference necessitating the use of background correction. It has become apparent to us that the presence of significant amounts of nitrate ions in the sample tends to suppress the instrument readings. Therefore, when analyzing samples that have been acidified with nitric acid, corresponding amounts of acid should be added to the standards.

1. Equipment and Reagents

An atomic absorption spectrophotometer equipped with suitable air-acetylene burner head and calcium hollow cathode lamp is used.

Most of the modern instruments have direct digital readout devices usable with or without strip chart recorders. It is customary for manufacturers of atomic absorption spectrophotometers to provide "cookbooks" providing information on instrument settings and outlining analytical procedures to be followed. A Technicon Autoanalyzer® proportioning pump or equivalent is required for the automated procedure.

Lanthanum solution (5% La): a weighed amount of 58.7 g of lanthanum oxide (La_2O_3) or 127 g lanthanum chloride ($LaCl_3 \cdot 6H_2O$) is moistened with a little deionized distilled water and 250 mℓ concentrated HCl is slowly added. After complete dissolution, the solution is diluted to 1000 mℓ with deionized distilled water.

Stock standard calcium solution, 400 mg/ℓ Ca, stock standard magnesium solution, 1000 mg/ℓ Mg, and mixed calcium/magnesium intermediate standard solutions are prepared as described previously in the procedure for the complexometric titration of calcium.

Blank solution: to 200 mℓ of lanthanum solution is added 1000 mℓ deionized distilled water. This solution is used as baseline water in the automated procedure.

2. Procedures

a. Manual

Prepare a series of calcium standards by diluting the stock standard solution to cover the expected sample concentration. If the samples have been acidified, add an equivalent amount of the same acid to each standard. Add lanthanum solution in the ratio of 1 mℓ for every 5 mℓ standard and mix thoroughly. This may be done either directly in the flask containing the working standards or in an extracted aliquot of appropriate volume. Prepare the samples by adding lanthanum in the same volume ratio as for the standards.

Set up the instrument in accordance with the manufacturer's instructions, and allow it to warm up for a few minutes. Use the wave-length selector to set the wavelength at exactly 422.7 nm. Check to ensure that all of the instrumental settings and gas flows are correct, and then aspirate the standards sequentially in order of lowest to highest and then in the reverse order. Record the instrument reading for each. Aspirate the

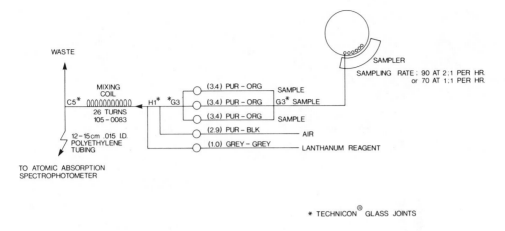

WASTE

MIXING
COIL
C5* 00000000000
26 TURNS
105 – 0083

H1* *G3

(3.4) PUR – ORG   SAMPLE
(3.4) PUR – ORG   G3* SAMPLE
(3.4) PUR – ORG   SAMPLE
(2.9) PUR – BLK   — AIR
(1.0) GREY – GREY — LANTHANUM REAGENT

SAMPLER
SAMPLING RATE : 90 AT 2:1 PER HR.
                  or 70 AT 1:1 PER HR.

12 – 15 cm .015 I.D.
POLYETHYLENE
TUBING

TO ATOMIC ABSORPTION
SPECTROPHOTOMETER

* TECHNICON® GLASS JOINTS

FIGURE 1.   Manifold for calcium and magnesium determination by automated atomic absorption spectrophotometry.

samples and record the instrument readings. Check for stability of analytical and instrumental conditions by aspirating a standard after every 10 or 15 samples and, finally, run all standards after all of the samples have been aspirated.

Plot a calibration curve using average peak heights or read the concentration directly off the instrument meter if the instrumental features permit. Read and record the calcium concentration for each sample.

b. Automated Procedure

The automated procedure is suitable when large numbers of samples are to be analyzed. Figure 1 shows the manifold used in our laboratory at Burlington, Ontario. When using the automated procedure, the rate of aspiration into the burner should be at its maximum. Also, the pump tubes should be checked and replaced at frequent intervals to ensure good sample-to-sample separation and overall satisfactory performance.

3. Precision and Accuracy

The repeatability of the manual method has been determined to be ± 0.05 and ± 0.18 mg/ℓ Ca at mean Ca levels of 8.7 and 43.1 mg/ℓ, respectively (Philbert & Traversy[7]).

The average between-run relative standard deviation obtained in a single lab (Water Quality Branch, Ontario Region) by duplicate sample analysis using the automated procedure, was ± 0.6% at the 10 mg/ℓ Ca level. Mean recovery from a variety of natural waters and precipitation samples was 96%.

C. Atomic Emission Spectrometry Method

Atomic emission spectrometry has long been established as a reliable analytical technique for metal analysis. The technique is based on the principle that when a sample is aspirated into a high energy source such as a flame, arc-spark discharge, or plasma, it is atomized, and the atoms, in absorbing energy from the source, are transformed into a state of excitation whereafter they re-emit energy of wavelength characteristic of the respective elements as they return to a less excited state. Calcium may be determined by means of flame atomic emission techniques, but the inductively coupled plasma atomic emission spectrometric technique, affording simultaneous multi-elemental analysis, is fast revolutionizing the conventional flame emission method.

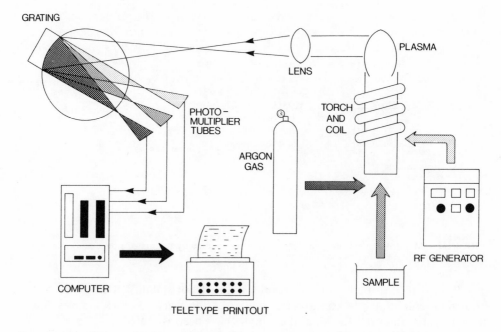

FIGURE 2.   Schematic representation of the basic components of the ICP-AES.

All in all, the inductively coupled plasma system has unique capabilities and, in general, has comparable sensitivity to flame emission and atomic absorption spectrometry.

During the last few years, the inductively coupled plasma atomic emission spectrometer (ICP-AES) has gained prominence and has evolved to be the principal contender with the atomic absorption spectrophotometer in their application to metal analysis. The advent of the ICP-AES has, without doubt, added new dimensions to the field of water analysis and analytical chemistry in general. The sample is introduced into a plasma via an ultrasonic or pneumatic nebulizer chamber system similar to that used in the conventional atomic absorption spectrophotometer. In the plasma, the sample is subjected to temperatures of about 7,000 to 10,000°K, and electrons of the atoms of elements contained in the sample undergo a transition to an excited state by absorbing thermal energy from the plasma source which they then re-emit at a characteristic frequency (atomic spectra) as they return to a lower energy state. Thus, the elements present may be determined by resolving individual spectral lines of the element of interest and choosing and measuring the intensity of one or more of the carefully selected lines for each element. The basic components of the ICP-AES is shown schematically in Figure 2.

The ICP-AES technique has a number of advantages over the atomic absorption and atomic emission spectroscopic techniques. For example, the high temperature and inert nature of the plasma eliminate most of the chemical interferences that are known to occur in flames and electric arcs. Its most outstanding feature is its potential for simultaneous multi-element analysis. This is particularly attractive in cases where the amount of sample available is limited, such as generally the case for precipitation samples. It is now possible to analyze a single sample for up to about 50 or more different elements simultaneously. Matrix correction is facilitated by means of suitable computer programs. Measurements are not necessarily restricted to ground state transitions, but could be made from the first and even second ionization state lines as well. The large linear dynamic range of concentration (about six orders of magnitude) afforded, makes the method ideal for the automated analysis of samples containing variable concentrations of constituents. Although the ICP-AES method is best suited for the analysis of liquids, the high temperature will also vaporize most solid samples introduced in the form of a fine powder. The technique permits, where necessary, sample preconcentration using conventional methods such as solvent extraction. Thus the ICP-AES system has the potential of becoming an attractive analytical tool.

The main disadvantages of the ICP-AES technique include possible interferences from stray light and the formation of refractories in the plasma. The very high temperatures naturally lead to complex spectral line interferences due to matrix. The latter results specifically from the presence of elements such as calcium, magnesium, iron, and aluminum. The analysis of soft water by ICP-AES is free from serious interferences but, in the case of hard water, the stray light interference due to calcium and magnesium presents problems. The use of a good spectrometer with a properly baffled optical system incorporating narrow band interference filters coupled with background correction will generally minimize most of these problems.

The main factors to be considered in the selection of a particular analytical method may be enumerated as: (1) the method must have the desired sensitivity, accuracy and precision for the element or compound of interest; (2) it must be simple and fast and have the desired specificity; (3) it must be free from interelemental and other interferences; (4) the sample, preferably, should require little or no treatment before it could be introduced into the analytical system; (5) it must be applicable over a wide enough range and (6) it must be economical.

While no known analytical method meets all of the above criteria completely, the ICP-AES, potentially, most closely approximate this ideal for multi-elemental analysis. Nevertheless, while the prospect of the ICP-AES system eventually becoming a popular analytical tool in most of the more advanced water chemistry analytical laboratories appears good, the technique, in its present stage of development, is no panacea for elemental analysis (trace elements particularly) on a routine basis. There still remains the need for improved detection limit capability and the elimination of interelemental and matrix interferences for more efficacious application of the technique to water analysis. Also, the use of the system becomes cost-effective only in cases where there is a requirement for large numbers of samples for simultaneous multielemental analysis. Where the number of elements of interest are few, i.e., six or less, it is more economical to use atomic absorption spectrometry.

In setting up and using the ICP-AES system, the manufacturer's instructions should be carefully followed. The detection limit for calcium is typically of the order of 0.001 mg/ℓ Ca, and the relative standard deviation at about the 1 mg/ℓ level is expected to be well below the 1% mark for most waters.

D. Ion-Selective Electrode Method

The ion-selective electrode (ISE) method may also be employed for measuring the concentration of ionized or unbound calcium in water. The ISE is most commonly applied in direct potentiometric measurement in which the particular electrode is responsive to changes in the ionic activity of the element of interest. Thus, for a constant activity coefficient, the measured cell potential is directly related to the concentration of the element. Since the activity coefficient of an electrolytic solution varies with the total ionic strength of the solution, it is necessary to work with solutions of either low or constant ionic strengths.

The method requires a good expanded scale pH meter or specific ion meter equipped with a calcium ion-selective electrode and suitable reference electrode, a magnetic stirrer and a thermostatted bath. After calibration of the instrument and the electrodes, the calcium concentration can be read off directly on instruments with logarithmic scale or computed by means of a calibration curve. The measurement is best performed by the standard addition technique.

In general, the method is simple to apply, quick, and economical. The detection limit is in the region of about 0.2 mg/ℓ Ca. It may be usefully applied to routine continuous monitoring and process control work. When properly applied, the method is capable of yielding reasonably good analytical results for most waters. The main drawback of the method is the fact that the commercially available calcium electrodes are particularly sensitive to ions such as iron and zinc. Other ions, including magnesium, barium, sodium, copper and nickel may also interfere.

E. Ion Chromatographic Method

The application of ion chromatography (IC) to the determination of anions and cations in liquid samples is relatively new, but is already proving to be useful in some applications. The IC method has comparatively high sensitivity and specificity and has potential for simultaneous multi-parameter analysis of liquids.

The ion chromatograph uses a technique based on ion-exchange chromatography incorporating a conductimetric detector and affords detection of virtually all of the ions in water or aqueous solutions that can be separated by ion-exchange. The IC method enables direct measurements of the concentrations of ions that do not absorb in the ultraviolet or visible range. The technique is particularly suited for determining non-chromophoric ions such as inorganic anions and/or cations including ammonia, as well as aliphatic amines and carboxylic acids present in simple or even complex matrices. The method is relatively interference-free, and sample pretreatment is seldom required. The practical working range is normally from about 0.010 to about 1000 mg/ℓ, and the volume of sample aliquot injected is determined by the size of the sample loop used. According to Crowther et al.,[8] injection of too high a volume may overload the system, and thus it is necessary to determine the optimum sample size to be injected for a given set of conditions.

The instrument basically comprises a system consisting of one or more cation or anion separator and suppressor (stripper) columns with loop-type sample injection valves; a series of four or more reservoirs for the eluent, regenerant, and water; an automatic timer, constant-volume pumps for controlling the flow rates of the eluent and regenerant, a conductimetric detector, and a recorder/integrator combination. The basic flow scheme of the Dionex® Ion Chromatographic system is shown in Figure 3. Other more sophisticated accessories such as automatic sampling and electronic data-handling devices may be easily incorporated to facilitate data capture/management. Anion and cation concentrator columns are also available for the measurement of low concentrations of ionic species in solution.

In principle, the sample is injected through the loop-type sample injection device into the separating column in a suitable background eluent solution. The ionic species are separated by conventional ion-exchange/elution chromatographic means in the separator column, after which the eluent enters the suppressor column, where it is stripped or neutralized allowing only the ions of interest to leave the suppressor column. These ions are carried in a stream of deionized water, or some other suitable low-conducting background solution, to the conductimetric detector which, in conjunction with the recorder/integrator, permits measurement of the ionic concentration. The combination of eluent-separator/suppressor columns selected will be governed by the determinands to be measured; the subject is dealt with in greater detail by Small et al.[9] When analyzing for anions, a low-capacity anion-exchange resin is used in the separator column, while a strong cation-exchange resin in the hydrogen form is employed in the suppressor column. The eluent normally consists of a carbonate/bicarbonate solution whereby the anions are eluted in a background of CO_3^{--}/HCO_3^{-}

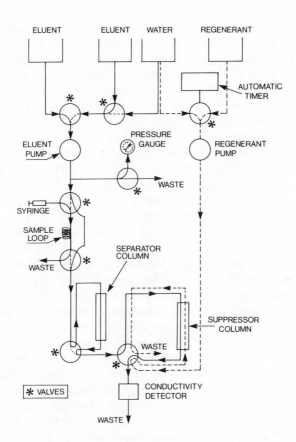

FIGURE 3.   Basic flow scheme of the Dionex® ion chromatograph.

through the separating column. In the suppressor column, carbonate and bicarbonate ions are removed by the hydrogen ions present in the resin $R-H^+$ to form carbonic acid as illustrated by the equations: $2R-H^+ + Na_2CO_3 \rightarrow 2R-Na + H_2CO_3$ and $R-H^+ + NaHCO_3 \rightarrow R-Na + H_2CO_3$. The sample ions present are then converted to their respective acids, according to the equation: $R-H^+ + Cation - A \rightarrow R-Na + H-A$, and reach the detector in this form in the low-conducting carbonic acid background.

The instrument, currently marketed by Dionex Corporation,[10] is a high-performance liquid chromatograph and is reputed to have the capability of separating and detecting alkali and alkaline earth elements or inorganic anions at concentrations down to about 0.010 mg/ℓ, with relative standard deviations better than ±3%. Lower detection limits are attainable with the use of concentrator columns. The work of Crowther et al.[8] shows that when the system is operated in the manual mode, the time required for a single parameter analysis is approximately 3 to 5 min, while it requires about 12 min to determine 8 anions simultaneously. Judging from information currently available, the emphasis is mainly on the determination of anions such as Cl, Br, F, NO_3, SO_4, and PO_4 in water and clinical samples.

III. DETERMINATION OF MAGNESIUM

Magnesium, like calcium, occurs abundantly in nature and is also a common constituent of natural waters. The amount found in a given water depends largely on the type of terrain traversed by that water. Entry into water systems is primarily through dissolution of ferromagnesium minerals from igneous rocks and dolomitic sedimentary rocks. Generally, dissolved magnesium concentration is lower than calcium for the majority of natural waters. Because of the high solubility of magnesium salts, however, the metal tends to remain in solution and is less readily precipitated than calcium. It is second only to calcium in contributing to water hardness.

Water containing excessive amounts of dissolved magnesium salts may have serious adverse effects, similar to those outlined for calcium, relating to certain domestic and industrial applications. Magnesium imparts an unpleasant bitterish taste to water when present in large amounts, and its salts, the sulfate particularly, may have purgative and diuretic effects if large quantities are ingested. Nevertheless, the element is essential for human life and the approximate daily adult requirement is believed to be about 200 to 300 mg. Magnesium poses no major concern with regard to public health or the aquatic environment, and limits of concentration set for water are based mainly on palatability, corrosion, and incrustation criteria, the criteria depending on the intended use.

Most of the methods described above for calcium analysis apply equally well to magnesium, and it is usually convenient to determine the two elements together. Apart from the economical benefits that would be derived therefrom, the use of mixed standards is advantageous from an analytical standpoint since the standards matrix is then brought one step closer to that of the samples to be analyzed. The titrimetric, atomic absorption, and atomic emission spectroscopic methods are the ones best suited for magnesium determination. These procedures have been dealt with in sufficient details under calcium, and only the necessary supplementary information is given below. Magnesium may be determined gravimetrically by precipitating it in an ammoniacal solution as the magnesium ammonium phosphate after first removing the calcium present. The precipitate is then washed with dilute ammonium hydroxide solution, dried, ignited and then weighed as the resulting magnesium pyrophosphate. The method is much too lengthy for all practical purposes and is not favored for routine sample analysis.

A. EDTA Complexometric Total Hardness Method

The method is applicable to the determination of magnesium in natural surface and ground waters, drinking water, precipitation, and wastewaters. The method entails the determination of total hardness in the form of calcium and magnesium salts, and the remarks made for the calcium method apply. Assuming that a suitable sample volume is titrated, the detection limit of the total hardness titrimetric method is about 1 mg/ℓ as $CaCO_3$.

The sum of calcium and magnesium concentrations is determined by titrating the sample with standard EDTA solution at a pH of 10.0 ± 0.1 using Eriochrome® Black T (EBT) as indicator. The indicator, when added to samples containing calcium and/ or magnesium, produces a wine-red colored solution. The EDTA titrant, when added, chelates the calcium and magnesium stoichiometrically to form the soluble metal complex. When all of the magnesium and calcium ions are chelated, the solution changes from red to blue. Thus we have Ca (Mg) + EBT → Ca (Mg) − EBT (wine-red color) and Ca(Mg) − EBT + EDTA → Ca (Mg) − EDTA + EBT (blue color). The sample must contain enough magnesium ions in solution in order to obtain a satisfactory endpoint. This is ensured by the addition of a small quantity of the magnesium salt of EDTA to the buffer solution. The concentration of calcium alone is determined by titrating another sample aliquot as described for calcium determination. The difference between the two values is the value of the magnesium concentration. The back-titration method described by Szekeres[4] may also be used, but it is not favored because of the lengthy procedure involved.

Waters such as seawater and brine that contain extreme quantities of salts may cause interferences. Certain heavy metals, including iron, manganese, aluminum, cadmium, copper, lead, cobalt, nickel, zinc, barium, and strontium may react with the EDTA or may interfere by causing faded or indistinct end-points. The interference may be minimized by the addition of hydroxylamine and cyanide solutions. Some of the substances, such as polyphosphates, organic phosphates, and EDTA/NTA compounds, that are commonly used for water treatment may interfere also. The effect of polyphosphates and heavy metals is overcome by addition of sodium sulfide. This solution should be used judiciously, however, as it forms a sulfide precipitate in the presence of appreciable amounts of heavy metals which tends to obscure the end-point. Addition of two or three crystals of potassium ferrocyanide suppresses interferences from any divalent manganese present. Highly colored waters, including those containing suspended or colloidal organic matter, may also interfere with the end-point detection. Before proceeding with the determination, such samples should either be filtered or centrifuged or evaporated to dryness, followed by ignition at about 600°C in a muffle furnace. The residue is then dissolved in about 20 ml of 1 M HCl, and neutralized to pH 7 with 1 M NaOH and then made up to the original volume with deionized distilled water. Some analysts have also removed the color from samples by first passing them through a column containing activated carbon. It should be noted, however, that some types of samples, groundwaters particularly, may give spurious results due to precipitation of iron and calcium upon exposure to the air. Furthermore, filtration may remove calcium present as colloidal $Ca(OH)_2$ or $CaCO_3$ from the sample.

1. Equipment and Reagents

The apparatus required is the same as previously described for the EDTA complexometric titration method for calcium.

Buffer solution: ammonium chloride (NH_4Cl), 67.6 g, is dissolved in 200 ml of deionized distilled water. Concentrated ammonium hydroxide (NH_4OH), 570 ml, is added, followed by 5.0 g of the magnesium salt of EDTA. The solution is diluted to 1000 ml and stored in a tightly stoppered plastic bottle. One ml of this buffer should bring the pH of a neutral sample to 10.0 ± 0.1. If not, the solution should be replaced. Alternatively, the solution is prepared by dissolving 4.716 g of analytical reagent grade disodium salt of EDTA and 3.12 g magnesium sulfate ($MgSO_4 \cdot 7H_2O$) or 2.576 g magnesium chloride ($MgCl_2 \cdot 6H_2O$) in about 200 ml of deionized distilled water. Then 67.6 g ammonium chloride and 570 ml of concentrated NH_4OH are added. The solution is mixed well to dissolve the salt and then made up to 1000 ml. For best results, the magnesium concentration should be adjusted to the exact equivalence by addition of an appropriately small amount of the disodium EDTA salt, or magnesium sulfate or chloride. Alternatively, a more stable glycine-potassium hydroxide buffer of pH 10.4 ± 0.1 may be prepared by dissolving 93.8 g analytical reagent grade glycine (NH_2CH_2COOH) in about 500 ml deionized distilled water. A solution of potassium hydroxide is added to adjust the pH of the solution to 10.4 ± 0.1, after which 5.0 g of the magnesium salt of EDTA is dissolved. The volume of the resulting solution is then made up to 1000 ml. A well-stoppered polyethylene bottle is used for solution storage.

The following reagents prepared for the calcium EDTA filtration procedure are required in addition to the standard 0.010 EDTA solution: mixed standard calcium/magnesium solution, hydroxylamine hydrochloride solution, sodium cyanide solution, and potassium ferrocyanide crystals.

Eriochrome® Black T Indicator solution: Eriochrome® Black T ($C_{20}H_{12}N_3NaO_7S$), 0.4 g is dissolved in 100 ml deionized distilled water. This solution is stable for only about a week. A much more stable indicator mixture with a shelf life of about one year may be prepared by grinding 0.5 g of the dye with 100 g of sodium chloride crystals and storing in a dark colored, wide-mouth bottle. About 0.2 g of the mixture is used for each titration.

Sodium sulfide solution: sodium sulfide ($Na_2S \cdot 9H_2O$), 5.0 g, or ($Na_2S \cdot 5H_2O$), 3.7 g, is dissolved in 100 ml deionized distilled water. The solution is stored in a bottle tightly stoppered with a rubber stopper. The solution deteriorates on exposure to air and should be replaced at such intervals as determined by the frequency with which it is used.

2. Procedure

Pipette 50.0 ml of the sample into a 250 ml Erlenmeyer flask. Use a smaller volume and dilute to 50 ml if the sample contains high Ca and Mg concentration or if the total alkalinity is greater than about 300 mg/l $CaCO_3$. It is best to use a sample aliquot containing between 5 to 10 mg Ca, plus Mg for titration. Seawater and brines should be diluted in order to reduce the salt content.

Place the flask on the magnetic stirrer and stir while adding 1 ml of hydroxylamine hydrochloride, followed by 1 ml of the buffer solution. Check the sample pH to ensure it is at 10.0 ± 0.1. If not, adjust it appropriately.

Add 2 ml of NaCN solution, and if manganese is present in significant amount, add a few drops of a 5% ascorbic acid solution or a few crystals of potassium ferrocyanide.

Add about 1 ml of sodium sulfide solution if titrating wastewater, effluent, or polluted water samples and particularly where the sample is known to contain significant amounts of polyphosphates.

Add about 0.2 g of the powdered indicator mixture or 3 to 4 drops of the solution, and titrate immediately with standard EDTA to a distinctive blue color end-point.

Record the volume of titrant used to reach the end-point and determine a reagent blank by repeating the above procedure with 50.0 ml of deionized distilled water in place of the sample.

Repeat the titration on another sample aliquot to check the readings, and use the average value to calculate the results. The titrations should be performed immediately after addition of the buffer solution and should not take more than about 5 min to be completed. This helps to minimize the chances of calcium precipitating out as insoluble calcium carbonate.

3. Calculation

Calculate the sum of the calcium and magnesium concentration from the formula:

$$\text{mg/}l \text{ Ca + Mg} = \frac{1000 \text{ Vf}}{\text{m}l \text{ sample titrated}}$$

where V is the average volume (ml) of EDTA titrant used minus the volume (ml) used for the blank titration, and f is a calculation factor based on the EDTA standardization and denoting the amount (mg) of Ca that is equivalent to each 1.0 ml of standard EDTA solution.

Subtract the value determined for the calcium concentration from the above value to obtain mg/l magnesium.

Use of the following equation is faster and more convenient for calculating the magnesium concentration: mg/l Mg = 12.16 (total hardness ($CaCO_3$) × 0.01998 — calcium as Ca × 0.0499). The numerical values are the respective factors used for converting mg/l to meq/l or vice versa.

4. Precision and Accuracy

The relative standard deviation of the total hardness (Ca + Mg) method is of the order of ±2 to 3% at the 200 mg/ℓ $CaCO_3$ level. The relative error is approximately ±1%.

The precision and accuracy of the magnesium results would, logically, be governed by the precision and accuracy with which both the calcium and total hardness determinations were made.

B. Atomic Absorption Spectrophotometric Method

Except for the instrumental conditions, the atomic absorption spectrophotometric method for calcium and magnesium is identical. Both determinations could be made simultaneously on a dual-channel instrument or on two single-channel instruments if available. Otherwise, the same aliquot may be analyzed sequentially for Ca and Mg on a single instrument.

The detection limit of the method is about 0.0005 mg/ℓ Mg and the practical upper working limit is about 10 mg/ℓ. Calcium, sodium, and potassium may interfere if their concentrations exceed about 400 mg/ℓ.

1. Equipment and Reagents

An atomic absorption spectrophotometer, equipped with an air-acetylene burner head and magnesium hollow-cathode lamp, is used with a strip chart recorder. An autoanalyzer proportioning pump may be employed if desired.

Lanthanum solution (5%) and the blank solution are prepared as described for the calcium atomic absorption method.

Stock standard magnesium solution: An acidified stock magnesium solution is prepared by dissolving 0.829 g analytical reagent grade magnesium oxide (MgO) in 10 mℓ of metal-free concentrated HNO_3. The solution is diluted to 1000 mℓ in a volumetric flask with deionized distilled water. It would be convenient to use the mixed intermediate magnesium standard solution prepared for total hardness determination if available.

2. Procedure

Prepare a series of magnesium or, preferably, calcium plus magnesium working standards to cover the expected range of sample concentration. If the samples have been acidified, add an equivalent amount of the same acid from the same lot and bottle if possible, to each standard.

Add lanthanum solution in the ratio of 1 mℓ for every 5 mℓ standard and mix thoroughly. This may be done either directly in the flask containing the working standards or in an extracted aliquot of appropriate volume. Prepare the samples by adding lanthanum solution in the same volume ratio as for the standards.

Set up the instrument in accordance with the manufacturer's instructions and allow it to warm up for a few minutes. Set the wave-length at exactly 285.2 nm. The less sensitive 202.5 nm line may also be used if considered appropriate. Check to ensure that all of the instrumental settings and gas flows are correct, and aspirate the standards sequentially in order of lowest to highest and then in the reverse order. Record the instrument reading for each. Check for stability of analytical and instrumental conditions by aspirating a standard after each 10 or 15 samples and, finally, run the series of standards after all samples have been aspirated.

Plot a calibration curve using peak heights, or read the concentration directly off the instrument meter if feasible on that particular instrument. Read and record the magnesium concentration for each sample.

The automated procedure described for calcium is applicable to magnesium determination as well. It is best suited for cases where large numbers of samples are to be analyzed at once.

3. Precision and Accuracy

The average between-run relative standard deviation obtained in a single laboratory (Water Quality Branch, Ontario Region) by duplicate sample analysis using the automated procedure was $\pm0.5\%$ at the 2.0 mg/ℓ Mg level. Mean recovery from a variety of natural waters and precipitation samples was 101%.

C. Atomic Emission Spectrometry

Magnesium may also be conveniently determined by the ICP-AES technique (see section describing ICP-AES for calcium determination). Calcium may interfere in this method, and it is best to use mixed standards containing Ca and Mg in amounts that are comparable to that of the sample.

A detection limit of about 0.0002 mg/ℓ Mg is attainable. The relative standard deviation is about 0.5% over a range of 1 to 10 mg/ℓ Mg.

IV. TOTAL HARDNESS

Water hardness is caused primarily by the presence of calcium and magnesium salts in the water. Hydrogen ions and polyvalent metals such as iron, manganese, aluminum, strontium and zinc, contribute to water hardness also. Apart from calcium and magnesium, other hardness-causing metals are seldom present in large enough quantities in water to have any significant effect on water hardness. Thus, in general, the sum of the calcium and magnesium ions present is the principal determining factor of water hardness. Because hardness-causing substances react and form insoluble precipitates with soap before a lather is formed, hard water causes soap to lose its detergent power with the attendant reduction of its foaming capacity. Water hardness, therefore, is sometimes broadly defined as a measure of the capacity of the water to precipitate soap.

Although most of the calcium and magnesium salts dissolved in water are found in the form of bicarbonates, sulfates, and to a lesser extent as chlorides and nitrates, water hardness is customarily expressed as an equivalent of calcium carbonate, the units being mg/ℓ hardness as $CaCO_3$.

Hardness may be categorized as being temporary or permanent. Temporary hardness is the hardness that is removable by boiling, while the remaining hardness is said to be permanent. The latter results mainly from the presence of calcium sulfate. Also, when the numerical value of the total hardness exceeds the sum of the carbonate and bicarbonate alkalinities, the fraction of the total hardness equivalent to the total alkalinity of the sample is termed carbonate hardness, while the excess hardness is called noncarbonate hardness. Carbonate hardness is attributed primarily to the presence of calcium and magnesium carbonates and bicarbonates while noncarbonate hardness results from other salts including sulfates, chlorides and nitrates, with sulfates being generally the most prominent. Sekerka and Lechner[11] have successfully used an automated programmable system for the direct simultaneous determination of total hardness, carbonate hardness, and noncarbonate hardness in water by using bivalent ion-selective electrodes.

The numerical value of total hardness (TH) of natural and wastewaters may vary from about zero to several hundred mg/ℓ, depending on the source and several other factors. In the case of natural water, for example, the type of terrain with which the water comes in contact is of prime importance. Waters in areas of carbonate bedrock are generally hard, while, on the other hand, waters draining from igneous rocks are relatively soft. Carbon dioxide entrained in surface or groundwater may cause the dissolution of calcium from soils and rocks, resulting in varying degrees of water hardness. Broadly speaking, water may be classified from as being very soft to very hard in accordance with criteria similar to those found in the Water Encyclopedia,[12] as follows:

> For very soft water, total hardness as $CaCO_3$ is in the approximate range of 0 to 30 mg/ℓ.
> For soft water, total hardness as $CaCO_3$ is in the approximate range of 31 to 60 mg/ℓ.
> For moderately hard water, total hardness as $CaCO_3$ is in the range of 61 to 120 mg/ℓ.
> For hard water, total hardness as $CaCO_3$ is in the range of 121 to 180 mg/ℓ.
> For very hard water, total hardness as $CaCO_3$ exceeds 180 mg/ℓ.

Because of the varying hardness characteristics of most natural waters and the possible wide ranges of water hardness encountered, the ranges of total hardness values used to classify water are variable. It is not uncommon, for instance, to find some waters with total hardness concentrations in the range of 150 to 300 mg/ℓ and above 300 mg/ℓ being classified as "hard" and "very hard", respectively.

Normally, water hardness is of interest mainly in relation to the intended use of the water. The one undesirable feature of hard water that affects virtually everyone is its relatively high soap consumption power with the attendant formation of the familiar curd and bathtub ring. This obviously has serious economic implications both in the home and in industrial activities, including cleaning and laundering. Hard water is corrosive to pipes and causes the formation of scales in boilers, radiators, pipes, and cooking utensils, as well as having diverse adverse effects in, for example, industries such as textiles, plating, and canning. High calcium and magnesium content in water may hamper proper cooking of some vegetables. It is noteworthy to point out, however, that results of recent epidemiological studies indicate that the incidence of cardiovascular diseases is appreciably higher in areas where municipal water supplies consist of soft water. While levels of acceptability will vary with local hydrogeological conditions and consumer acceptance, water with total hardness as $CaCO_3$ of 120 mg/ℓ or less is generally considered to be satisfactory for most domestic, agricultural, and industrial uses. Water with hardness in excess of 500 mg/ℓ as $CaCO_3$, may be unsuitable for such uses and may require softening before use.

Total hardness may be determined in a number of ways of which the following are more common: (1) palmitate (soap) method; (2) soda-reagent method; (3) EDTA complexometric titration method; (4) calculation method; (5) automated colorimetric method; (6) ion-selective electrode method.

The EDTA complexometric titration and automated colorimetric methods measure essentially hardness due to calcium and magnesium. The calculation method, which takes into account other hardness-causing cations, is the more accurate and, therefore, preferred method. The description of the EDTA complexometric titration method is already given for magnesium determination.

The palmitate or soap method involves precipitation of calcium and magnesium ions from the samples with an alcoholic standardized solution of potassium or sodium palmitate. After boiling to expel CO_2, the sample is titrated with the palmitate solution such that when all the calcium and magnesium ions present have been precipitated, the excess palmitate in solution is hydrolyzed, resulting in an end-point color change.

In the soda reagent method, sulfuric acid is added to decompose the carbonates and bicarbonates and the sample is boiled to expel CO_2. The calcium and magnesium ions present are then removed by precipitation with a mixture of sodium carbonate and sodium hydroxide solution. The solution is filtered and the excess soda reagent in the filtrate is determined by titrating against a standard sulfuric acid solution.

A. Calculation Method

In determining total hardness, it is far more convenient to use the faster and more accurate calculation method if the individual hardness-causing cations have already been determined separately. Apart from being faster than the other methods, the calculation method may also be more accurate and precise depending on the degree of precision and accuracy with which the concentrations of the individual cations were measured in the first place. The factors for converting the concentration in mg/ℓ of each cation to the equivalent calcium carbonate concentration is derived from the ratio of the respective equivalent weights. Thus the conversion factors for calcium and magnesium are

$$\frac{50.045}{20.04} \text{ and } \frac{50.045}{12.156} = 2.497 \text{ and } 4.117, \text{ respectively.}$$

Similarly, the conversion factors for Fe, Sr, Al, Zn and Mn are respectively 1.792, 1.142, 5.564, 1.531, and 1.822. Summation of the computed $CaCO_3$ equivalence for the individual cation gives the total hardness value in mg/ℓ as $CaCO_3$. In practice, if the concentrations of the above cations, apart from calcium and magnesium, are insignificant, summation of the converted calcium and magnesium values only will suffice.

B. The Automated Colorimetric Method

The method being described is applicable to natural surface and ground waters, precipitation, and wastewaters over a practical working range of about 5 mg/ℓ to 250 mg/ℓ total hardness as $CaCO_3$. The upper limit may be extended by suitable dilution of the sample through variations in the sample to dilution water ratio and/or by suitable adjustment of the standard calibration setting on the colorimeter.

The Technicon® automated colorimetric method[13] is essentially an adaptation of the EDTA complexometric titration method. The disodium magnesium salt of ethylenediamine tetraacetic acid (EDTA) or the magnesium salt is used to exchange, or an exact equivalent basis, magnesium ions for calcium and/or other cations that form a more stable EDTA complex than magnesium. The free magnesium then reacts with a calmagite solution at pH 10 to form a red-violet colored complex. The total hardness concentration is thus determined by measuring the magnesium concentration in the final sample-reagent stream. The absorbance of the final solution is measured at 520 nm.

Aluminum, cobalt, copper, iron, nickel, and zinc may interfere. Sample turbidity and color may interfere. If necessary, remove particulate matter from turbid samples by filtering before analysis.

1. Equipment and Reagents

The equipment used consists of a Technicon AutoAnalyzer® or equivalent system consisting of a sampler, proportioning pump, manifold, colorimeter equipped with 15 mm × 2.0 mm flow cell and 520 interference filters, recorder, digital printer/reader (optional), and linearizer if a digital printer/reader is used.

Buffer solution, pH 10.0: analytical reagent grade sodium hydroxide (NaOH), 3.59 g, and 6.18 g boric acid (H_3BO_3) are dissolved in approximately 800 mℓ deionized distilled water. The solution is diluted to 1 ℓ with more water. The pH of the solution must be adjusted to 10.0, using a 1 M solution of NaOH or HCl if necessary. Alternatively, an ammonia buffer is prepared by dissolving 67.6 g ammonium chloride in 572 mℓ of ammonium hydroxide (NH$_4$OH). The solution is made up to 1 ℓ with deionized distilled water.

Disodium magnesium EDTA solution, 0.00376 M: disodium magnesium ethylenediamine tetraacetate (disodium magnesium EDTA; C$_{10}$H$_{12}$MgN$_2$Na$_2$O$_8$), 1.35 g, is dissolved in approximately 800 mℓ of the pH 10.0 buffer solution. The solution is diluted to a 1 ℓ with more buffer and 0.5 mℓ of Brij-35® (polyoxyethylene 23 lauryl ether — available from Technicon Instruments Corporation, New York) is added.

Stock calmagite indicator solution: calmagite (3-hydroxy-4 [(2-hydroxy-5-methyl-phenyl)azo]-1-napthalnesulfonic acid), 0.25 g, is dissolved in 1 ℓ deionized distilled water. The solution is stirred for about ½ hr and then filtered.

Working calmagite indicator solution: a 35 mℓ aliquot of the stock calmagite solution is mixed with 515 mℓ of the pH 10.0 buffer solution and 450 mℓ of the EDTA solution; 0.5 mℓ of Brij-35® is then added. This solution is good for approximately one week, after which it should be prepared afresh.

Stock standard solution: the 1000 mg/ℓ stock CaCO$_3$ solution prepared for the calcium EDTA complexometric titration method is used.

Working standard solution: a series of working standard solutions are prepared by making appropriate dilutions of the stock standard solution.

2. Procedure
Set up the manifold as shown in Figure 4, and follow the recommended procedure for the automated procedure such as previously described for chloride or sulfate. Set the reagent baseline absorbancę with reference to water to approximately 0.72 absorbane units.

C. The Ion-Selective Electrode Method
Sekerka and Lechner,[14] recently described a method for the determination of total hardness by an automated direct potentiometric procedure using a bivalent ion-selective hardness electrode. The authors[11] later reported that because of carbonate ion interference, the direct potentiometric method was not suitable for total hardness determination in water samples. The problem is reported to have been overcome by employing a standard addition technique, and results reported therefrom agreed satisfactorily with the conventional manual titration method. Relative standard deviation at 1, 50, and 500 mg/ℓ total hardness as CaCO$_3$ are reported to be ±2.5%, ±1.2%, and ±4.5%, respectively.

V. SODIUM

Sodium is present in varying amounts in virtually all natural waters. Practically all sodium compounds are water-soluble and tend to remain in aqueous solution. Water in contact with igneous rocks will dissolve sodium from its natural source. Other potential natural sources include clay minerals, feldspars, and minerals such as halite and mirabilite. Sodium may also enter water through industrial and municipal waste discharges and run-offs from diffuse sources such as oil fields drainage. In general, highly mineralized waters usually contain high concentrations of sodium. Seawater normally contains about 10,000 mg/ℓ Na, while brines may contain sodium ranging in concentration from about 25,000 mg/ℓ to about 100,000 mg/ℓ. Hard waters that have been softened by the sodium cation exchange process are likely to contain a considerable amount of sodium.

The ratio of sodium in the sum of cations is an important factor in considering water

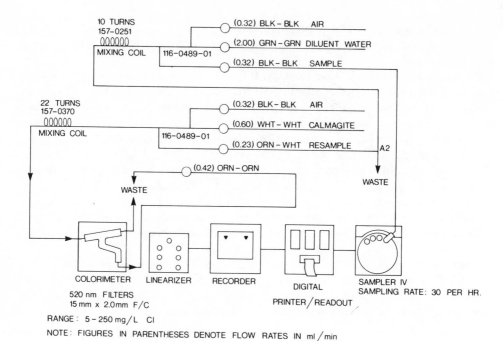

FIGURE 4.   Total hardness manifold.

for agricultural uses. It is important in regard to plant and animal pathology. For example, relatively large concentrations of sodium may adversely affect soil structure and permeability, resulting in alkaline soils which may restrict agricultural uses. Excessive sodium uptake by plants may prove detrimental to fruit crops particularly. Dilute waters that have been softened by base exchange or which have traversed through siliceous igneous terrain, may acquire a high percentage of sodium relative to the total cation concentration. The amount of intake of sodium must be carefully controlled by those suffering from diseases such as coronary thrombosis, cirrhosis of the liver, and hypertension.

Excessive amounts of sodium in drinking water would normally affect mainly the palatability of the water, and water containing up to about 1000 mg/ℓ may generally be physiologically tolerable. The presence of large amounts of sodium in water intended for industrial and agricultural uses is of greater importance, however. Sodium contributes to corrosion and incrustation in pipes and boiler systems and may cause foaming in boiler water if present in excess of 2 mg/ℓ.

The sodium concentration of irrigation water is of prime importance and plays a part in determining the permeability of the soil to water. The reaction that a particular water type could have towards soil may be predicted by the numerical value of an expression known as the sodium adsorption ratio (SAR). The SAR is an estimate of the extent to which sodium ions present in the water would be adsorbed by the soil and is generally expressed as follows:

$$SAR = \frac{Na\ (epm)}{\dfrac{Ca + Mg\ (epm)}{2}}$$

where epm is equivalents per million. Basically, the SAR provides an indication of whether or not the proportion of sodium ions in the soil is likely to exceed certain desired limits due to base exchange between the sodium ions in water in contact with the soil and the exchangeable calcium and magnesium ions present. The SAR also serves as a good indicator of the stability of clay soil with respect to its potential for breakdown and particle dispersion.

Because of the adverse effect that water with high sodium content may have on soil structure and permeability, the SAR is a useful index for monitoring and determining the suitability of water required for irrigation. In general, a maximum SAR limit of six is allowable for irrigating most crops, but there are certain sensitive crops that should not be subjected to water having SAR values higher than about four.

Two other expressions, one giving the percent sodium in a given water and the other giving the percent sum of sodium and potassium relative to the sum of the major cations, are sometimes useful indicators also.

Thus:

$$\text{percent Na} = \frac{\text{Na (epm)} \times 100}{\text{Sum of major cations in epm}}$$

$$\text{percent Na + K} = \frac{(\text{Na + K}) \times 100}{\text{Ca + Mg + Na + K}}$$

Sodium may be measured by a variety of methods including gravimetric, colorimetric, flame photometric, atomic absorption or atomic emission spectrophotometric, and ion-selective electrode procedures. Of these, the flame photometric, atomic absorption, and atomic emission spectrophotometric methods are considered to be the best. They are rapid, sensitive, precise and relatively free of interferences.

The gravimetric method entails the precipitation of sodium as the hydrated sodium zinc uranyl acetate by means of zinc uranyl acetate. The method is long, imprecise, and lacks the sensitivity required for most natural fresh water analysis. It is susceptible to interferences from lithium, potassium, sulfate, phosphate, and organic acids such as oxalic, citric and tartaric. Another potential source of error stems from possible partial dehydration and precipitation of silica upon evaporation of the sample. The method is, therefore, not favored.

The colorimetric method entails following the same general procedure used for the gravimetric determination of sodium; except that the precipitated sodium salt, after being washed successively with an ethanol-acetic acid mixture and diethyl ether, is dissolved in a few mℓ of ammonium thiocyanate solution. The optical density of the resulting yellow-colored solution, the intensity of which is proportional to the sodium concentration, is measured at 420 nm.

A. The Flame Photometric Method

The method is applicable to natural surface and ground waters, drinking water, precipitation, and wastewaters. A detection limit of about 0.05 mg/ℓ is attainable, and the practical upper working limit is about 100 mg/ℓ Na. Higher Na concentrations can be conveniently measured by making appropriate sample dilutions.

When the sample containing sodium ions is introduced into a gas flame, the element is atomized, and the excited atoms emit light at a specific wavelength characteristic of the element. The intensity of the light is proportional to the concentration of the element and is measured spectrometrically. The method is ideally suited for the simultaneous determination of sodium and potassium. Sodium is measured by 589 nm and potassium at 768 nm. A direct-reading on an internal-standard type flame photometer may be used. There are various types of flame photometers commercially available, and the manufacturer's operating instructions should be followed for the particular instrument being used.

The most prominent source of interferences is the radiation interference resulting from other elements, the other alkali metals particularly, that are present. For example, potassium, calcium, and magnesium interfere if the potassium to sodium, calcium to sodium, or magnesium to sodium ratios equal or exceed 5:1, 10:1 and 100:1, respectively. Aluminum, if present in sufficient amount, interferes by suppressing the intensity of the emitted light. Chloride, sulfate, and bicarbonates may interfere also if present in large enough amounts. Suspended particulate matter may affect the determination by clogging the burner. This effect could be overcome by aspirating a filtered or centrifuged sample. Also, the addition of a little nonionic detergent such as Brij-35® (polyoxyethylene 23 lauryl ether) to the lithium standard solution generally assures proper solution aspiration. The interelemental interference effect may be minimized by (1) working at as low a sodium and potassium concentration range as is conveniently possible; (2) using the internal standard or standard addition technique; (3) prior removal of the interfering elements; (4) addition of radiation buffers; or (5) by adding equivalent amounts of the interfering elements, if known, to each of the calibration standards. It is inadvisable to make direct determinations at too high a concentration, because accuracy may be reduced at such levels due to the effect of self-absorption, which causes the intensity vs. concentration curve to decrease in slope. The presence of extraneous light, drafts, airborne dusts, or tobacco smoke may lead to erroneous results. Careless handling of glassware and equipment could easily lead to contamination. Perspiration is an obvious source of contamination that should be guarded against. Because of the high risk of contamination by cigarette smoke, smoking should not be permitted in the immediate area of the analysis.

1. Equipment and Reagents

A flame photometer of either the direct reading or internal-standard type is used, the dual channel instrument being preferred.

All glassware must be cleaned thoroughly using a good detergent, then rinsed with 1 + 15 nitric acid, followed by several rinses with deionized distilled water. Glassware should be alkali metal-free and of resistant glass.

Stock standard sodium solution, 1000 mg/ℓ Na: a few grams of analytical reagent grade sodium chloride (NaCl) is dried at 105°C for about 1 hr and then cooled in a desiccator. The salt, 2.5418 g, is dissolved in deionized distilled water in a 1 ℓ volumetric flask and the solution made up to the mark with deionized distilled water. The solution is stored in a polyethylene or similar container.

Standard lithium solution, 1000 mg/ℓ Li: analytical reagent grade lithium chloride (LiCl) or lithium nitrate (LiNO$_3$) is dried overnight at 105°C and then cooled in a desiccator. Rapidly, 6.109 g of the nitrate or 9.935 g of the chloride is weighed and dissolved in deionized distilled water in a 1 ℓ volumetric flask. The solution is made up to the mark with deionized distilled water, and stored in a polyethylene or similar container.

2. Procedure

Set up the flame photometer and related apparatus in accordance with the manufacturer's operating instructions.

Prepare a series of standards to cover the required working range. Working ranges of 0 to 1.0, 0 to 10.0, and 0 to 100.0 mg/ℓ Na may be the most convenient, depending on the sample concentration. In preparing the working standards for 0 to 1.0 mg/ℓ and 0 to 10.0 mg/ℓ ranges, it is best to make dilutions from a 10 mg/ℓ and 100 mg/ℓ

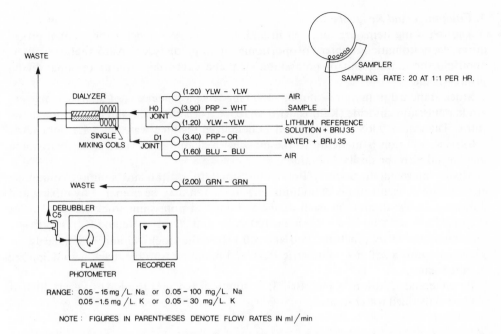

RANGE: 0.05 – 15 mg /L. Na or 0.05 – 100 mg /L. Na
0.05 –1.5 mg /L. K or 0.05 – 30 mg /L. K

NOTE : FIGURES IN PARENTHESES DENOTE FLOW RATES IN ml / min

FIGURE 5.   Sodium and potassium manifold.

intermediate standard solution, respectively. Prepare enough standards (about 11) in each series to bracket the expected concentration ranges of the samples.

Pipette equal volumes of each working standard, a reagent blank, and each sample into marked beakers or test tubes. Add exactly the same measured volume of lithium solution to each in the ratio of approximately 1.0 ml lithium solution to every 4.0 ml of standard, blank, or sample.

Mix well and aspirate the blanks and standards, starting first with the lowest and progressing up to the highest, and then in the reverse order. Record the instrument emission intensity reading at 589 nm. Aspirate the samples with a suitable standard interspersed at intervals of about 10 to 15 samples. Aspirate all of the standards again at the end of the sample aspiration.

Check the peak heights of the respective standards for consistency, and plot a calibration curve, taking into account the blank reading. Read the sodium concentration of each sample from the calibration curve.

B. Automated Flame Photometric Procedure for Sodium and Potassium

Sodium and potassium can be conveniently measured simultaneously by an automated procedure using the manifold shown in Figure 5. The automated method is particularly beneficial when large numbers of samples are to be analyzed at once. The sample is mixed with the lithium internal standard solution, dialyzed, and then introduced into the flame photometer equipped with interference filters to isolate the spectral lines of sodium and potassium. The method described below is similar to that given in the Water Quality Branch Analytical Methods Manual,[5] and is actually an adaptation of the manual procedure described above. Its practical working ranges are 0.05 to 100 mg/l Na and 0.05 to 30 mg/l, but more concentrated samples could be easily analyzed by adding a dilution line to the manifold or by diluting the samples manually.

1. Equipment and Reagents

The following items are required in addition to those listed for the manual procedure: An automatic sampler, proportioning pump, dialyzer, AutoAnalyzer®-type manifold for sodium and/or potassium, strip chart recorder, and an optional digital printer/reader.

Stock standard potassium solution, 1000 mg/ℓ K: a few grams of analytical reagent grade potassium chloride (KCl) is dried at 105°C for about 1 hr and cooled in a desiccator. The salt, 1.9068 g, is dissolved in deionized distilled water in a 1 ℓ volumetric flask. The solution is made up to the mark with deionized distilled water and stored in a polyethylene or similar container.

Mixed Intermediate Sodium/Potassium Standard: a standard solution containing appropriate concentrations of sodium and potassium may be prepared by mixing and diluting suitable aliquots of each of the sodium and potassium stock standards. For example, a mixed standard containing 100 mg/ℓ and 20 mg/ℓ Na and K, respectively, may be prepared by pipetting 25.0 and 5.0 mℓ of the stock Na and K standards, respectively, into a 250 mℓ volumetric flask and making up to the mark with deionized distilled water.

Baseline and Dilution Water: Brij-35® solution, 0.5 mℓ, is added to every liter of deionized distilled water needed.

2. Procedure

The manifold shown in Figure 5 is based on the Technicon Auto-Analyzer® System.[15] Analysts are expected to follow the instruction manuals supplied with the particular instrument being used. The following is an outline of the procedure generally followed.

Open both oxygen and natural gas main tank valves to maximum. Slowly open fuel gas (in this case natural gas), valve on module and apply lighted match to burner tip. A yellow flame will appear. Slowly turn this gas valve until flame reaches chimney height.

Open oxygen pilot valve, and small blue flames will appear on the tips of the three pilot tubes. Continue oxygen input until one or more of the three blue flames rise off the pilot tubes. At this point, slowly close the oxygen valve until the three blue flames return to the pilot tubes. Turn the main oxygen valve on module counter-clockwise until the flame turns blue and its inner cone reaches a height of about 2 cm. Allow about ½ hr for warm-up and stabilization.

Place the standards and samples in the sample tray in order of lowest to highest standards, immediately followed by the same standards placed in the reverse order, and followed by the samples. Insert a suitable standard after every 15 to 20 samples and, finally, place a series of standards at the very end of the run. It is best not to load the tray with too many samples at once in order to minimize changes due to evaporation.

After aspiration of samples and standards is completed, wash out the system with deionized distilled water for 15 min. Shut off pump and allow approximately 5 min for the burner to expel liquids. Shut off the oxygen and then the gas supply.

If only a small number of samples have higher concentrations of sodium and/or potassium than the highest standard, dilute the samples and re-run using the same manifold. If many samples have high sodium and/or potassium concentrations, add a suitable pump tube to dilute the samples with deionized distilled water, and re-cycle the diluted samples to the original manifold assembly. In this case, suitable higher concentrations of sodium and potassium standards should be used.

Prepare calibration curves derived from the average peak heights obtained for each of the sodium and potassium standards. Determine the concentration of sodium and potassium in the samples by comparing sample peak heights against the calibration curves.

3. Precision and Accuracy

The relative standard deviation obtained within a single lab (Water Quality Branch, Ontario Region) was ±0.67% and ±1.3% at sodium levels of 12 mg/ℓ and 67 mg/ℓ respectively. That for potassium was ±1.1% and ±1.8% at levels of 6.4 mg/ℓ and 36 mg/ℓ K, respectively.

The repeatability of the method at the 2 and 5 mg/ℓ Na concentration levels was determined to be ±0.11 and ±0.05 mg/ℓ, respectively. That at the 0.2 and 0.5 mg/ℓ K levels was ±0.04 and ±0.02 mg/ℓ K, respectively. Average recovery from a variety of natural waters has been determined to be 100%.

C. Atomic Absorption Spectrophotometric Method

The atomic absorption spectrophotometric method has been used successfully for the determination of sodium, particularly at low concentrations. Some analysts have expressed a preference for the atomic absorption method over the flame emission technique in analyzing certain water types such as river, estuarine and sea water. Measurements could be made on the acidified portion of the sample intended for trace metal analysis using an oxidizing air-acetylene flame at a wavelength of 589.0 nm. The less sensitive 589.6 nm or 330.2 nm lines could also be used where necessary. This may obviate the need for dilution of samples containing relatively high sodium concentrations.

The instrumental settings would depend on the make and model of instrument being used, and the analyst should refer to the manufacturer's instruction manual for operating procedures. In general, however, the optimum working range under standard conditions is from about 0.005 mg/ℓ to 1.0 mg/ℓ Na with a sensitivity of 0.015 mg/ℓ, although much higher concentrations may be measured directly. The method is subject to ionization interferences, which could be minimized by the addition of another alkali metal such as lithium or cesium as an internal standard. The ionization effect may be reduced substantially by using the less hot air-hydrogen flame. It should be borne in mind though, that the chances of possible chemical interferences are increased with the lower temperature flame.

Philbert and Traversy[7] reported a repeatability of ±0.04 and ±0.06 mg/ℓ Na, at the 2 mg/ℓ and 5 mg/ℓ Na concentration levels, respectively, using a Perkin-Elmer® Model 403 atomic absorption spectrophotometer.

D. Atomic Emission Spectroscopic Method

When an inductively coupled plasma emission system is available, sodium could be determined by multielemental analysis using this technique. The detection limit is in the range of about 0.0002 mg/ℓ Na with a relative standard deviation of about ±1% at the 1 mg/ℓ Na level.

E. Ion-Selective Electrode Method

The principle of the method is as outlined in the section dealing with calcium. Because the sodium electrode is also responsive to other ions, including lithium, potassium, silver, and hydrogen, the sample pH must be adjusted to ensure best electrode specificity. This may be accomplished by the addition of anhydrous ammonia, or a secondary amine such as dimethyl amine or diisopropyl amine, or a buffer solution containing morpholine. Under the conditions of the test, the method is relatively free from interferences and has a detection limit of approximately 0.1 mg/ℓ Na. As is the case for the calcium method, it is best suited for continuous monitoring and process control work, or when interferences from sources such as color, turbidity, colloidal matter, oxidants and reductants preclude application of the other methods. The simplicity, speed, and inexpensiveness of the method are some of its attractive features. Another attractive feature is the amenability of the method to on-line monitoring or to automation. Sekerka and Lechner,[16] for example, demonstrated the comparability of sodium and potassium results derived from the flame photometric method and an automated electrode method.

VI. POTASSIUM

Potassium is the seventh most abundant element in, and constitutes some 2.5% of, the earth's crust. Significant amounts of potassium may be found in feldspar, micas, and clay minerals. Contact with potassium-bearing soils, industrial discharges, and agricultural run-offs into receiving waters are some possible ways for potassium to enter surface waters.

Despite its abundance in nature, potassium is found in relatively small concentrations in most natural waters mainly because of it being reconstituted into insoluble secondary minerals formed in the process of weathering. Its concentration in natural surface waters is generally below the 10 mg/ℓ K level. Some hot springs contain up to about 100 mg/ℓ K, and seawater and brines are likely to contain several times more. There are many natural waters that are weakly radioactive and the presence of potassium-40 may contribute partially to that. For instance, it is believed that most of the radioactivity of seawater is due to the presence of K-40.

Potassium is an essential nutritive element for both plant and animal life. However, ingestion of excessive amounts (>2,000 mg/ℓ) may prove detrimental to the human nervous and digestive systems. Fortunately, potassium is seldom, if ever, a problem in potable water. On the other hand, water intended for certain industrial uses and containing potassium and sodium together in sufficiently large amounts (about 50 to 100 mg/ℓ total) may be conducive to foaming corrosion and incrustation.

Both potassium and sodium belong to the family of alkali metals and have generally similar chemical characteristics. Accordingly, the same analytical methodologies are more or less applicable to both elements. As in the case of sodium, the flame photometric, atomic absorption and atomic emission spectroscopic methods are preferred. The Ion-Selective electrode method is useful for certain special applications.

When determined gravimetrically, potassium and sodium salts present in the sample are converted first to the chlorides by addition of concentrated HCl after removal of sulfate, iron, aluminum, calcium and magnesium. A few drops of 1 + 3 H_2SO_4 are added to the solution containing the mixed chlorides, followed by the addition of a suitable volume of 10% platinic chloride solution. The potassium is precipitated as potassium chloroplatinate (K_2PtCl_6) which is filtered, washed, dried, and weighed. The method is much too long, insensitive, and imprecise for any consideration.

The colorimetric method entails precipitating the potassium with sodium cobaltinitrite in the form of the dipotassium sodium cobaltinitrite complex. Upon oxidizing with standard potassium dichromate or thiocyanate solution under acidic conditions, it reacts quantitatively with the formation of a characteristic color, the intensity of which is proportional to the amount of potassium originally present. The color may be measured either photometrically or visually. Ammonium ions and possibly silica are the only known significant sources of interference. Apart from being relatively long, the method lacks the sensitivity and precision required for the determination of this constituent and is not favored. It is my opinion, however, that automation of the procedure would result in considerable overall improvements.

A. Flame Photometric Method

As stated earlier, the flame photometric method described for sodium determination is equally applicable to potassium. Sodium, calcium, and magnesium may interfere if the sodium to potassium, calcium to potassium, and magnesium to potassium ratios equal or exceed 5:1, 10:1, and 100:1, respectively. A detection limit of 0.05 mg/ℓ K is attainable.

1. Equipment and Reagents

The equipment used is similar to that described for sodium. A stock standard potassium solution containing 1000 mg/ℓ K and a standard lithium solution containing 1000 mg/ℓ Li are prepared as previously described for sodium.

2. Procedure

Follow the procedure described for either the manual or automated method for sodium determination.

B. Atomic Absorption Spectrophotometric Method

The comments made for sodium are generally applicable to potassium in applying the atomic absorption spectrophotometric method. Potassium is usually measured at 766.5 nm, but other less-sensitive wavelengths such as 769.9 nm and 404.7 nm may be used if necessary. The optimum working range is generally from about 0.005 to 2 mg/ℓ K, with a sensitivity of about 0.04 mg/ℓ K, but considerably higher concentrations may be measured directly without major problems. The ionization interference effect is minimized by addition of another alkali metal such as lithium or cesium. An air-hydrogen flame could also be used if conditions warrant. Sodium is liable to interfere if the Na:K ratio is substantially high. In such a case, it is advisable to match approximately the sodium content of the sample by appropriate addition of a sodium standard to the potassium standard. The use of the sodium/potassium mixed standard may suffice, however.

The repeatability of the method at the 0.2 and 0.5 mg/ℓ K concentration levels was determined to be ±0.004 and ±0.001 mg/ℓ K, respectively.

VII. ACIDITY/ALKALINITY

Acidity and alkalinity are best considered together because of their interrelationship. Standard methods[6] defines the acidity of a water as "its quantitative capacity to neutralize a strong base to a designated pH". Acidity, then, is evidently dependent on the pH value and buffering capacity of the water under consideration. The presence of free carbon dioxide, strong mineral acids, weak acids of organic or inorganic composition such as carbonic, acetic, silicic, and boric acids, and hydrolyzable cations such as iron and aluminum, governs the acidity value, which itself depends on the extent to which the acids or acid salts are ionized in the sample.

Alkalinity, conversely, is a quantitative measure of the capacity of a water to neutralize a strong acid to a designated pH. Salts of weak acids or strong basic compounds are the principal contributors, and alkalinity of natural waters is attributed primarily to the presence of bicarbonates, carbonates, and hydroxides in the water. Alkalinity is thus reported as hydroxide, carbonate, or bicarbonate alkalinity, depending on certain specific relationships to be determined in making the measurements. It is dependent on sample pH, mineral composition, temperature, ionic strength and the overall buffering capacity of the water.

Basically, acidity and alkalinity give a measure of the total resistance to pH change or the buffering capacity of the given water. The measured acidity and alkalinity are expressed in terms of an equivalent concentration of calcium carbonate in mg/ℓ. It should be noted, however, that apart from carbonates, bicarbonates, and the hydroxyl ion, constituents such as silicates, borates, and phosphate may also contribute to alkalinity. Dissolved gases such as CO_2, H_2S, and NH_3 may affect acidity or alkalinity values also.

Acidity and alkalinity values are important factors to be considered in studying water chemistry and its uses. For instance, high acidity, (or low alkalinity) is a prime contributor to water corrosiveness. Alkalinity measurements are useful in studying water quality characteristics and in the evaluation and control of water and wastewater treatment processes. Too little or too much alkalinity may create problems in the treatment of water and wastewaters. Waters containing excessive alkalinity are undesirable for most uses because of the associated excessive hardness or high mineralization. In general, it is desirable to maintain water alkalinity values at such a level to ensure proper chemical balance of the water so that it is neither too corrosive nor encrusting and to ensure its potability. For example, gastrointestinal discomfort or irritation may result from consumption of excessively high-alkalinity water. When high-alkalinity water is boiled over an extended period, it may result either in the formation of a deposit or the creation of an unpleasant taste depending on its composition. Domestic and industrial use of such water invariable leads to incrustation on cooking utensils, service pipes and boilers. On the other hand, low alkalinity may promote reactions resulting in iron pick-up, and consequently, development of the so-called red water. Alkalinity in the range of 30 to 500 mg/ℓ $CaCO_3$ is generally acceptable for drinking water, while limits ranging somewhere between 85 to 150 mg/ℓ $CaCO_3$ apply generally to water required for most industrial uses.

Optimum pH, acidity, and alkalinity values are essential for healthy aquatic life. Variations in alkalinity values may result from variable uptake of CO_2, depending on aquatic productivity, and other related biological activities. According to Bricker and Garrels,[17] it is possible for aquatic organisms to produce local conditions that are conducive to the precipitation of $CaCO_3$ compounds even when the aquatic medium is undersaturated with such compounds. Acidity/alkalinity determination is valuable in studying precipitation chemistry. Recent studies have revealed that aerial transportation of pollutants such as oxides of sulfur and nitrogen results in the generation and deposition of sulfuric and nitric acids into surface waters. This may have marked adverse effects on the water chemistry, depending on the neutralizing capacity of the receiving water. Increased acidity in certain lakes, for example, may enhance the release of toxic mercury compounds in that water. All in all, the great significance that the $CaCO_3$ — H_2O — CO_2 equilibrium system has with respect to the chemical, physical and biological properties of water, makes alkalinity and acidity very important water quality parameters.

By definition, measurement of acidity/alkalinity obviously implies a choice of a variable end-point pH, depending on the water type and the purpose for which the data are required. It is customary, however, to titrate to the standard pH value of 3.7 (Standard Methods[6]), 4.5 (Analytical Methods Manual[5]), or 8.3 when determining acidity in the majority of natural waters and wastewaters. The value obtained by titrating to pH 3.7 is equivalent to what is termed methyl orange acidity, while that for an end-point of pH 8.3 is equivalent to the phenolphthalein acidity/alkalinity. It should be noted, however, that according to Kemp,[18] the methyl orange and phenolphthalein end-points in acidity/alkalinity titrations may occur anywhere between pH values of 4 to 5, and 7 to 9, respectively, depending on the molar concentration of carbonic acid in the solution. The relationships and calculation of the different alkalinity groups are given at the end of this section.

A. Acidity

Acidity is measured titrimetrically, and the end-point is detected either electrometrically or colorimetrically. The hydrogen ions present in the sample are neutralized by a standard alkaline solution. The colorimetric end-point titration method uses methyl orange or phenolphthalein indicator to denote the point at which a persistent color change indicates the equivalence point. A bromocresol green/methyl orange indicator is commonly used to denote the equivalence point pH of 4.5. This procedure is not recommended for colored or turbid samples. Also, residual chlorine and waste material present in some waters may interfere with the indicator. The electrometric end-point indicator method is more precise and accurate and is the one of choice.

A coulometric titration method described by Liberti et al.[19] may be used for the determination of strong acid concentration in precipitation. The acid is titrated at a constant current with hydroxyl ions liberated at a platinum electrode with a silver-silver bromide electrode functioning as a counter electrode. By recording the emf of a glass-calomel electrode assembly at specific time intervals until the potential changes sign, and plotting the corresponding Gran function against the electrolysis time in seconds, the end-point is determined by extrapolating the straight part of the curve. The equivalence point is denoted by the point at which the plot intercepts the abscissa.

The electrometric titration method is applicable to surface and ground waters, wastewaters and precipitation. There is no fixed applicable range, and the practical working range is governed by the acidity concentration and the volume of sample taken for titration. A detection limit of about 0.5 mg/ℓ $CaCO_3$ acidity is attainable. The method may also be used to generate a titration curve for any given sample.

The sample is titrated with standard sodium hydroxide solution to a designated pH (usually 3.7 or 4.5 and 8.3). Suspended matter, soaps, oily substances or other waste matter as well as precipitates that may be formed during the titration may affect the glass electrode. The net result is sluggish response, which should be offset by pausing for about 15 to 20 seconds between successive additions of the titrant or by adding the titrant slowly in small increments as the end-point is approached. The presence of oxidizable or hydrolyzable ions such as iron, aluminum or manganese may result in drifting end-points. Changes liable to occur in the concentrations of dissolved gases such ad CO_2, H_2S, and NH_3 during sampling, storage, or titration, will be reflected in the measurement. It is possible for metal ions contained in acid samples to neutralize the titrant. Samples from mine drainage and receiving streams and other waters containing waste acids, their salts, or substantial amounts of ferrous iron and other polyvalent cations in a reduced state must be first treated with hydrogen peroxide as recommended by the ASTM.[20]

1. Equipment and Reagents

An automated electrometric titration assembly or just simply a suitable pH meter is required.

CO_2-Free Water: several liters of deionized distilled water is boiled for about 15 min and allowed to cool to room temperature without the water getting in contact with atmospheric CO_2. The water should have a pH value of 6.0 or greater and a specific conductance below $2\mu s$/cm. It is best to prepare fresh just before use.

Standard sodium hydroxide solution, 0.10 M: sodium hydroxide (NaOH), 11 g, is dissolved in about 10 mℓ of CO_2-free water. The solution is cooled and filtered through a Gooch crucible using a pre-washed glass filter disc. Then 10.9 mℓ of the filtrate is diluted to 2.0 liters with CO_2-free water and is stored in a plastic bottle. The solution is protected from atmospheric CO_2 by keeping the bottle tightly capped or by attaching a soda-lime tube. The solution is standardized by titrating against 0.05 M standard potassium biphthalate before use.

Standard sodium hydroxide solution, 0.02 M: 200 ml of the 0.10 M solution is diluted to 1.0 l with CO_2-free water. The solution is stored as described above for the 0.10 M standard and is standardized before use with 0.010 M standard potassium biphthalate solution.

Standard potassium biphthalate solution, 0.010 M: a few grams of analytical reagent grade potassium biphthalate crystals ($KHC_8H_4O_4$) are crushed and then dried at approximately 120°C for about 2 hr. The salt is cooled in a desiccator, after which 2.0422 g is dissolved in CO_2-free water in a 1 l flask and then made up to the mark.

2. Procedure

Standardization of 0.020 M NaOH solution: pipette at least three 20.0 ml aliquots of 0.010 M potassium biphthalate into 150 ml breakers or 125 ml Erlenmeyer flasks, if they would accommodate the electrode assembly, and dilute with approximately 25 ml CO_2-free water. Place one of the flasks or beakers on the magnetic stirrer, and immerse the pH electrode assembly so as not to touch the magnetic bar and/or the sides of the container. Stir gently while titrating with the NaOH solution from a burette to an end-point of pH 8.3, taking care to add the titrant slowly as the end-point is approached. Repeat the titration using the other aliquots. The volume of titrant used should agree to within ±0.2 ml.

Calculate the molarity M_x of the NaOH from the equation:

$$M_x = \frac{0.01 \times 20}{V}$$

where V is the average volume (ml) of NaOH titrant used.

Measurement of Sample Acidity: pipette a suitable sample aliquot into a beaker or an Erlenmeyer flask. The sample size should be selected such that no more than 25 ml of the titrant is required. Samples with acidity less than about 1000 mg/l as $CaCO_3$ should be titrated against 0.02 M NaOH, while samples with acidity in excess of 1000 mg/l as $CaCO_3$ are better titrated with 0.10 M NaOH. If necessary, add CO_2-free water to the measured sample aliquot in order to ensure adequate immersion of the pH electrodes.

Immerse the washed electrode assembly into the sample and, while stirring gently, titrate with the standardized NaOH solution to pH 3.7 if the pH of the sample was initially below that value. Then continue titrating to pH 8.3 or to some other predetermined pH. Record the volumes of titrant required to reach both end-points, and repeat the determination on another sample aliquot. The volume of titrant used in each case should not differ by more than 0.2 ml.

The procedure for the H_2O_2 treatment is as follows: pipette a suitable aliquot of the sample into the beaker or flask and measure the pH. If the pH is above 4.0, add 0.10 M standardized H_2SO_4 stepwise in increments of 5 ml to bring the pH to 4 or lower. Remove the electrodes, add 5 drops of 30% H_2O_2 and boil for 2 to 4 minutes. Cool to room temperature, and titrate electrometrically to pH 8.3 as described above.

3. Calculation

Calculate the acidity from the following expressions:

$$\text{Acidity to pH 3.7 as mg/} l \text{ } CaCO_3 = \frac{50,000 \text{ } VM_x}{m l \text{ of sample titrated}}$$

where V = ml of NaOH used to titrate to pH 3.7, and M_x = molarity of NaOH titrant, or

$$\text{Acidity to pH 8.3 as mg/}\ell\text{ CaCO}_3 = \frac{50{,}000 \, V_1 \, M_x}{\text{m}\ell \text{ of sample titrated}}$$

where V_1 = mℓ of NaOH used to titrate to pH 8.3.

When the peroxide procedure is employed, the sample acidity to pH 8.3 as mg/ℓ CaCO$_3$ is calculated from the expression:

$$\text{Acidity} = \frac{[(V_1 \, M_x) - (V_2 \, M_y)] \times 50{,}000}{\text{m}\ell \text{ of sample titrated}}$$

where V_2 = volume of standard H_2SO_4 used to adjust sample pH, and M_y = molarity of the H_2SO_4 solution. A negative result in indicative of alkalinity.

4. Precision and Accuracy

A repeatability of about 2% or less is considered reasonable. The accuracy of the test is governed to a large extent by the sample composition, sample handling, and the time lapse between sample collection and the acidity determination. The sample container should be filled to the top and kept tightly capped until commencement of the titration. Titration should be done as soon as possible after the container is opened. It is advisable that the sample temperature be kept constant during the measurements.

B. Determination of Alkalinity

The methods available for making alkalinity measurements include potentiometric (electrometric) titration, colorimetry, and a CO_2 evolution procedure. The potentiometric method gives all-round good performance, but its major drawback is its slowness relative to the automated colorimetric and CO_2 evolution methods. The CO_2 evolution method results in carbonate alkalinity only, which may not necessarily equate to the total alkalinity value. The manual or automated colorimetric method involves the use of phenolphthalein or methyl orange solution for end-point detection. The method, however, is not suited for colored or turbid samples in which case the potentiometric titration procedure is preferred.

1. Potentiometric Titration Method

The method is applicable to natural surface and ground waters, precipitation and wastewaters. There is no fixed applicable range, and the practical working range is governed by the alkalinity concentration and the volume of sample taken for titration.

The results are reported as either phenolphthalein alkalinity or as total alkalinity, depending on the end-point of the titration. A detection limit of about 0.5 mg/ℓ alkalinity as CaCO$_3$ is attainable.

Phenolphthalein alkalinity is determined by titrating a sample aliquot with a standard solution of a strong acid to pH 8.3. The total (methyl orange) alkalinity is determined by titrating to a predetermined end-point, depending on the total alkalinity value and the sample type. The recommended end-point for most waters is pH 4.5, but titration to a pH of 3.7 is recommended for industrial wastes or waters of a complex nature. Standard Methods[6] recommends end-point pH values of even 5.1 and 4.8 for alkalinity values of about 30 and 150 mg/ℓ CaCO$_3$, respectively. The two end-point technique used by Thomas and Lynch[21] is employed to enable determination of the actual inflection point of the titration, i.e., the true equivalence point of the bicarbonate-carbonic acid dissociation. Furthermore, the need for precise standardization of the pH meter is eliminated. Detection of the true end-point could be affected by rough and too vigorous stirring of the sample during titration, which would be conducive to loss of CO_2 to the air.

As in the case of acidity determination, the electrode response may be rendered sluggish due to the presence of suspended matter, soaps, oily substances or other waste matter, as well as precipitates that may form during the titration. Titration of samples with high total alkalinity and hardness are likely to show a positive bias due to possible supersaturation and precipitation of $CaCO_3$ before the end-point is reached.

a. Equipment and Reagents

An automated electrometric titration assembly or just a suitable pH meter is required.

CO_2-free deionized distilled water: this is prepared as described under Acidity, Section VI.A.

Standard sodium carbonate solution, 0.010 *M*: a few grams of primary standard anhydrous sodium carbonate (Na_2CO_3) is dried at about 250°C for 4 or 5 hr and allowed to dry in a desiccator. Quickly, 1.0599 g is weighed and dissolved in CO_2-free water in a 1 *l* volumetric flask. The solution is then made up to the mark.

Standard sulfuric acid solution, 0.05 *M*: 3 m*l* of concentrated H_2SO_4 is diluted to 1.0 *l* with CO_2-free water. A 0.05 *M* HCl solution may be used instead, in which case 8.3 m*l* of concentrated HCl is made up to 1.0 *l*. The solution is standardized against a standard sodium carbonate solution as described below.

Standard sulfuric acid solution, 0.010 *M*: one hundred milliliters of the 0.05 *M* H_2SO_4 solution is diluted to 500.0 m*l* with CO_2-free water. The solution is standardized by titrating electrometrically against standard Na_2CO_3 solution.

b. Procedure

Standardization of H_2SO_4: pipette at least three 20.0 m*l* aliquots of the standard carbonate solution into 150 m*l* beakers of 125 m*l* Erlenmeyer flasks, if they would accommodate the electrode assembly, and dilute with approximately 25 m*l* CO_2-free water. Place on the magnetic stirrer, immerse the pH electrodes so as not to touch the magnetic bar or the container walls. Stir gently while titrating with the H_2SO_4 solution from the burette. Titrate to an end-point of pH 4.5, record the burette reading, and then continue titration to pH 4.2. Record the burette reading again. Repeat with the other aliquots. The volume of titrant used should agree to within ±0.2 m*l*.

Calculate the acid molarity (M_1) from the equation:

$$M_1 = \frac{M_2 V_2}{V_1}$$

where M_2 = molarity of the Na_2CO_3, and V_1 and V_2 = average m*l* of acid and carbonate solution, respectively, used for titration. The calculation of alkalinity of an unknown sample may be simplified by the determination and use of a calculation factor (f) based on the following illustrative example:

A 0.010 *M* Na_2CO_3 solution is equivalent to 1060 mg/*l* Na_2CO_3 or 1000 mg/*l* $CaCO_3$. Thus each 1.0 m*l* of the standard Na_2CO_3 is equivalent to 1.0 mg $CaCO_3$.

The volume of acid needed for the titration is given as $x_1-(x_2-x_1)$ where x_1 = m*l* of acid required to reach pH 4.5 end-point and x_2 = m*l* of acid used to titrate to 4.2 pH, i.e., the total volume of acid used. Assume that the burette reading at the start of titration was 0.00 m*l*, and readings at the pH 4.5 and 4.2 end-points were 20.94 and 21.42 m*l*, respectively.

Thus, volume of acid used to titrate the alkalinity to the equivalence point of the bicarbonate-carbonic acid dissociation

$$= 20.94 - (21.42 - 20.94)$$

$$= 20.46 \text{ m}\ell$$

So 20.46 mℓ of the acid is equivalent to 20.0 mℓ of Na_2CO_3 solution or 20.0 mg $CaCO_3$.

Thus, each mℓ of acid used

$$= \frac{20.0}{20.45} \text{ mg CaCO}_3$$

$$= 0.978 \text{ mg CaCO}_3$$

The factor to be applied in calculating the sample alkalinity will depend on the volume of sample titrated. Thus for a 20.0 mℓ sample the factor would be:

$$f = \frac{1000}{20} \times 0.978 = 48.9, \text{ whereas for a 100 m}\ell \text{ sample } f = \frac{1000}{100} \times 0.978 = 9.78$$

Sample Measurement: pipette a suitable sample aliquot into the beaker or Erlenmeyer flask. Wash the electrode assembly, and immerse it into the sample and stir gently. If the sample pH is over 8.3, titrate to 8.3, and record the volume of acid used. Continue titrating to pH 4.5 and 4.2, and record the burette readings each time. Repeat, using another aliquot.

c. Calculations

Phenolphthalein Alkalinity: let V mℓ be the volume of acid used to titrate sample to pH 8.3. Then phenolphthalein alkalinity as mg/ℓ $CaCO_3$ = Vf, where f is the calculation factor determined for the specific volume of sample titrated.

Total Alkalinity: as illustrated in the calculation example above, the volume of acid required for sample titration to methyl orange equivalence end-point is:

$$x_1 \ (x_2 \ x_1)$$

whence the total alkalinity as mg/ℓ $CaCO_3$ is = $f[x_1 - (x_2 - x_1)]$ where f is determined as illustrated above.

d. Precision and Accuracy

The same precautions outlined for acidity determination apply here also. The repeatability of the test is of the order of ±1.5% at the 35 mg/ℓ $CaCO_3$ level. Average recovery from spiked natural samples was determined to be 98% at the 95% confidence level. A positive bias was observed for certain samples containing relatively high amounts of borates and silicates. The U.S. EPA Analytical Methods Manual[22] reports interlaboratory relative standard deviations of 15.9% and 4.5% for total alkalinity measurements on synthetic water samples at the 8 and 119 mg/ℓ $CaCO_3$ levels, respectively. The same study revealed a positive bias of 10.6% and 22.3% at the 8 mg/ℓ and 9 mg/ℓ $CaCO_3$ levels, respectively, while a negative bias of 7.4% was seen at the 119 mg/ℓ $CaCO_3$ level.

e. Procedure for Precipitation Samples

The following simple procedure is being used in some laboratories for the determination of acidity/alkalinity in precipitation samples:

A suitable aliquot of sample is pipetted into a flask or beaker and titrated to pH

4.0 with standard H_2SO_4 or HCl if the sample pH is above 4.0. The volume (Va) of the acid used is recorded. The excess CO_2 formed is removed by bubbling air or an inert gas through the sample which is then titrated with standard NaOH or KOH to pH 5.6. The volume of base (Vb) is recorded. The 5.6 pH end-point is taken as being equivalent to the pH of distilled water in equilibrium with atmospheric CO_2.

The acidity or alkalinity concentration is calculated from the equation:

$$\text{Acidity as Mg/}\ell\text{ CaCO}_3 = \frac{(V_b T_b - V_a T_a)1000}{\text{m}\ell\text{ of sample titrated}}$$

where T_b and T_a are the titer (mg $CaCO_3/m\ell$) of the standard base and acid solutions, respectively. A negative value signifies alkalinity.

The between-run relative standard deviation determination for this procedure from duplicate sample analysis is ±7.6% and ±3.9% at mean acidity and alkalinity values of 1.83 mg/ℓ and 3.55 mg/ℓ as $CaCO_3$, respectively.

f. The Gran Technique

The fixed end-point titration is based on the assumption that, as far as alkalinity is concerned, the water system is defined by the carbonate/bicarbonate dissociation only and that the total carbonate concentration in the system is known and remains constant for different waters. Strictly speaking, however, this is not the case. Thus, while the fixed end-point titration is satisfactory for waters with relatively high acidity or alkalinity values (> about 20 mg/ℓ $CaCO_3$), the precision with which the true end-point could be determined and, hence, the accuracy of the results is generally unsatisfactory for waters of low alkalinity acidity such as, for example, precipitation, steam condensate, or other very soft waters. The carbonate-bicarbonate components may not necessarily be the predominant or only groups in such waters; and other buffering species, including organics and metal ions, may be of major significance. Accurate detection of the actual inflection point is essential in such cases.

Thomas and Lynch[21] demonstrated how the equivalence point pH varies with total alkalinity values. The variation is paticularly pronounced over the lower concentration range of 0 to about 50 mg/ℓ $CaCO_3$ with the corresponding equivalence end-point pH covering a range of approximately 5.5 to 5.0. Accordingly, even if the titration end-point were taken at, say, ±0.1 pH unit of the true equivalence point, this may result in considerably large Percentage errors in the total alkalinity results. In other words, the error resulting from titration to a fixed end-point pH increases with decreasing alkalinity values.

Titration of low alkalinity water to a fixed end-point pH of less than 5 will normally bias the total alkalinity results high. The preferred and recommended approach is to determine the actual inflecton pint by constructing a Gran Plot using the procedure developed by Gran[23] and now generally known as the Gran titration technique. The procedure is based on the principle that the hydrogen ion concentration of the sample increases linearly with successive increments of the strong acid added during titration. Thus the inflection point is established from a ttration curve obtained by plotting the cumulative volume of acid (titrant) added vs. the measured pH or emf reading after each successive acid increment. This requires taking a series of burette and pH meter readings for each sample titrated. It is advisable to titrate to a final pH of about 3.5 to allow for the presence of organic acids which may contribute to the sample alkalinity. For best results, the sample should be titrated in a closed vessel out of contact with the atmosphere and suitable amounts of a salt solution (e.g., NaCl or KCl) should be added to obtain a constant ionic strength. After addition of each increment of the titrant, the solution pH should become stabilized before it is recorded. In order to ensure accurate identification of the inflection point, enough data points should be obtained around that point by limiting successive titrant additions to relatively small volumes.

The calculations may be done either graphically or, more conveniently, by a linear regression analysis of the titration curve. A linear least squares analysis of a series of about 5 to 10 data points is normally adequate, but up to about 50 points may be required for a complete analysis of the entire titration curve if desired. By calculating the second order difference coefficient for each successive titration point, the inflection point is denoted by the point at which the coefficient changes sign. The Gran function (F) is determined from the equation:

$$F = (Vs + V) \ [H^+]$$

where Vs is the original sample volume, and V is the volume of titrant added. Then the F values are plotted against V for only the data points that are well enough removed from the inflection (equivalence) point. The volume of titrant corresponding to the equivalence point, and hence the sample alkalinity, is calculated according to the regression coefficients.

The desired accuracy and precision may be attained through the use of a pH meter and electrodes with a linear response over the pH range of interest, an accurately standardized acid solution, and suitable precision titration burettes, such as a micrometer screw burette, or syringes. When performed manually, the procedure is somewhat time-consuming, making it unattractive for routine sample measurements. There is now available, however, a number of automated modular titration systems, incorporating programable calculators or microprocessors, through which the measurements could be made fairly rapidly. Consequently, the technique is now considered to be operationally feasible for routine determinations of low alkalinity values. The procedure is applicable also to the determination of low sample acidity by alkalimetric titration.

2. Colorimetric Method

The phenolphthalein and total alkalinity of a sample may be determined titrimetrically, using phenolphthalein and methyl orange indicators instead of a pH meter to denote the respective end-points. An automated adaptation of the colorimetric method is now in general use for total alkalinity determination. The method has been used successfully for several years for field analysis of waters of the Laurentian Great Lakes and inter-connecting channels (Philbert and Traversy).[7] The manifold used is shown in Figure 6.

The method is applicable to natural surface and ground waters, precipitation, and wastewaters. It is not applicable to colored or turbid samples nor samples with pH lower than 3.1. The method typically covers a total alkalinity range of from 0.5 to 100 mg/ℓ CaCO$_3$. The upper concentration range may be extended by appropriate dilution of the sample.

The sample is segmented with air and then mixed with a weakly buffered solution of methyl orange. The resulting color change is measured at 550 nm by means of a continuous flow colorimeter. The pH of the methyl orange is adjusted to just below the equivalence point, such that any addition of alkalinity results in a proportional loss in color intensity.

Sample color and turbidity may interfere. Turbidity should be removed by centrifugation or filtration, but in samples where color persists, the potentiometric titration method should be used instead.

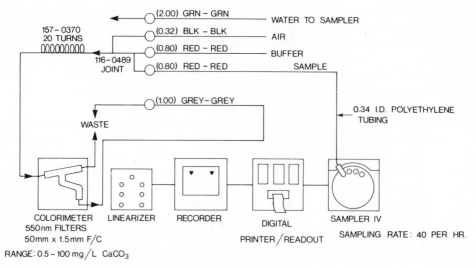

FIGURE 6.   Total alkalinity manifold.

a. Equipment and Reagents

Technicon AutoAnalyzer®-type system consisting of a sampler, proportioning pump, and alkalinity manifold is used. A colorimeter equipped with 50 mm × 1.5 mm flow cell and 550 nm filters, a strip chart recorder, linearizer, and an optional digital printer/readout, are required also.

Buffer solution, pH 3.1: potassium biphthalate ($KHC_8H_4O_4$), 10.2 g, is dissolved in 500 mℓ deionized distilled water. One molar HCl is used to adjust the solution pH to 3.1, and then the solution is diluted to 1 ℓ. This solution is stable for about 1 week.

Methyl orange indicator: methyl orange powder, 0.5 g, is dissolved in deionized distilled water and made up to 1 ℓ. The solution is filtered if necessary.

Buffered methyl orange indicator: a solution containing 85 mℓ of methyl orange solution and 250 mℓ of the buffer solution is diluted to 2 ℓ. Brij.-35® (polyoxyethylene 23 lauryl ether, available from Technicon Corp), 1.0 mℓ, is then added.

Standard alkalinity solution: this solution is prepared as described or the potentiometric titration method previously described.

b. Procedure

Set up the automated analyzer system in accordance with the manufacturer's instructions. Prepare a series of alkalinity standards to cover the expected sample range. When the baseline is steady, place the standards and samples in the sample cups. Run the standards from lowest to highest and then in the reverse order. Run the samples with a suitable standard interspersed between every 15th to 20th sample, and finally run the complete series of standards at the end. Do not leave the sample or standard aliquots exposed unnecessarily to the air.

Check for consistency in the peak heights, in which case the average peak heights are used to draw a calibration curve. If there is evidence of a drift, that must be taken into account This may be due to a number of factors of which coating of the colorimeter flow cell is the most common. This effect may be minimized by washing out the system at frequent intervals. Read the sample alkalinity from the calibration curve or directly from the digital printer readout.

c. *Precision and Accuracy*

The repeatability of the method is about 1% at the 50 mg/ℓ CaCO₃ total alkalinity level. The average between-run relative standard deviation calculated from duplicate analysis of natural samples was ±0.70% at the 90 mg/ℓ CaCO₃ level. Recoveries from natural surface water samples ranged from 90 to 99%.

3. *The Carbon Dioxide Evolution Method*

According to Thomas and Lynch,[21] the most accurate procedure for alkalinity determination is by the determination of total CO_2 content. The automated CO_2-evolution method is based on this principle and is essentially an adaptation of a similar method described by Goulden.[24] Our experiences with the method, which has been in use for the past three years, have been very good indeed. The analytical system being used is shown in Figure 7 with detail of the gas separator shown in Figure 8.

Although comparisons of the automated carbon dioxide evolution method with the conventional electrometric titration method using a variety of sample types indicated that there were no significant differences between results, any non-carbonate alkalinity present would, understandably, remain undetected by the CO_2 evolution method. The results thus obtained are better reported as "carbonate" alkalinity instead of total alkalinity. The method described enables analysis of much smaller sample volumes at an increased rate than is possible with the potentiometric method. It is applicable to the determination of alkalinity in non-turbid natural waters in the range of 2 to 200 mg/ℓ alkalinity as $CaCO_3$. The upper range can be extended by dilution of the original sample.

Inorganic carbonates are converted to carbon dioxide by reacting an aliquot of sample with acid in a closed system. The CO_2 produced is swept from the solution in a stream of oxygen and measured in a nondispersive infrared analyzer. Thus:

$$H^+ + CO_3^{--} \rightleftharpoons HCO_3^-$$
$$H^+ + HCO_3^- \rightleftharpoons H_2CO_3 \rightleftharpoons CO_2 + H_2O$$

There are no known significant interferences, but alkalinity caused by non-carbonate substances will go undetected.

a. *Equipment and Reagents*

A Technicon AutoAnalyzer® or similar system consisting of sampler, proportioning pump, and manifold is used. A non-dispersive infrared analyzer such as a Beckman® Model IR 215A, 865, or equivalent, a recorder (Beckman® Model 1005 or equivalent), and a flowmeter are also required. In this system, the oxygen supply is regulated by the flowmeter in the Beckman® 915 TOC Analyzer. The oxygen used should be very dry.

Nitric acid solution, 5% v/v: 25 mℓ of concentrated nitric acid, (HNO_3) is diluted to 500 mℓ with deionized water.

Stock standard alkalinity solution, 0.020 *M*: a few grams of primary standard anhydrons sodium carbonate is dried at about 250°C for 4 or 5 hr and then cooled in a desiccator. The salt, 2.120 g, is dissolved in CO_2-free water and the solution diluted to 1000 mℓ. This solution is equivalent to 2000 mg/ℓ alkalinity as $CaCO_3$. Changes in concentrations of this standard have been observed upon its exposure to the air, for example, when aliquots placed in the sample cups are run successively. A better-buffered and more stable standard solution, comprising a mixture of sodium carbonate and sodium bicarbonate, has been used with better success. This is prepared by mixing proportionate amounts of a 0.020 *M* sodium carbonate solution (2.120 g Na_2CO_3/ℓ) and a 0.040 *M* sodium bicarbonate solution (3.360 g $NaHCO_3$/ℓ) and making up to

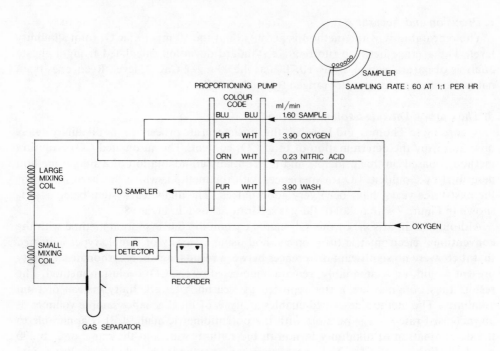

FIGURE 7.   Carbonate alkalinity manifold.

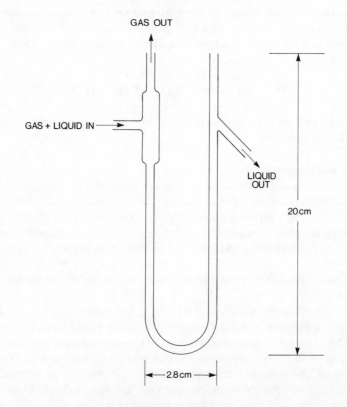

FIGURE 8.   Gas separator.

appropriate volume to yield an intermediate standard solution equivalent to 1000 mg/
ℓ CaCO$_3$. For example, 700.0 mℓ of the NaHCO$_3$ standard plus 300.0 mℓ of the

Na₂CO₃ standard would result in a mixed standard solution equivalent to 1000 mg/ℓ $CaCO_3$. The proportions of the two mixed standards used will be determined by the relative concentrations of CO_3^{--} and HCO_3^- ions in the samples to be analyzed. Apart from the stability factor, this mixed standard has the added advantage of having a composition somewhat more representative of the sample.

Alkalinity working solutions: a series of standards is prepared daily by appropriate dilution of the mixed stock solution.

b. Procedure

The following procedure is applicable if the Technicon Auto-Analyzer® components and Beckman infrared analyzer are used. In setting up the analytical system, the analyst should follow the instruction manuals supplied with the components making up the system.

Open the oxygen tank and set the operating gas pressure on the secondary gauge to about 3 to 4 psi. A flowmeter should be employed between the oxygen supply and the manifold. Set the gas flow at 275 mℓ per minute. Place the reagent lines in the proper bottles and start the proportioning pump. Turn on the infrared analyzer selection Range 1. Allow the system thirty minutes of warm-up time before proceeding to the next step. Turn on the recorder selecting the 100 mV setting. Zero both the infrared analyzer and the recorder on a deionized distilled water baseline.

Run a series of standards to cover the concentration range of the samples to be analyzed first from the lowest to the highest and then in the reverse order. Run the samples with appropriate standards interspersed between every 15th or 20th sample and then run the standards again at the end. The standards and samples should not be left standing for too long in the sample cups before being analyzed. Ater completion of samples and standards, wash out the system with distilled water for 15 min. Shut off the equipment and the gas supply.

c. Calculations

Check the respective peak heights of the standards for consistency, and use the average peak height to draw a calibration curve. If replicate peak heights of the standards differ significantly from one another, the cause should be checked and eliminated. Determine the concentration of alkalinity in the samples by comparing peak heights with the calibration curve.

d. Precision and Accuracy

In a single laboratory (Water Quality Branch, Ontario Region), the relative standard deviation, based on replicate sample analysis at concentrations of 11.3, 32.7, and 96.0 mg/ℓ $CaCO_3$, were ± 1.4%, ± 0.64%, and ± 0.58%, respectively. Recoveries from spiked samples at concentrations of 11.3, 32.7, and 96.0 mg/ℓ $CaCO_3$ averaged 102%, 107%, and 100%, respectively.

4. Relationship and Interpretation of Alkalinity Results

Theoretically, there is no rigorous direct procedure to individually determine the carbonic acid, bicarbonate, or carbonate concentration of all waters by alkalimetric or acidimetric titration. The following computational approach is commonly used in determining the concentrations of each of the basic components (i.e., H_2CO_3, HCO_3^-, CO_3^{--}, OH^-, and H^+) making up the simple carbonate equilibrium systems under consideration.

As indicated above, the predominant contributors to alkalinity are the carbonates, bicarbonates, or hydroxides present in the water. The phenolphthalein and total (methyl orange) alkalinity values provide a means for the stoichiometric classification of these three forms of alkalinity, based on the assumption that other contributing species are absent.

At the end-point of the phenolphthalein titration, the reaction between the acid and hydroxide and carbonate present would have been completed, whereas that involving the bicarbonate present is just about to start. Thus, for example, the following equations represent the reaction state at the phenolphthalein end-point:

$$Ca(OH)_2 + H_2SO_4 = CaSO_4 + 2H_2O \text{ and}$$

$$2CaCO_3 + H_2SO_4 = Ca(HCO_3)_2 + CaSO_4$$

From the above equations, it is evident that the phenolphthalein titration measures all of the hydroxide and one half of the carbonate alkalinities.

At the total (methyl orange) alkalinity end-point, we have the reactions between the acid and the hydroxides, carbonates, and bicarbonates in the sample all completed as exemplified by the following reactions:

$$Ca(OH)_2 + H_2SO_4 = CaSO_4 + 2H_2O$$

$$CaCO_3 + H_2SO_4 = CaSO_4 + H_2CO_3$$

and

$$Ca(HCO_3)_2 + H_2SO_4 = CaSO_4 + 2H_2CO_3$$

Therefore, the total alkalinity determination is essentially a measure of the total content of the hydroxides, carbonates, and/or bicarbonates present in the sample. The above conditions present five sets of possibilities centered around the following criteria:

> Carbonate alkalinity is present if the phenolphthalein alkalinity (P) is not zero but is less than the total alkalinity (T).
> Hydroxide alkalinity is present if P is greater than ½ T.
> Bicarbonate alkalinity is present if P is less than ½ T.

Thus:

1. When $P = O$, which is generally the case for most natural samples, T is equivalent to the bicarbonate alkalinity only, and we have bicarbonate concentration $= (T \times 1.219)$ mg/ℓ.
2. When $P = T$, the T value is equivalent to hydroxide alkalinity only, and hydroxide concentration $= (T \times 0.3399)$ mg/ℓ.
3. For $P < 1/2T$, the carbonate alkalinity is equivalent to 2P, and bicarbonate alkalinity makes up the rest, i.e., $T - 2P$. Thus the carbonate and bicarbonate concentrations are equal to $(T \times 1.2)$ mg/ℓ and $1.219 (T - 2P)$ mg/ℓ, respectively.
4. For $P = 1/2T$, all of the alkalinity is in the form of carbonate alkalinity and the carbonate concentration is equal to $(2P \times 0.6)$ or $0.6T$ mg/ℓ.
5. When $P > 1/2T$, hydroxide alkalinity is equal to $(2P - T)$ and carbonate alkalinity is equal to $2(T - P)$. The hydroxide and carbonate concentrations are then given as $0.3399 (2P - T)$ mg/ℓ and $1.2 (T - P)$ mg/ℓ, respectively.

In forms of ionic concentrations, we have, $[CO^{--}_3] = 2P - 2[OH^-]$ and $[HCO^-_3] = T - 2P + [OH^-]$.

The carbon dioxide content of a water is comprised of the sum of the free CO_2 and the CO_2 available in the form of carbonate and bicarbonate ions present in that sample. Thus, if in addition to the total and phenolphthalein alkalinities, the free CO_2 content is known, knowing the above relationships enables the total CO_2 value to be determined from the formula:

$$mg/\ell \text{ total } CO_2 = mg/\ell \text{ free } CO_2 + 0.88 \text{ (A + B)}$$

where A $=$ mg/ℓ bicarbonate alkalinity, and B $=$ 1/2 (mg/ℓ carbonate alkalinity). Furthermore, if the total alkalinity, pH, temperature, and mineral content of the sample are known, the carbonate, bicarbonate and hydroxide alkalinity, and the free CO_2 concentration may be determined nomographically using a series of nomographs such as given by Dye[25] or as found in Standard Methods.[6] It should be noted, however, that because of certain limitations in their applications, the nomographs must be used judiciously. For best results, the pH and temperature measurements must be done in the field immediately after sample collection.

VIII. CHLORIDE

Chloride is an anion found in variable amounts in natural waters and wastewaters. The origin of chloride in surface and ground water may be from diverse sources such as weathering and leaching of sedimentary rocks and soils, infiltration of seawater, windblown sea salt in precipitation, domestic and industrial waste discharges, municipal storm drainage, household cleaners, and run-offs from roads treated with salts in winter.

In general, the concentrations of chloride in surface waters vary according to climatic conditions and geographical location. The chloride concentration in surface waters in humid regions removed from the sea coast is normally below 10 mg/ℓ, whereas waters in semi-arid and arid regions may contain up to several hundred mg/ℓ Cl. The chloride concentration of seawater is of the order of some 20,000 mg/ℓ, whereas brines may contain as much as 10 times that amount.

Excessive chloride in potable water is not particularly harmful, and criteria set for this anion are based primarily on palatability and its potentially high corrosiveness. The distinctive salty taste that waters with relatively high chloride concentrations bear, varies depending on the accompanying cation. Thus sodium chloride gives a considerably more objectionable taste than the equivalent amount of calcium or magnesium chloride. The corrosiveness of large amounts of chlorides is enhanced in the presence of calcium and magnesium, in which case it is particularly damaging to metallic equipment and structures. High chloride concentration may adversely affect certain crops necessitating its control in irrigation water supplies. Excessive chloride content may also have undesirable effects on household and commercial food and beverage preparation. The limit set for most domestic and industrial water uses is 250 mg/ℓ Cl, but the limits for irrigation waters may range from 150 mg/ℓ to <20 mg/ℓ Cl depending on the type of crop being grown.

The concentrations of chloride in natural waters generally bear a strong correlation with the sodium content and specific conductance. Chloride determinations may serve to indicate the intrusion of waters of different compositions or to trace and measure rates and volumes of water mass movements. From an environmental standpoint, chloride is basically a conservative parameter and may serve as an index of pollution occurring in natural fresh waters from primary sources such as industrial and municipal outlets in a given watershed.

At least eight different methods, encompassing titrimetric, colorimetric, potentiometric, electroanalytical, and ion chromatographic techniques, are available for chloride determination in water. These are: the argentometric (Mohr's), the thiocyanate-titrimetric, the potentiometric titration, the mercuric nitrate titrimetric, the coulometric, the colorimetric, the ion-selective electrode, and the ion chromatographic methods.

The first three methods are based on the reaction of the chloride ion with aqueous silver nitrate to form the white silver chloride precipitate, with the main difference being in the method used for detecting the end-point. In the Argentometric method, potassium chromate is used to indicate the end-point, while the thiocyanate method entails back-titrating with a standard potassium or ammonium thiocyanate solution in the presence of ferric alum after precipitation of the chloride present by excess amounts of standard silver nitrate solution. The argentometric, mercuric nitrate, and potentiometric titration methods are the preferred procedures. The quantity of the silver chloride precipitate formed is proportional to the chloride concentration in the sample and is sometimes determined by turbidity measurements. This procedure is not favored, notwithstanding claims that chloride concentration as low as 0.005 mg/ℓ with an accuracy of $\pm$ 0.005 mg/ℓ Cl are measurable by this technique.

The mercuric nitrate method is based on the reaction of chloride ions with mercuric nitrate to form the soluble slightly dissociated mercuric chloride. The indicator used is diphenylcarbazone-bromophenol blue, which in the presence of excess H_g^{++} ions, forms the purple or violet-blue mercury diphenylcarbazone complex. The determination is highly pH-dependent, however, and requires careful control of the solution pH. There is not much to choose from between the argentometric and the mercuric nitrate titrimetric methods, except that the mercuric nitrate method is better suited for samples of relatively low mineral content; and the end-point is easier to detect compared to the argentometric method. However, methods requiring the use of the mercury salt should be used with reservation in view of the likelihood of adding to the already serious problem of mercury pollution in the aquatic environment. The potentiometric, coulometric, and ISE methods are suitable for turbid or colored waters to which the colorimetric and titration methods are inapplicable. Over the past several years, we have been using the automated mercuric thiocyanate/ferricyanide colorimetric method with very good success in analyzing a wide range of natural waters, wastewaters, and precipitation samples. The ion chromatographic application is relatively new, and we have so far had no experience with this method for chloride determination. Performance data available so far show the method to be promising for chloride in addition to a variety of other anions and cations.

A. Argentometric (Mohr's) Method

The method is applicable to natural surface and ground waters, precipitation, and wastewaters. The method is best suited for waters that are free of color or turbidity, and where the aliquot to be titrated contains about 0.1 to 10 mg Cl. The applicable range of the method depends on the volume of sample titrated. The upper range may be extended by suitable sample dilutions.

The chloride ion reacts quantitatively with a standard silver nitrate solution to form the white silver chloride precipitate as exemplified by the equation:

$$NaCl + AgNO_3 = AgCl + NaNO_3$$

Under neutral or slightly alkaline conditions, the presence of excess silver ions is detected by means of potassium chromate due to the formation of the colored silver chromate (i.e., $2AgNO_3 + K_2CrO_4 = Ag_2CrO_4 + 2KNO_3$).

Sample color or turbidity would interfere by obscuring the end-point. The presence of other halides (bromides and iodides specifically), cyanides, sulfides, sulfites, and thiosulfate would bias the results high. Orthophosphates and polyphosphates, if present in large amount (about 25 mg/ℓ or more), and iron in excess of 10 mg/ℓ Fe interfere.

1. Equipment and Reagents

Standard titration equipment is required.

Potassium chromate indicator solution: analytical reagent grade potassium chromate (K_2CrO_4), 50 g, is dissolved in about 100 mℓ of deionized distilled water. Silver nitrate solution is added until a red precipitate forms. The solution is allowed to stand away from light for about 12 to 24 hr after which it is filtered and diluted to 1 ℓ with deionized distilled water.

Standard silver nitrate solution, 0.0141 M: a few grams of analytical reagent grade silver nitrate ($AgNO_3$) crystals are crushed and heated at 40°C for about 2 or 3 hr, followed by cooling in a desiccator. The salt, 2.395 g, is dissolved in deionized distilled water in a 1 ℓ volumetric flask and the solution made up to the mark. The resulting solution is standardized against a standard sodium chloride or potassium chloride solution. This standard is stored in a dark-colored bottle away from light and is standardized each time before use.

Standard chloride solution, 0.0141 M: a few grams of analytical reagent grade sodium chloride (NaCl) crystals are dried at 140°C for a few hours and cooled in a desiccator. The salt, 0.8241 g, is dissolved in deionized distilled water in a 1 ℓ volumetric flask and the solution made up to the mark. This solution contains 500 mg/ℓ Cl.

Sodium hydroxide solution, 1 M: analytical reagent grade sodium hydroxide, 40 g, is dissolved in deionized distilled water and made up to 1 ℓ.

Sulfuric acid solution, 0.5 M: 30 mℓ of concentrated H_2SO_4 is slowly added to about 500 mℓ of deionized distilled water. The resulting solution is mixed, cooled and then made up to 1 ℓ.

Phenolphthalein indicator, 10 g/ℓ: 1g of phenolphthalein is dissolved in 100 mℓ ethanol.

Analytical reagent grade hydrogen peroxide is required.

2. Procedure

a. Standardization of AgNO₃ Solution

Pipette 25.0 mℓ of the standard chloride solution into a 250 mℓ Erlenmeyer flask or other suitable container such as a white casserole. The pH of the solution should be within the range of 7 to 10; if not, use the sodium hydroxide or sulfuric acid to adjust to approximately pH 8.3, using the phenolphthalein solution as indicator.

Place the flask on an illuminated white surface with a white background. Add about 1.0 mℓ of K_2CrO_4 indicator solution, and stir while adding the silver nitrate solution from the burette. Titrate until a faint reddish-brown or pinkish yellow color persists, and record the volume of $AgNO_3$ used. Repeat at least twice ensuring that the end-point detection is consistent. The volumes of titrant used in each case should not vary by more than about 1%. If a deep reddish-brown color forms, it indicates that the end-point has been overstepped. Determine the titration blank by repeating the above with deionized distilled water instead. Calculate the titer (mg Cl/mℓ) of the silver nitrate solution as follows:

25.0 mℓ of NaCl contain 12.5 mg Cl. Let V mℓ be the average volume, corrected for blank, of $AgNO_3$ used for titration. Then each mℓ of $AgNO_3$ titrant is equivalent to 12.5/V mg Cl = a factor f, for example.

b. Measurement of Sample

Sample color, if marked, may be removed by treatment with an aluminum hydroxide suspension. In such cases, however, it is advisable to use instead another analytical method such as the potentiometric titration or ion-selective electrode procedure. The effect of sulfide, sulfite, and thiosulfate, is present, may be overcome by adding 1 mℓ of H_2O_2 and stirring for about one min. Pipette a suitable sample aliquot, and repeat the procedure described above for the $AgNO_3$ standardization.

3. Calculation

Calculate the chloride concentration in the sample from the formula:

$$mg/\ell\ Cl = \frac{(V_S - V_B)\ 1000f}{\text{volume of sample titrated}}$$

where V_s = average volume of $AgNO_3$ required for sample titration, V_B = average volume required for blank titration, and f = calculation factor determined by $AgNO_3$ standardization.

4. Precision and Accuracy

According to Standard Methods,[6] the between-laboratory relative standard deviation and relative error at the 241 mg/ℓ Cl level were determined to be 4.2% and 1.7%, respectively, for 41 participating laboratories.

B. Automated Colorimetric Method

This method may be applied to natural surface and ground waters, wastewaters, and precipitation with a chloride content in the range of about 0.1 to 100 mg/ℓ. Higher concentrations can be brought within this range by proper sample dilution and/or adjustment of the standard calibration control on the colorimeter. The manifold and concentration of color reagent may vary depending on the range of interest. Figure 9 shows a typical manifold in use. The practical working range may be varied by including a dilution water line as appropriate and/or by suitable adjustment of the standard calibration control knob on the colorimeter.

The method is based on the displacement of thiocyanate ion (SCN^-) from mercuric thiocyanate by chloride ion and the subsequent reaction of the liberated thiocyanate ion with ferric ion to form the red colored complex, ferric thiocyanate. Thus:

$$Hg\ (SCN)_2 + 2Cl^- = HgCl_2 + 2SCN^-$$

$$SCN^- + Fe^{+++} = Fe\ (SCN)^{++}$$

The color intensity of this complex is proportional to the original chloride concentration and is measured in a colorimeter at 480 nm. The ferric ion is furnished by either a ferric alum of ferric nitrate solution. The current AutoAnalyzer® II methodology employs ferric nitrate. Natural sample color or turbidity may interfere. Bromides, iodides, cyanides, nitrites, and thiosulfates may interfere if present in significant amounts.

1. Equipment and Reagents

A Technicon AutoAnalyzer® unit or equivalent equipment consisting of a sampler, proportioning pump, chloride manifold, and a colorimeter equipped with a 15 mm flow cell and 480 nm filters, a strip chart recorder, and an optional digital printer/reader is used.

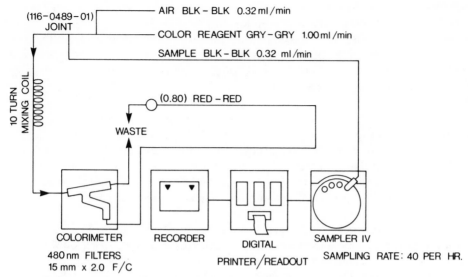

RANGE : 0.1–10 mg/L AND 0.1–30 mg/L Cl DEPENDING ON COLOR REAGENT MIXTURE.

NOTE : FIGURES IN PARENTHESES DENOTE FLOW RATES IN ml / min

FIGURE 9.   Chloride manifold.

Ferric nitrate solution, 20.2%: ferric nitrate (Fe(NO$_3$)$_3$·9H$_2$O), 202 g, is dissolved in about 500 mℓ deionized distilled water. Concentrated HNO$_3$, 22.5 mℓ, is added and the solution diluted to 1 ℓ with deionized distilled water.

Mercuric thiocyanate solution, 0.417%: mercuric thiocyanate [Hg(SCN)$_2$], 4.17 g, is dissolved in 1 ℓ of methanol, and the solution is filtered if necessary.

Color reagent: for a working range of 0.1 to 10 mg/ℓ Cl, 30 mℓ of the ferric nitrate solution is mixed with 30 mℓ of the mercuric thiocyanate solution. The resulting solution is diluted to 500 mℓ with deionized distilled water, and 0.5 mℓ of Levor® IV wetting agent or Brij-35® (available from Technicon Corp.) is added.

For a working range of 0.1 to 30 mg/ℓ Cl, mix 75 mℓ of each of the two solutions, and make up to 500 mℓ. Add 0.5 mℓ of Levor IV or Brij-35®.

Manifold cleaning solution, 4% H$_2$SO$_4$ solution: 40 mℓ of concentrated H$_2$SO$_4$, is mixed with 960 mℓ of distilled water, and Levor® IV is added while shaking until the solution becomes cloudy.

Stock chloride solution: a few grams of analytical reagent grade sodium chloride (NaCl) is dried at about 140°C for a few hr and cooled in a desiccator. An amount of 1.6482 g is dissolved in deionized distilled water in a 1 ℓ volumetric flask and the solution made up to the mark with deionized distilled water. The chloride concentration of this solution is 1000 mg/ℓ Cl.

2. Procedure

Prepare a series of working standards to cover the expected sample range. Set up the manifold and adjust the AutoAnalyzer® modules in accordance with the manufacturer's instructions. When the baseline is steady, place a series of working standards in the sampler tray starting first from the lowest to the highest, then in the reverse order and followed by the samples to be analyzed. Also, intersperse an appropriate standard between every 15th or 20th sample. Finally, place a series of standards at the end in the sampler tray. Run the standards and samples in the order that they are placed in the tray.

It is sometimes advisable to compensate for natural color effect in samples where color is prominent. This is done by running all of the reagent and sample lines with the exclusion of the color-producing reagent. The decision on whether or not this step is necessary will depend largely on the judgment and experience of the analyst. After completion of analysis, wash out the system by placing the reagent lines in the cleaning solution and running for several minutes.

When the digital printer-reader is not being used, prepare a calibration curve from the peak heights recorded. The peak heights of the respective standards run should be checked for consistency and an average height used in plotting the curve. If there are signs of drifts and/or erratic behavior, check and correct the cause before proceeding with the analysis. It may be necessary to wash out the system using the cleaning solution before performance is satisfactory. Determine the concentration of chloride in the samples by comparing peak heights against the standard calibration curve.

3. Precision and Accuracy

The within-run and between-run relative standard deviations, based on duplicate sample analyses, were determined to be 0.53% and 0.45% at levels of 15.0 mg/ℓ and 26.5 mg/ℓ Cl, respectively. The average recovery from natural water samples has been 95%.

C. Potentiometric Titration Method

The potentiometric titration method is useful, particularly in cases where prominent sample color and/or turbidity preclude determining the chloride concentration by the titration methods described above. The method is also applicable to direct chloride measurements in the presence of otherwise interfering constituents such as phosphate, chromic ions, ferrous iron, and some other heavy metals. The method is suitable for practically all water types. The range covered is similar to that given for the argentometric method.

The method is based on the measurement of the changes occurring in the potential between two electrodes of a pH meter or electronic voltmeter, equipped with glass and Ag/AgCl electrodes, while a standard solution of silver nitrate is slowly added to a measured volume of the sample. The $AgNO_3$ reacts quantitatively with, and precipitates out, the chloride present as AgCl. The end-point is reached when the meter records the greatest change in voltage for small and constant increments of $AgNO_3$. Iodides, bromides, and ferricyanides interfere and bias the results high. Ferric iron interferes if its concentration exceeds the chloride concentration by a substantial amount. Chromate and dichromate ions interfere also. Acidification with a few drops of concentrated HNO_3 is generally sufficient to counteract interferences in most natural waters, but samples containing relatively high concentrations of organic compounds and other interfering substances should be pretreated by following the procedure given in Standard Methods.[6] The potentiometric titration method is comparatively long, and, for this reason, the ISE method has an advantage over it.

D. Coulometric Titration Method

The coulometric method uses the principle of generating silver ions in solution coulometrically and determining the "dead-stop" end-point. Jacobsen and Tandberg[26] describe a simplified method for determining chloride in natural water. The sample is contained in an electrolysis cell, and a constant current of about 1.1 μA is applied between two silver wire electrodes coated with silver chloride and immersed in the sample to which is added a HNO_3-methanol mixture. The potential/time curve is traced by means of a suitable strip chart recorder. Thus by noting the time of electrolysis, i.e., the time from the start of the titration to the point at which there is a sudden decrease in the recorded potential, the chloride concentration may be determined from a calibration curve drawn for standard chloride solutions subjected to the same conditions as the sample. The method is reported to be fast and is applicable to samples containing chloride concentrations between the range of 0.1 to 100 mg/ℓ Cl.

E. Colorimetric Method

The most often used colorimetric method for chloride determination in water is the mercuric thiocyanate/ferricyanide method. However, there are several other colorimetric methods available. For example, Afghan et al.[27] describe an automated method covering a range of 0.15 to 100 mg/ℓ Cl based on the catalytic effect of chloride in the conversion of nitrate to nitrite and the subsequent measurement of the nitrite by chromotropic acid. The absorbance of the resulting colored solution is measured at 505 nm. Bertolacini and Barney[28] also describe a method involving the use of mercury chloranilate which, when reacted with chloride in solution maintained at a pH of about 7.0, liberates the reddish-purple acid chloranilate ion according to the equation:

$$HgC_6Cl_2O_4 + 2Cl^- + H^+ \rightarrow HgCl_2 + HC_6Cl_2O_4^-$$

The color intensity of the resulting solution is proportional to the chloride concentration, and absorbance may be measured at 530 nm or 305 nm. The method has a practical working range of about 0.2 mg/ℓ to about 100 mg/ℓ Cl, but the detection limit can be improved if absorbance is measured at the lower wavelength. As in the case of sulfate determination, most cations interfere and must first be removed. Anions such as bromide, iodide, fluoride, phosphate, and thiocyanate interfere also.

F. Mercuric Thiocyanate/Ferricyanide Method

This method permits both manual and automated determinations of chloride to be made. However, because it is felt that the titrimetric methods are good alternative manual procedures, and in keeping with current trends towards complete automation of analytical procedures in the laboratory, only the automated procedure will be considered here. Readers interested in the manual procedure are referred to the Annual Book of ASTM Standards.[20]

G. Ion-Selective Electrode Method

The determination of chloride concentration by means of a chloride ISE has gained a considerable amount of popularity in present-day environmental monitoring. The method is particularly suited for turbid or colored samples or in cases where continuous on-line monitoring is desired. There is a variety of chloride ion-selective electrodes commercially available. In general, these electrodes are of the silver halide-silver sulfide membrane type permitting detection of chloride concentration down to about 1 mg/ℓ Cl. Sekerka et al.[29,30] describe an improved and more sensitive solid state electrode that uses a mercuric sulfide-mercurous chloride membrane which permits measurement of chloride concentrations down to as low as about 0.020 mg/ℓ Cl.

The ISE chloride method suffers primarily from the potential interferences from sulfide, iodide, bromide, and thiocyanate ions. The sulfide ions tend to poison the electrode membrane with marked adverse effects. This effect, however, can be overcome by oxidizing any sulfide present with H_2O_2, or by adding a solution of bismuth nitrate to the sample aliquot in order to precipitate the sulfide as the bismuth sulfide before proceeding with the measurements. The other interfering ions may bias the chloride results high.

The method is quite simple and entails the addition to the sample of an ionic strength adjusting buffer (ISAB) solution, comprising a saturated potassium nitrate solution, the pH of which has been adjusted to two with nitric acid, before measuring the chloride concentration. The relative standard deviation of the method, based on replicate analysis of a simple sample is reported to be 4.2%, 0.7% and 0.14% at chloride concentrations of 0.010 mg/ℓ, 1.0 mg/ℓ, and 98 mg/ℓ, respectively. Recovery from a variety of natural waters and wastewaters ranged from 95 to 110%. In general, the results from the ISE method compared satisfactorily with those obtained by the conventional standard method.

The method can be easily automated if required. All in all, the ISE chloride method, because of its simplicity, sensitivity, speed, inexpensiveness, and its ability to generate good quality data, is considered to be an attractive and economical method of determining chloride concentrations in most, if not all, water types.

H. Ion Chromatographic Method

The ion chromatographic (IC) method for chloride determination in water and other aqueous solutions is gaining a great deal of popularity. The method has a detection limit of about 0.010 mg/ℓ Cl or better, depending on whether or not a concentrator column is used. It is of particular advantage in cases where interferences from sample color and other constituents such as bromide and iodide preclude the application of other chloride methods. The method is simple, specific, sensitive, and relatively fast and economical, particularly when used for multiple-anion analysis. The eluent in common use comprises a mixture of 0.003 M sodium bicarbonate and 0.0024 M sodium carbonate solutions. A second eluent comprising of a 0.0015 M $NaHCO_3$ solution is also required. The separator column is made up of a low capacity pellicular anion-exchange resin, while the suppressor column consists of a high capacity cation-exchange resin. The sample loop size will depend on the concentration range of interest.

Standard chloride solutions are injected to enable identification on the basis of retention time and for comparison of sample peak heights whereby the chloride concentration may be determined.

IX. SULFATE

Sulfate, being a stable, highly oxidized and soluble form of sulfur, is the form in which the element is generally present in natural surface and ground waters. Sulfate may enter water through weathering of sulfide-bearing rocks or by direct dissolution of evaporation deposits. It may be leached from sedimentary rocks and particularly from sulfate deposits such as gypsum and anhydrite. Effluents from certain industries may also be a major source of sulfate to the receiving waters. Another significant source to water systems is air-borne industrial pollutants containing oxides of sulfur which convert to sulfuric acid in precipitation (acid-rain). Sulfate can be produced also by bacterial or oxidizing action as, for example, in the oxidation of organosulfur compounds.

Sulfates in water are generally bound to alkali and alkaline earth metals and are readily soluble. Concentrations in natural waters, precipitation, and wastes will vary depending on the source. The amount present in natural water will depend largely on the type of terrain to which it has been exposed. High concentrations may be found in some well waters and surface waters in arid regions where sulfate minerals abound. Brines may contain up to about 200,000 mg/ℓ SO_4.

Sulfate is believed to play a role in the metabolism of some species of aquatic plants. Concentrations of about 200 to 300 mg/ℓ SO_4 in drinking water may prove objectionable, depending on the physiological tolerance of the consumer. In particular, consumption of excessive amounts of magnesium and sodium sulfate may have cathartic and gastrointestinal irritation effects on the consumer. However, although concentrations in excess of about 500 mg/ℓ SO₄ may impart a bitter taste to water, waters with concentrations of up to about 500 mg/ℓ and 1,000 mg/ℓ SO₄ are considered acceptable for human and livestock consumption, respectively. Sulfate is one of the secondary contributors to water hardness. Sulfate concentration is natural water seldom reaches high enough levels to adversely affect aquatic life, but concentrations reaching several hundred mg/ℓ SO₄ in irrigation water may cause problems. Excessive amounts in water intended for certain industrial and food manufacturing uses may also have undesirable effects. High sulfate concentrations may cause, in certain cases, corrosion of concrete and may accelerate corrosion and incrustation effects of water. It is not uncommon, for example, to find a hard deposit of calcium sulfate in pipes, condensers and boilers. Under certain conditions, sulfate may be reduced to the sulfide resulting in water odor and, possibly, corrosiveness.

The objective and acceptable limit set for drinking water in Canada are <250 mg/ℓ and 500 mg/ℓ SO₄, respectively. Where limits are set for industrial applications, the desirable limit is generally about 250 mg/ℓ, but a limit of 100 mg/ℓ SO₄ is considered desirable in the brewing industry.

Methods in common use for the determination of sulfate concentration in water include measurements by gravimetric, turbidimetric, titrimetric, conductimetric, potentiometric titration, colorimetric and ion chromatographic techniques. The gravimetric, turbidimetric and titrimetric (volumetric) methods are generally considered to be standard methods for sulfate determination (Standard Methods,[6] ASTM[20]) but in the present author's view, they each have certain features which make them somewhat unattractive from a practical standpoint.

The gravimetric method is based on the quantitative precipitation of sulfate by barium chloride in a hydrochloric acid medium, after removal of silica and suspended solids (e.g., $CaSO_4 + BaCl_2 = BaSO_4 + CaCl_2$). The precipitated barium sulfate is digested, filtered, washed with hot water until free of chloride, and then ignited or dried, after which it is weighed. The method is suited for a working range of about 10 to 100 mg/ℓ SO₄, although the range could be extended by working with a suitable sample volume. The method is subject to interferences from sulfite, sulfide, silica, nitrate, certain heavy metals such as chromium and iron, and suspended matter. According to Standard Methods,[6] the gravimetric method is the most accurate method for determining sulfate concentrations above 10 mg/ℓ SO₄. However, the method is time-consuming and normally requires relatively large volumes of samples. Moreover, the quality of results depends to some large measure on the skill and experience of the analyst. The relatively high detection limit of the method precludes its application to waters containing less than 10 mg/ℓ SO₄. Standard Methods[6] reports a between-laboratory relative standard deviation and relative error of 4.7% and 1.9%, respectively, for 32 laboratories analyzing an unknown synthetic sample with 259 mg/ℓ SO₄.

The turbidimetric method is based on the same principle as the gravimetric method, except the precipitated barium sulfate is measured by a nephelometer or transmission photometer. The resulting turbidity in the solution is proportional to the sulfate concentration, which is determined by converting the measured turbidity through a calibration curve prepared by following the same procedure with a series of sulfate standards. The method is much more rapid than the gravimetric procedure, but is usually less accurate than the latter at concentrations above 10 mg/ℓ SO₄. According to the Hach Company,[31] some nephelometric-type turbidimeters enable trace amounts of sulfate in solution to be measured. The precision and accuracy with which this could be accomplished is highly dependent on the skill and experience of the analyst, however. In addition to the above interferences, natural sample color may also interfere.

Standard Methods[6] reports a between-laboratory relative standard deviation and relative error of 9.1% and 1.2%, respectively, for 19 laboratories analyzing an unknown synthetic sample with 259 mg/ℓ SO$_4$. The U.S. EPA[22] reports a between-laboratory relative standard deviation of 26.7% and 5.9% for 16 laboratories analyzing a synthetic sample containing 8.6 mg/ℓ and 199 mg/ℓ SO$_4$, respectively. The bias observed at those levels were -3.7% and -1.7%, respectively.

The conductimetric method entails the titration of sulfate ions present in the sample with a standard barium chloride solution in the presence of a barium sulfate adsorption (precipitation) support. The latter is prepared by mixing solutions of H$_2$SO$_4$ and BaCl$_2$ and allowing the BaSO$_4$ precipitate to stand for about 24 hr, after which the supernatant liquid is decanted. The precipitate is then washed several times to get rid of all sulfate ions. The sulfate concentration in the sample is determined by measuring the solution conductivity with successive additions of BaCl$_2$ titrant and plotting a conductivity curve, which normally consists of two straight lines intersecting at the equivalence point. Knowing the titer of the BaCl$_2$ solution from titration of a standard sulfate solution, the concentration of the sample may be calculated.

Vogel[32] describes a potentiometric method in which sulfate ions in solution are titrated with a standard lead nitrate solution after addition of potassium ferrocyanide and potassium ferricyanide. Titration is done potentiometrically in a 1 + 1 water-ethanol solution with a platinum wire electrode and a silver reference electrode. The end-point is detected by a sharp change in the electrode potential due to the precipitation of lead ferrocyanide. Cortellessa and Napoli[33] have used an adaptation of this procedure for the determining of sulfate ions at considerably low concentrations in solution.

Already, ion chromatography (IC) is, seemingly, a good choice of method for determining sulfate concentration in some waters. The normal operating conditions are more or less the same as for chloride and facilitate simultaneous determination of these and other anions as well. A detection limit of 0.001 mg/ℓ SO$_4$ is supposedly attainable with the use of concentrator columns.

A. Titrimetric (Volumetric) Method

Sulfate concentration may be determined titrimetrically in a number of different ways. One procedure involves addition of a few milliliters of a 1% solution of hydroxylamine hydrochloride to a suitable volume of sample to overcome potential interference from ferric iron. A solution of benzidine hydrochloride is then added and the solution stirred vigorously. The benzidine hydrochloride reacts with the sulfate ions in a hydrochloric acid medium to form a slightly soluble benzidine-sulfuric acid complex, as exemplified by the equation C$_{12}$H$_8$(NH$_2$)$\cdot$2HCL + CaSO$_4$ = C$_{12}$H$_8$(NH$_2$)$_2$$\cdot$H$_2SO_4$ + CaCl$_2$. Thus the benzidine-sulfuric acid complex precipitates out. The precipitate is filtered, washed to get rid of excess HCl, and then transferred quantitatively into an Erlenmeyer flask or beaker. The precipitate is dissolved by heating to boiling in deionized distilled water. The solution is then titrated against a standard NaOH solution in the presence of phenolphthalein indicator to determine the amount of sulfuric acid, and hence sulfate ion, present. This procedure is obviously fairly time-consuming and is not favored for routine work.

A more rapid and popular method entails titration of the sulfate ions in solution with standard barium chloride solution in an alcoholic medium and under controlled conditions. The sample is titrated at a pH of 3.8 to 4.0 with thorin (disodium salt of (0[(2-hydroxy-3, 6 disulfo-1-naphthyl)azo] benzenearsonic acid) or sodium alizarinsulfonate being used to indicate the end-point. The end-point is reached when the solution color changes from yellow to pink. The solution takes on a reddish-pink color as the end-point is overstepped. The color change is relatively sharp under controlled conditions, but the analyst may experience difficulties in end-point detection when titrating some types of samples. For example, chloride in excess of about 1000 mg/ℓ Cl obscure the end-point especially at low sulfate concentrations (<10 mg/l SO$_4$). Metallic ions interfere by forming colored complexes with the indicator. Certain cations such as potassium, aluminum, and ion and anions such as phosphate, nitrate, sulfite, sulfide, and fluoride interfere by coprecipitating with the barium sulfate. Natural sample color and turbidity may interfere also. The method is applicable to natural surface and ground waters, precipitation, and wastewaters over a practical working range of about 5 to 1000 mg/l SO$_4$.

The sulfate ions in solution are titrated with a standard solution of barium chloride in an alcoholic medium under controlled conditions. Excess barium ions are detected by thorin indicator.

$$\text{Thus: } CaSO_4 + BaCl_2 = BaSO_4 + CaCl_2$$

Interfering cations are removed by first passing the sample through a cation exchange column similar to that shown in Figure 10. Ortho-metaphosphates in excess of 2 mg/l should be removed by mixing the sample aliquot to be titrated with a solution of magnesium carbonate and filtering out the precipitated phosphate. Sulfides may also be removed by precipitating it out as zinc sulfide. Sulfites must be compensated for by determining the sulfite concentration in the sample and subtracting the equivalent sulfate concentration from the titration results. The procedure given by the ASTM[20] for the determination of sulfate concentration in the presence of sulfite, phosphate, and chromium (including chromates and dichromates) should be followed where the concentrations of these constituents are significant.

1. Equipment and Reagents

A titration assembly including a burette or microburette and an ion exchange resin column are required. The ion exchange column should be regenerated when about two thirds exhausted by passing 1 + 4 HCl through the resin column and then washing thoroughly with deionized distilled water. This may be required each time after passage of about two to six samples depending on their cation content. White porcelain dishes or casseroles of approximately 100 to 125 ml capacity are used.

Alcohol: isopropanol, or 95% ethanol, or methanol is used.

Solutions of 1 + 99 ammonium hydroxide, 1 + 99 hydrochloric acid and 1 + 4 hydrochloric acid are required.

Thorin indicator solution, 0.2%: thorin [(HO)$_2$ As OC$_6$H$_4$N:NC$_{10}$H$_4$ (OH) (SO$_3$Na)$_2$], 0.2 g, is dissolved in 100 ml deionized distilled water.

Cation exchange resin: Strong cation exchange resin such as Amerlite® IR-120, Dowex® 50W - X8, Bio-Rex® 70, or equivalent is used.

Stock standard sulfate solution, 0.0104 M: analytical reagent-grade anhydrous sodium sulfate (Na$_2$SO$_4$) is dried at 105°C and, when cooled, 1.479 g is dissolved in deionized distilled water in a 1 l volumetric flask. The solution is then made up to the mark with deionized distilled water. This solution contains 1000 mg/l SO$_4$.

Working standard sulfate solution: a series of standards is prepared to cover the expected concentration range of the samples by making appropriate dilutions of the stock standard sulfate solution.

Standard barium chloride solution, 0.005 M: barium chloride (BaCl$_2 \cdot$ 2H$_2$O), 1.221 g, is transferred quantitatively into a 1l volumetric flask and is dissolved in deionized distilled water, which has been adjusted to a pH of 3.8 to 4.0 with dilute HCl or, if necessary, ammonium hydroxide solution. The solution is made up to the mark with more of the same water. Each 1.0 ml of this solution is equivalent to 0.500 mg of sulfate. The titer of the solution is checked with working standard sulfate solutions of suitable concentration each time before use. Each standard is titrated at least three

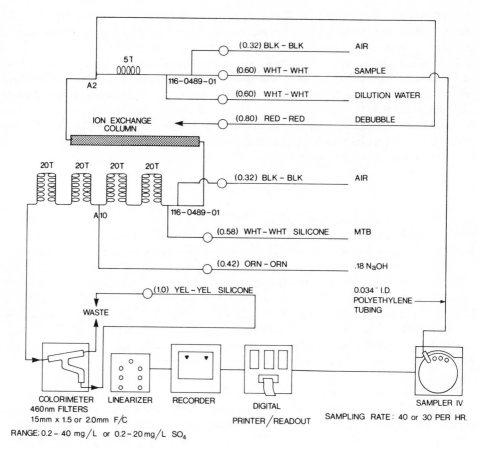

FIGURE 10.   Sulfate manifold.

times following the procedure described below for the sample and blank determination.

2. Procedure

Filter the sample if turbid, and pass about 50 mℓ of it through the ion-exchange column. Pass approximately 10 to 15 mℓ aliquots through each time, and discard the first two or three effluents. Pipette 10.0 mℓ of the next effluent into a white porcelain dish or casserole. Add 40 mℓ alcohol and 2 drops thorin indicator. Adjust the pH to 3.8 to 4.0 by carefully adding, drop by drop, NH_4OH (1 + 99 solution) until the solution just turns pink. Then add HCl (1 + 99 solution), drop by drop, until the pink color disappears: 1 drop is usually sufficient. If the NH_4OH is added too fast, it is possible to overrun the color change from yellow to pink and the sample continues to be yellow. It is then impossible to develop the pink color by addition of NH_4OH. Using an untitrated yellow blank solution in another porcelain dish or casserole as reference, titrate the sample with $BaCl_2$ solution until it just turns pink. Check the titration results by repeating with a second 10.0 mℓ sample aliquot, and then determine a titration blank by repeating the procedure with deionized distilled water instead of the sample.

3. Calculation

The sulfate concentration of the sample could be calculated in one of two ways as follows:

1. Titrate a series of the sulfate standards with the barium chloride solution and, after correcting for the titration blank, plot a calibration curve of titer (mℓ) vs. equivalent sulfate concentration in mg/ℓ. Read the sulfate concentration in the sample directly from the calibration curve.
2. Determine a calculation factor (f) from the barium chloride standardization, and calculate the sulfate concentration in the sample by the equation:

$$\text{mg/}\ell\ SO_4 = f\,(V_S - V_B)$$

where

$$V_S = \text{m}\ell \text{ of BaCl}_2 \text{ solution used to titrate the sample}$$

$$V_B = \text{m}\ell \text{ of BaCl}_2 \text{ solution used to titrate the blank}$$

and

$$f = \frac{C}{x - V_B}$$

where

$$C = \text{concentration in mg/}\ell\ SO_4 \text{ of the sulfate standard titrated}$$

and

$$x = \text{m}\ell \text{ of BaCl}_2 \text{ solution used to titrate the standard.}$$

4. Precision and Accuracy

In a single laboratory (Water Quality Branch, Ontario Region), the within-laboratory relative standard deviation was $\pm\ 3.0\%$ at 25 mg/ℓ SO_4. The method has an accuracy of the order of $\pm\ 1.5$ mg/ℓ SO_4 over the practical working range.

B. Colorimetric Method

Sulfate concentration may be determined colorimetrically in a number of ways. Application of the principle of competitive precipitation described by Bertolacini and Barney,[28,34] is the one commonly used. For example, the sample, after being passed through a cation exchange column to get rid of interfering cations, may be reacted with barium chloranilate, whereby barium sulfate is precipitated with the formation of the colored acid chloranilate ion, i.e., $BaC_6Cl_2O_4 + SO_4^{--} + H^+ \rightarrow BaSO_4 + HC_6Cl_2O_4^-$. The color intensity of the resulting solution is proportional to the sulfate concentration and is measured at 530 nm or 332 nm. Absorbance in the ultraviolet region is much greater than in the visible range, and measurements made at 332 nm enable detection of sulfate concentrations of about 0.06 mg/ℓ or lower. An automated adaptation of the method is generally used. The reaction takes place under controlled pH conditions, since unreacted barium chloranitate will hydrolyze to chloranilic acid in the presence of excess H^+ ions such as may be released upon passage of the sample through some cation exchange resins. This causes a positive bias in the sulfate results, the extent of which depends on the filterable residue content of the sample. This positive error may become apparent at filterable residue concentrations in the range of about 1000 to 1500 mg/ℓ and may be eliminated by conducting the reaction in the presence of a suitable buffer solution, by suitable sample dilution and/or by using a cation exchange resin in the sodium form. An acetate buffer solution of pH 4.63 is normally used.

MacKellar and Wiederanders[35] describe another method for indirect determination of sulfate concentration in natural waters involving spectrophotometric titration of excess barium ion with EDTA in aqueous solution at pH 10. The sample is first passed through a cation exchange resin column and then boiled gently for a few minutes. A measured amount of a standard barium chloride solution is added to ensure excess barium ions remaining after reacting with the sulfate ions present. Boiling is continued for a few minutes longer, after which the solution is allowed to cool to room temperature. The barium sulfate precipitate is filtered and washed and the filtrate plus washings quantitatively transferred to a flask. A few milliliters of an ammonia buffer solution of pH 10 is added, and the solution is made up to a specific volume. The solution is then titrated spectrophotometrically and absorbance read at 240 nm. The method is reported to have a detection limit of 5 mg/ℓ SO_4 with an average relative standard deviation of 2.2% over a concentration range of 20 to 1000 mg/ℓ SO_4.

Another method using methylthymol blue and developed by Lazrus et al.,[36] is now widely used. Our experience with this automated procedure during the past decade has been quite good, and the method is recommended for use where the sample number warrants the use of an automated procedure. Apart from its speed, the method is far more sensitive than the gravimetric or titration methods. Participation in numerous interlaboratory comparison studies have also demonstrated the accuracy and reliability of the method. The method was included about two years ago in the United States EPA Federal Register (Guidelines establishing test procedures for analysis of pollutants) as an alternative test method for compliance monitoring for the National Pollutant Discharge Elimination System (NPDES) in the United States.

The automated methylthymol blue method described below is applicable to natural surface and ground waters, precipitation, and wastewaters over a practical working range of 0.2 mg/ℓ to 300 mg/ℓ SO_4. The upper limit may be extended by suitable sample dilution and/or adjustment of the standard calibration knob on the colorimeter. Sulfate concentrations in the range of 0.2 to about 40 mg/ℓ are best measured with the manifold shown in Figure 10, while higher concentrations may be measured by appropriate reduction of the sample to dilution water ratio.

The sample is passed through a cation exchange column to remove cations that may interfere. Then sulfate ions present in solution are reacted with an equimolar alcoholic solution of barium chloride and methylthymol blue ($C_{37}H_{44}N_2O_{13}S$) at a pH of 2.5 to 3.0. The pH is then raised to 12.5 to 13.0, and the unreacted barium ions react with the methylthymol blue (MTB) to form a blue complex whereby the amount of uncomplexed MTB remaining gives a measure of the sulfate concentration. That is, the greater the concentration of sulfate originally present, the less intense is the blue color formed.

The reactions taking place may be represented by the following equations:

$$CaSO_4 + BaCl_2 + MTB \text{ (gray color)} \rightarrow BaSO_4 , + CaCl_2 + MTB \text{ (gray color) at pH 2.5 to 3.0}$$

$$2Ba^{++} + MTB \rightarrow Ba_2 MTB \text{ (blue complex)} + \text{unreacted MTB (gray color) at pH 12.5 to 13.0}$$

The absorbance of the colored solution is measured at 460 nm. There are no known significant interferences. However, samples having pH values below two should be neutralized before being passed through the cation exchange resin to prevent the elution of cations, calcium particularly, from the resin. Also sodium in concentrations exceeding about 2000 mg/ℓ Na may scrub any accumulated Ca ions from the exchange column with a resulting negative error. Anions such as sulfite, phosphate, nitrate, and fluoride that may coprecipitate with barium sulfate may cause errors. Natural sample color or turbidity may interfere.

1. Equipment and Reagents

A Technicon AutoAnalyzer® II or equivalent system consisting of colorimeter, sampler, proportioning pump, manifold, a linearizer, a recorder, and an optional digital printer/reader is used. The colorimeter is equipped with a 15 mm × 1.5 or 2.0 mm flow cell and 460 nm interference filter.

Barium chloride solution, 0.00625 M: barium chloride dihydrate ($BaCl_2 \cdot 2H_2O$), 1.5267 g, is dissolved in 500 mℓ of deionized distilled water and the solution diluted to 1 ℓ.

Methylthymol blue: methylthymol blue (3'3"-bis-N,N-bis (carboxymethyl) - amino methylthymolsulfone-phthalein monosodium salt, i.e., $C_{37}H_{43}N_2O_{13}S \cdot Na$, 0.1217 g, or 0.1286 g of the trisodium salt ($C_{37}H_{41}N_2O_{13}S \cdot Na_3$) or 0.1354 g of the pentasodium salt ($C_{37}H_{39}N_2O_{13}S \cdot Na_5$) is dissolved in a 25 m$\ell$ barium chloride solution contained in a 500 mℓ volumetric flask. Four milliliters of 1.0 M hydrochloric acid is added, upon which the solution color changes to bright orange. Then 71 mℓ of deionized distilled water is added, followed by 0.5 mℓ Brij-35®. The resulting solution is diluted to 500 mℓ with 95% ethanol. The pH of this solution should be 2.6. The solution should be prepared fresh daily and stored in a dark glass bottle. For better convenience, a stock MTB solution may be prepared by following the alternative procedure given below.

The monosodium salt, 0.6085 g, or 0.6430 g of the trisodium salt, or 0.6770 g of the pentasodium salt of methylthymol blue is dissolved in 125 mℓ of the barium chloride solution contained in a 500 mℓ volumetric flask. Twenty milliliter of 1.0 M HCl is added, followed by sufficient deionized distilled water to make up to the 500mℓ mark. The solution is stored in a dark glass bottle. This solution is stable for about 5 to 7 days when stored in the dark at about 5°C. A working solution is prepared immediately before using by pipetting 50.0 mℓ of the stock solution into a 250 mℓ volumetric flask, 0.5 mℓ of Brij-35® is added, and the solution is made up to the mark with 95% ethanol.

Buffer solution, pH 10.5 ± 0.5: ammonium chloride, 6.75 g, is dissolved in 500 mℓ of deionized distilled water, followed by the addition of 57 mℓ of concentrated ammonium hydroxide (NH_4OH). The solution is diluted to 1 ℓ with deionized distilled water.

Buffered EDTA solution: tetrasodium EDTA (ethylene-diaminetetra-acetic acid tetrasodium salt), 40 g, is dissolved in about 500 mℓ of the above buffer solution. The solution is diluted to 1 ℓ with more buffer solution.

Sodium hydroxide, 0.18 M: sodium hydroxide, 7.2 g, is dissolved in 800 mℓ of deionized distilled water. The solution is cooled and then diluted to 1 ℓ with deionized distilled water.

A solution of 1 M hydrochloric acid is prepared by diluting 83.3 mℓ concentrated HCl to 1 ℓ with deionized distilled water.

Ion exchange resin: a strong cation exchange resin such as Amberlite® IR-120, Dowex® 50W-X8, Bio-Rex® 70, or equivalent is used. Dowex® is preferred since it tends to compact less so as not to cause excessive back-pressure in the system. Fines are removed by stirring with several portions of deionized distilled water followed by decantation of the supernatant before settling is complete. The ion exchange column is prepared by putting a slurry of the resin into a piece of glass tubing about 20 cm long, 2.0 mm ID and 3.6 mm OD., or into a piece of gray or purple AutoAnalyzer® pump tubing. This is conveniently done by using a pipette and a loosely fitted glass wool plug at one end of the tubing. There should be no air bubbles trapped within the column. The column can exchange the equivalent of about 35 mg of calcium and should be regenerated when about half ot two thirds exhausted. The frequency with which this should be done would depend on the number and cationic concentration of the samples being analyzed. A 1000 mg/ℓ SO_4 stock standard sulfate solution is prepared as described in Section XVII.A.1.

Working standard sulfate solution: a series of standards is prepared to cover the expected concentration range of the samples by making appropriate dilutions of the stock standard sulfate solution.

2. Procedure

Set up the low- or high-level manifold as required. Refer to the manufacturer's instruction manual in setting up and adjusting the system modules. When the baseline is steady, place a series of working standards in the sample tray, starting first from the lowest to the highest, then in the reverse order, and followed by the samples to be analyzed. Also, intersperse an appropriate standard between every 15th or 20th sample. Finally, place a series of standards at the end of the samples in the tray.

Run the standards and samples in the order that they are placed in the tray. If it is necessary to compensate for sample color effect, run all the reagent lines as usual with the MTB line in deionized distilled water and note the response on the recorder or digital printer/readout. After completion of the analysis, wash out the system by placing the MTB and NaOH lines first in wash water and then in the buffered EDTA solution, and allow the system to run for about 15 min. Then wash all lines with distilled water for about another 15 min.

When a digital printer/readout is used in conjunction with the linearizer, the sulfate concentration of the samples may be read directly from the printout. If not, check the peak heights of the respective standards for consistency, and use average peak heights to plot a calibration curve. Compute the sample concentration by comparing peak heights against the standard calibration curve.

3. Precision and Accuracy

An average within-lab relative standard deviation of $\pm 0.75\%$ at 20 mg/l SO$_4$ has been obtained by duplicate analysis of natural samples, using the low-level manifold. The relative standard deviation at the 100 mg/l SO$_4$ level is of the order of $\pm 1.0\%$, using the high-level manifold. The mean recovery from a variety of natural and waste waters was 98.0%.

REFERENCES

1. **Philbert, F. J.**, The effect of sample preservation by freezing prior to chemical analysis of Great Lakes waters, in *Proc. 16th Conf. Great Lakes Res.*, Internat. Assoc. Great Lakes Res., 1973, 282.
2. *Guidelines for Canadian Drinking Water Quality, 1978*, Department of National Health and Welfare, Ottawa, 1979.
3. **Brittain, H. G.**, The use of hydroxynapthol blue in the ultramicrodetermination of alkaline earth and lanthanide elements: an improved method, *Analytica Chimica Acta*, 96, 165, 1978.
4. **Szekeres, L.**, Determination of calcium, magnesium, and phosphate in the presence of iron and each other, *Microchemical Journal*, 19, 330, 1974.
5. *Analytical Methods Manual*, Water Quality Branch, Inland Waters Directorate, Environment Canada, Ottawa, 1979.
6. *Standard Methods for the Examination of Water and Wastewater*, 14th ed., APHA-AWWA-WPCF, American Public Health Association, Washington, D.C. 20036, 1975.
7. **Philbert, F. J. and Traversy, W. J.**, Methods of sample treatment and analysis of Great Lakes water and precipitation samples, in *Proc. 16th Conf. Great Lakes Res.*, Int. Assoc. Great Lakes Res., 1973, 294.

8. Crowther, J., McBride, J., and Villard, S., personal communication, Ontario Ministry of the Environment, Toronto, 1978.

9. Small, H., Stevens, T. S., and Bauman, W. C., Novel ion exchange chromatographic method using conductimetric detection, *Analytica Chimica Acta*, 47 (11), 1801, 1975.

10. Dionex Corp., Operation and Maintenance Manual for AutoIon® System 12 Analyzer, Dionex Ion Chromatography Systems, Dionex Corp., 1228 Titan Way, Sunnyvale, Calif. 94086, 1979.

11. Sekerka, I. and Lechner, J. F., Simultaneous determination of total, non-carbonate, and carbonate water hardness by direct potentiometry, *Talanta*, 22, 459, 1975.

12. *The Water Encyclopedia*, Todd, D. K., Ed., Water Information Center, Inc., Port Washington, N.Y., 1970, 306.

13. Technicon Corporation, AutoAnalyzer Industrial Method No. 251-73 WM and 165-71W, Technicon Instruments Corp., Tarrytown, N.Y.

14. Sekerka I. and Lechner, J. F., Automated simultaneous determination of water hardness, specific conductance, and pH, *Analytical Letters*, 7, (6), 399, 1974.

15. Technicon Corporation, AutoAnalyzer Industrial Methodologies, Technicon Instrument Corp., Tarrytown, N.Y.

16. Sekerka, I. and Lechner, J. F., Simultaneous determination of sodium, potassium, and ammonium ions by automated direct potentiometry, *Analytical Letters*, 7 (7), 463, 1974.

17. Bricker, O. P. and Garrels, R. M., Mineralogic factors in natural water equilibria, in *Principles and Applications of Water Chemistry*, Faust, S. D. and Hunter, J. V., Eds., John Wiley & Sons, N.Y., 1967, 449.

18. Kemp, P. H., Chemistry of natural waters — 1, Fundamental relationships, *Water Research*, 5, 297, 1971.

19. Liberti, A., Possanzini, M., and Vicedomini, M., The determination of the non-volatile acidity of rain water by a coulometric procedure, *Analyst*, 97, 352, 1972.

20. *Annual Book of ASTM Standards, Part 31, Water*, American Society for Testing and Materials, Philadelphia, Pa. 19103, 1977.

21. Thomas, J. F. J. and Lynch, J. J., Determination of carbonate alkalinity in natural waters, *J. Am. Water Works Assoc.*, 52, 259, 1960.

22. *Manual of Methods for Chemical Analysis of Water and Wastes*, United States Environmental Protection Agency, Office of Technology Transfer, Washington, D.C. 20460, 1974.

23. Gran, G., Determination of the equivalence point in potentiometric titrations, Part 11, *Analyst*, 77, 661, 1952.

24. Goulden, P. D., Automated determination of carbon in natural waters, *Water Research*, 10, 487, 1976.

25. Dye, J. F., The calculation of alkalinities and free carbon dioxide in water by use of nomographs, *J. Am. Water Works Assoc.*, 36, 895, 1944.

26. Jacobsen, E. and Tandberg, G., A simple coulometric method for the determination of chloride in natural waters, *Analytica Chimica Acta*, 64, 280, 1973.

27. Afghan, B. K., Leung, R., Kulkarni, A. V., and Ryan, J. F., New automated colorimetric method for the determination of chloride using chromotropic acid, *Anal. Chem.*, 47, (3), 556, 1975.

28. Bertolacini, R. J. and Barney II, J. E., Ultraviolet spectrophotometric determination of sulfate, chloride, and fluoride with chloranilic acid, *Anal. Chem.*, 30, (2), 202, 1958.

29. Sekerka, I., Lechner, J. F., and Wales, R., Determination of chloride in water with a HgS/Hg_2Cl_2 electrode, *Water Research*, 9, 663, 1975.

30. Sekerka, I., Lechner, J. F., and Harrison, L., Analysis for chloride ion in high purity water, and heavy water of pressurized reactory and cooling systems by ion selective electrode, *Journal of the AOAC*, 60, (3), 625, 1977.

31. Hach Chemical Company, Instruction manuals for Hach Laboratory Model 2100A turbidimeter, Hach Chemical Co., Ames, Iowa 50010, 1970.

32. Vogel, A. I., *A Text-book of Quantitative Inorganic Analysis Including Elementary Instrumental Analysis*, 3rd ed., Longmans, London, 1961, 958.

33. Cortellessa, G. C. and Napoli, C. A., Potentiometric micro determination of the sulphate ion, *Analyst*, 93, 546, 1968.

34. Bertolacini, R. J. and Barney II, J. E., Colorimetric determination of sulfate with barium chloranilate, *Anal. Chem.*, 29, (2) 281, 1957.

35. MacKellar, W. J. and Wiederanders, R. S., Indirect determination of sulfate ion by spectrophotometric titration of excess barium (II) ion with ethylenediamine tetra-acetate, *Anal. Chem.*, 50, (1), 160, 1978.

36. Lazrus, A. L., Hill, K. C., and Lodge, J. P., Jr., A new colorimetric microdetermination of sulfate ion, in *Automation in Analytical Chemistry*, Technicon International Symposia, N.Y., 1966, 291.

Chapter 3

NUTRIENTS

B. K. Afghan

TABLE OF CONTENTS

I. INTRODUCTION*

Carbon, nitrogen, phosphorus, silica, and other trace elements and their compounds are widely distributed throughout the environment. These elements play an important role during the photosynthesis of aquatic macro- and microorganisms; hence, they are referred to as nutrients.

The most important nutrient elements essential to growth and maintenance of living organisms are phosphorus, nitrogen, and carbon. Silicon, iron and other elements play a lesser role and should not be considered limiting to plant growth except where diatoms (high silica content) are present. This can be illustrated by examining the phosphorus (P), nitrogen (N) and carbon (C) content in typical plant tissues relative to dry weight and fresh weight. The ratios of weight for an average community of algae are approximately 1P:7N:40C:100 dry weight:500 fresh weight. This means that if one of the three elements is growth-limiting in a lake and all other elements are present in excess of physiological needs, phosphorus can theoretically generate 500 times its weight in living algae; nitrogen, 71 (500:7) times; carbon, 12 (500:40) times.[1]

---

* Methods presented in this chapter are reproduced by permission of the Ministry of Supply and Services, Canada.[51]

All nutrients constantly undergo cycling processes which influence molecular forms by metabolism uptake, transformation, storage and release. The cycling process for each nutrient involves the incorporation of inorganic materials into organic matter (photosyntehsis) and the subsequent mineralization of organic matter to respective inorganic species (respiration) by bacteria and other respiring organisms which catalyze the redox processes and tend to restore chemical equilibrium. A simple model can be constructed to demonstrate the biochemical cycle in streams and predict chemical equilibrium conditions by fixed stoichiometry. In oxygenated water, this is represented by the following equation.[2]

$$106 \ CO_2 + 16 \ NO_3^- + HPO_4^{2-} + 122 \ H_2O + 18H^+ \ (+ \ trace \ elements, \ energy)$$

$$P \Big\uparrow \Big\downarrow \ R$$

$$[C_{106} \ H_{263} \ O_{110} \ N_{16} \ P_1] \ + \ 138 \ O_2$$
(algal protoplasm)

where photosynthesis, P, converts light into chemical energy, which results in the production of organic material-algal biomass; respiration, R, symbolizes the degradation of organic matter, resulting in the regeneration of nutrients and the release of energy.

In unpolluted waters, the steady state between photosynthetic production (algal biomass) and heterotrophic respiration (rate of destruction of organic matter) is maintained. The resultant balance between P and R is responsible for regulating the oxygen content in the water as well as maintaining an acceptable quality of water. However, this equilibrium can easily be altered by an increased supply of plant nutrients to water, resulting in increased productivity of aquatic macro- and microorganisms. This, in turn, results in oxygen depletion caused by the decay of massive growth of aquatic vegetation, resulting in the subsequent deterioration of the quality of water. This phenomena can ultimately result in the decreased use of such waters for recreation, fishing and industrial purposes. In fact, increases in the biomass resulting from intensified loadings of nutrients due to man's activities have been well documented and are familiar to limnologists.[3]

Scientists use the term ''cultural or man-made eutrophication'' to describe the process of enhancement of productivity and respiration due to over-enrichment of water by nutrients derived from human activities. Numerous cases of man-made eutrophication have been reported in the literature. Eutrophication has been observed in lakes, rivers and reservoirs in almost all highly developed countries which depend upon increased use of chemicals in order to upgrade their living standards, increase productivity and food supplies.

Since World War II, increased industrialization and urbanization of the North American continent also has affected the world's largest single fresh water body, i.e., the Great Lakes-St. Lawrence system. During the mid-sixties, the governments of the United States and Canada recognized the increased pollution of the Great Lakes, particularly Lakes Erie and Ontario. Both governments requested that the International Joint Commission (IJC), under the Boundary Water Treaty of 1909, inquire into and report upon the extent of pollution of the Great Lakes-St. Lawrence system and recommend remedial action.

Subsequently, during the period 1965 to 1969, many studies of eutrophication and other pollution problems were carried out in the Lower Great Lakes by scientists from Canada and the United States. These studies revealed that the increased loadings of nutrients, particularly phosphates, were major contributors to the process of eutrophication. Based on this, the IJC recommended an extensive program of nutrient discharge control, the most important step being control of phosphate discharges to the Lower Great Lakes.[4]

To date, numerous studies are being conducted throughout the world to assess and evaluate distribution, transformation, and effects of nutrients. In order to carry out these studies (to increase our knowledge of the behavior and effects of nutrients in the environment), it is important that the most appropriate analytical techniques be used. In order to select the most appropriate techniques for quantitative analysis of the various forms of nutrients, it is essential that the analyst possess sufficient relevant knowledge (the possible forms of each nutrient which may be present in the aquatic environment, the possible transformations the nutrients may undergo at air-water and water-sediment interfaces, the pathways of nutrients into the aquatic environment, etc.). Therefore, this chapter will initially cover the above aspects. Following this, details regarding the principles and analytical potentialities of analytical techniques that are used prior to analysis of nutrients in the aquatic environment will be discussed.

A. Transformations and Forms of Nutrients in Natural Waters

The precise and accurate chemical analysis of all forms of nutrients has always been a challenging target for analytical chemists, liminologists and oceanographers. In order to understand the magnitude of the problem, one must know what types of transformations take place when nutrients enter the aquatic system. These interactions normally occur at the air-water/water-sediment interfaces as well as in water and sediment. The transformations can be classified as physical, chemical and biological. All nutrients undergo continual changes in forms in the aquatic ecosystem and produce a wide variety of chemical species. This can be illustrated by describing the major nutrient cycles in natural wastes.

1. Carbon Cycle

Plant material and animal tissue is composed of carbon and other essential elements. Over 50% of the composition of the majority of micro- and macro-organisms is in the form of carbon. The main source of this carbon is carbon dioxide from the atmosphere which is incorporated into organic compounds chiefly by photosynthetic reactions. The organic matter produced by photosynthesis is subsequently assimilated and becomes a part of the protoplasm of plant material and/or biota in the aquatic environment.

Plant or tissue material, after the death of the cell, is decomposed by bacteria and other respiring organisms and results in the conversion of carbon to carbon dioxide. The carbon dioxide eventually is released into the atmosphere and is recycled again to form organic matter. Without this continual cycling of carbon, the small amount (0.03%) of carbon dioxide in the air would eventually be depleted by photosynthesis, and life would eventually disappear from the earth.[5]

An excellent review on the subject of the carbon cycle in the aquatic ecosystem is published by Kerr et al.[6] Figure 1 is the schematic representation of the carbon cycle showing the various components of the system. There are two chief forms of carbon involved in the cycle: inorganic and organic carbon. The main sources of inorganic carbon in the aquatic ecosystem are from atmosphere, carbonates, allochthanous inorganic carbon and biological cycling of autochthonous and allochthanous materials. The relative importance of these sources will vary with place, time, pH and chemical composition of water.

Autotrophic organisms convert carbon dioxide and/or bicarbonate into organic carbon. This autochthonous organic carbon plus any allochthanous organic material is converted by very complex interconversions by aerobic and anaerobic organisms into carbon dioxide which is then available as a carbon source for autotrophs. This results

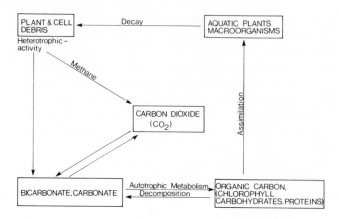

FIGURE 1.   Schematic of the carbon cycle.

in the constant change of the forms of carbon; the rates of these changes determine instantaneous availability of specific forms of carbon to organisms or plants. These forms in turn significantly affect and control the growth of the types of plants and/or organisms in the aquatic ecosystem (Figure 1).

Based on the above, the main forms of carbon may be classified as inorganic (carbonate-bicarbonate) and organic carbon. Furthermore, these forms may be present as dissolved, colloidal, or suspended forms. In water pollution analysis, these forms are present as "soluble" or "total" carbon. The distinction is based on the basis of the pore size of the filter used in the filtration of the sample prior to analysis.

2. Nitrogen Cycle

Nitrogen is the most abundant element in the atmosphere. It comprises about 78% (by volume) of the atmosphere. It can assume a number of oxidation states varying from minus 3 to plus 5 and thus is able to exist in many compounds. Atmospheric nitrogen can combine with oxygen, by lightning, to form various oxides which enter water bodies via rain. Nitrogen compounds also enter water fom sources such as precipitation, dustfall, runoffs, waste effluents, etc.

Microorganisms play a large role in transformations that occur in the nitrogen cycle in the aquatic ecosystem. The nitrogen cycle is mainly concerned with incorporation of gaseous nitrogen and the nitrogen from plants and animal tissues into forms that are usable by higher plants, whereas the carbon cycle is dependant upon the release of carbon dioxide into the atmosphere.

The transformations which play a large part in the nitrogen cycle are: fixation, ammonification, assimilation, nitrification and denitrification[7,8] (Figure 2). These reactions are carried out by microorganisms with either net gain or loss of energy. The nature and rates of reaction depend upon this net energy gain or loss. Principle compounds which are formed or assimilated are nitrogen gas, ammonia, organic nitrogen, and nitrate In natural waters, these forms are present as "soluble", "insoluble" or "suspended" nitrogen and are widely distributed in water and sediment.

3. Phosphorus Cycle

In the phosphorus cycle, many chemical physical and biological factors play a significant role. The inter-relationship between the various transformations and their relative importance is not well understood. Phosphorus does not occur as abundantly in nature as either carbon or nitrogen. It is generally accepted that phosphorus is the

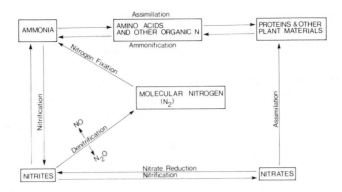

FIGURE 2.   Schematic of the nitrogen cycle.

main nutrient that triggers eutrophication, although it is required by algae in smaller quantities as compared to other nutrients. In 1970, Vollenweider[9] analyzed the data published by Thomas[10] on 46 Swiss lakes and related the decrease in carbon, nitrogen, and phosphorus in the lakes during the growing season to the initial concentrations of these elements in the spring. The seasonal decrease of these nutrients is largely due to the uptake of nutrients by the biomass. Significant correlation was found between storing concentration and decrease during the growing season for each nutrient. However, the highest correlation coefficient was found for phosphorus availability and phosphorus decrease. Phophate appeared to be the key substance governing the production of algae in these 46 Swiss lakes.

Similarly, in the fall of 1969, the IJC Report[11] identified eutrophication as a major problem in Lakes Erie and Ontario. The major study, coordinated by the IJC, of the pollution of Lakes Erie and Ontario and the International Section of the St. Lawrence River identified phosphorus as the controlling and controllable nutrient in the lake, and recommended a phosphorus control program to arrest and reverse the eutrophication problem. Subsequently, Canada and the United States passed appropriate legislations to control the phosphorus loading to the Great Lakes in order to reverse eutrophication and upgrade the quality of water in the Great Lakes.

In natural waters, phosphorus occurs principally as inorganic orthophosphate. During summer periods, the phosphate is split into two parts. In the warm waters, biological activity is intense, resulting in a depletion of orthophosphate phosphorus. However, the deeper and colder waters gain phosphate as phosphate-rich detritus falls and is decomposed by bacteria. In the deep water, some of the phosphorus precipitates as apatite, a calcium fluorophosphate mineral.[12] In water, phosphorus occurs in numerous forms such as particulate phosphorus, reactive phosphate, hydrolyzable and orthophosphate, and organic phosphate. It can occur in soluble and insoluble forms.

4. Loadings or Inputs of Nutrients to Water Bodies

The supply of materials containing nutrients essential for algal growth, such as phosphorus, nitrogen and carbon, enter the aquatic ecosystem in dissolved or suspended form. This input of materials into a water body is often referred to as "loading". These loadings may be from either man-made or natural sources. Concentrations and forms of nutrients in a given body of water depend upon the input rates from various sources as well as the chemical and biological reactions that occur within a water body in sediment and at the air/water or water/sediment interface.

Nutrient loading due to man's activities may originate from municipal, agricultural and industrial sources as well as urban runoffs. Natural sources may include precipitation, dust fall, non-urban runoffs, and biological fixation as well as sediment loading. The loadings of nutrients to a water body are estimated by sampling of wastes and streams and related discharge measurements. These measurements are normally made near the shoreline as well as at a selected network of sampling stations in a water basin. The input from tributaries is also considered by assuming that materials added to such tributaries would eventually reach the main water body.

Numerous studies are reported in the literature which relates to estimation of nutrient loadings due to man-made or natural sources. In a study on the mass contribution of the atmosphere to Lakes Huron and Superior, the atmosphere is shown to be a highly significant source of chemical materials to these lakes. The atmosphere contributes significantly to the total phosphorus and nitrogen loadings of Lakes Superior and Huron.[13] In another study on the pollution of Lake Erie, Lake Ontario, and the International section of the St. Lawrence River, it was estimated that the municipal waste discharged directly into Lake Ontario and the St. Lawrence River during 1966/67 amounted to approximately 2,100 and 13,300 tons (respectively) of total phosphorus and total nitrogen.[11] This report also gave figures for industrial waste discharges exceeding 4,000 tons/year of total nitrogen and phosphorus loadings. The nutrient loadings from other sources such as major tributaries, vessel wastes, dredging, stream bank erosion, and atmosphere are also calculated.

The material loads or wastes introduced into water bodies can produce several deleterious effects. These include excessive growth of macro- and microorganisms, an increase in the dissolved oxygen depletion, toxicity to fish life, and public health effects. The major problem created by excessive nutrients, such as carbon and phosphorus, is eutrophication (which results in nuisance blooms of algae, increased turbidity, depletion of deep-water oxygen, and changes in the composition of algae, invertebrates and fish). These symptoms were increasingly evident in the Great Lakes during the early 1960s and resulted in the deterioration of water quality.

The forms of nutrients vary depending upon location and use. For example, carbon is available in lakes from a large number of natural sources. Surface waters are usually saturated with carbon dioxide from the atmosphere. Bicarbonates and carbonates are present in abundance in most lakes due to natural chemical processes and are readily converted to carbon dioxide by well-known chemical reactions. Sewage and agricultural and industrial wastes are additional sources of carbon in polluted lakes. The types of compounds entering the water body vary enormously, depending upon the man-activity in a particular geographic location of the basin. Similarly, nitrogen enters lakes in the form of nitrates, ammonia, and organo-nitrogen compounds. These come from a variety of natural sources, including natural drainage from soils and precipitation from the atmosphere. In addition, farm fertilizers, manure, and organic wastes from municipalities contribute to the nitrogen load.

The analytical techniques and methods for the measurement of various forms of nutrients vary depending upon the form and the concentration range in a given sample. In addition, there are various other factors that must be considered prior to selection of an appropriate technique for the analysis.

B. Criteria for Selection of a Suitable Technique

In environmental analysis, the choice of a suitable analytical technique depends upon many factors. Those that are normally considered to be of prime importance are sensitivity, selectivity, precision, and accuracy. Other factors such as time of analysis, treatment of sample prior to quantitative data acquisition, sampling, cost of total analysis, and availability of a suitable standard and standard reference materials also play a significant role.[14]

1. Sensitivity

The term "sensitivity" of a technique may be defined as the lowest concentration of an ion, or substance of interest, which can be detected under well-defined and optimum conditions. It is very difficult to generalize about the relative sensitivities of numerous techniques and methods that are currently in use.

The term "lower detection limit" is used to describe the sensitivity of various methods. This may correspond to the sample signal (S) which is produced by a minimum concentration of analyte, as distinguished from the blank signal (N). In the majority of cases "lower detection limit" corresponds to a concentration which produces a S/N ratio of 2:1.

The ability of a given technique to detect the lowest amount of analyte depends upon many factors. For example, the sensitivity of atomic absorption depends on the intensity of the line emitted by the light source, on the feeding rate of the solution in the flame, atomizer efficiency, and on the spectroscopic and photometric condition. Similarly, in spectrophotometry the limit of detection is a function of intensity of color, the wavelength, and the thickness of the absorbing solution.

Sensitivity in spectrophotometric procedures is commonly expressed by two methods.[14] First, the "sensitivity index", which is the quantity of the analyte, converted to the colored product, which in a column of solution of cross-section 1 cm^2 gives an absorbance of 0.001; this is commonly referred to as Sandell's Expression. The alternate method of expressing sensitivity is based on molar absorptivity of the colored compound at the wavelength of measurement. The molar absorptivity, ε, can be calculated from the equation

$$\log \frac{I_o}{I} = A = \varepsilon \, I c, \varepsilon = \frac{A}{Ic}$$

where A = absorbance, c = concentration of colored species (mole/ℓ) and I = light path length (cm).

When the molecular formula of colored species is not known, the molar absorptivity may also be expressed with regard to one gram atom of the element determined per liter of a solution instead of one mole of the colored species. This expression of sensitivity is termed the "ionic molar absorptivity".

The knowledge of molar absorptivity or the sensitivity index enables the analyst to compare the sensitivity of various methods for determination of a particular species with several reagents.

The lower limit of detection of a method may be expressed in two ways: the absolute limit, i.e., the smallest detectable weight of substance expressed in micro- (10^{-6} g), nano- (10^{-9} g), pico- (10^{-12}) grams, etc., or the lowest detectable concentration in a sample expressed as parts per million (mg/ℓ), parts per billion (μg/ℓ), parts per trillion (ng/ℓ), etc.

2. Selectivity

The ability of a technique or a given method to provide a specific signal, under well-defined experimental conditions, may be defined as selectivity of a method. In the analysis of trace nutrients, it is important that quantitative measurement is based on the signal which is contributed by the species being measured and not other substances in the sample.

In order to obtain accurate results at trace levels, it is important that the signal to noise ratio be maximized in relation to the species being determined. This may be done by eliminating, minimizing or correcting the signal derived from interfering substances by pre-treatment of a sample. The alternate way of correcting interferences is by mathematical treatment of data, providing the response from interfering substances is not greater than that of the species of interest. In some cases, selectivity may be achieved by changing instrumental conditions so that the signal due to interfering substances is minimized or eliminated without significantly affecting the signal from the species of interest. One of the most practical ways of achieving selectivity is to obtain as much information about the chemical characteristics of the sample as possible, and select the technique that will respond to the analyte alone. In some cases, selectivity may be achieved by the use of masking agents.

3. Accuracy and Precision

In an ideal situation, the analytical result should always be equal to the true value of the analyte to be determined. In practice, it may not be possible, since the analytical result at trace and ultra-trace levels in a complex mixture is subject to absolute or relative errors from many sources. Therefore, the analyst must ensure that his results are reproducible and are as close as possible to the true value.

The term "accuracy" is defined as the correctness of measurement, and "precision" is a measure of the reproducibility of measurement. If μ = the true value, X = the value obtained experimentally, and E = the error, then

$$\mu = X \pm E$$

In routine analysis, μ is always known and is determined from $X \pm E$. In an ideal situation, if $E = 0$, then $\mu = X$, and the measurement is accurate. However, in practice, it is impossible to obtain zero error. The magnitude of the error in analytical results depends upon many factors, such as the skill and conscientiousness of the analyst, the chemical composition of the sample, the type of instrumentation, purity of reagents, contamination during sample treatment prior to analysis, etc. In the majority of cases, the analyst utilizes statistical techniques to measure the relative error, and these techniques usually measure precision rather than accuracy. However, statistical techniques are essential for the measurement of accuracy, because precision must be known before accuracy can be evaluated.

The term "bias" may be used to describe the accuracy of a measurement. In an ideal situation, an accurate method would have little or negligible bias. In practice, the bias is usually determined by analyzing a series of samples by a number of different methods, and the results are compared with the method known to give a negligible bias. If the results do not differ significantly, it is assumed that the value obtained represents a true analytical result. There are many ways of evaluating precision and accuracy and estimating analytical errors. These include calculation of standard deviation, the variance, bias, etc.[14-17]

4. Additional Considerations

When more than one method is available, the selection of an appropriate technique is usually based on other practical considerations. These include speed, versatility, ability of a technique to analyze a species on a multi-elemental basis, and the ultimate use to which the acquired data will be put.[18]

C. Factors Affecting Analytical Results

The analysis of water samples for various nutrients consists of four major steps:

1.   Sampling
2.   Pre-treatment and sample preparation (preservation and concentration)

3.   Derivation
4.   Quantitative analysis

To obtain valid results, emphasis should be given to all steps mentioned above.[19]

1. Sampling
Sampling and preservation are the most important steps in obtaining true results. Generally, the analyst is not involved in sample collection and preservation. The collection of a sample is often carried out by personnel with little knowledge of the type of analysis required and the factors that affect an analytical result. Therefore, it is essential that detailed procedures must be outlined and followed during collection. Analysis is not usually carried out on site, and most often the sample is collected, stored, and moved to a central location for analysis. The sample container can affect the concentration of many chemical parameters and, therefore, it is important to select an appropriate container for collection and storage of samples. The material used for a container should be chosen so that contamination does not occur, and the container must be washed, cleaned, and then rinsed two or three times with the sample prior to collection. For example, plastic containers are often used for sampling trace metals or inorganic chemical constituents. However, in the case of trace organics, glass containers are more often used to avoid contamination from organic materials.

In some cases, automated sampling systems are used to transport samples from the collection site to the location of analysis. These systems are thoroughly discussed by other authors.[20,21]

2. Sample Preservation and Preparation
The chemical composition of a water sample may vary from time of collection to analysis due to various physical, chemical and biological reactions. For example, it is well established that lead does plate out on the walls of glass bottles unless the sample is made distinctly acidic. Similarly, it is possible that some species of nutrients, when present at trace or ultra-trace levels, may be adsorbed onto the surface of the sample container resulting in significant loss. Condensed inorganic phosphates and polymeric silicic acids can depolymerize, resulting in increases in the concentration of ortho-phosphate and monomeric silicic acid. Biological activity can also alter the concentration of various chemical parameters, particularly nutrients such as phosphorus and nitrogen. Therefore, when water is taken for analysis, it should be treated to preserve its original composition. Three types of methods are generally used for this purpose:

1.   Storage in the dark at low temperatures, e.g., 4°C and below
2.   Immediate filtration of the sample using filters of fine pore-size (approximately 0.45 μm)
3.   The addition of preserving agents

Storage at 4°C or below and filtration are only effective in preventing or reducing biological activity.

Satisfactory sample preservation is obtained by the addition of preserving agents such as mineral acids, mercuric chloride, chloroform, etc. Jenkins[22] has investigated a number of techniques for the preservation of nitrogen and phosphorus forms in water. Methods for preserving nitrogen forms included: (1) storage at 4°C; (2) storage at -10°C; (3) storage at 4°C with the addition of 2 mℓ/ℓ of 5% sulfuric acid; and (4) storage at 4°C with the addition of 40mg of mercuric ion per liter. Mercuric ion was added to samples in the form of mercuric chloride. The best method was found to be storage at 4°C in the presence of 40 mg of mercuric ion per liter. Preservation of phosphorus for long periods (one month) was achieved by storing samples at $-10°C$ with 40 mg mercuric ion per liter.

Sample preparation usually varies depending upon the form of the nutrient to be determined, the concentration range, and desired limit of detection. In some cases, nutrients may be present in low concentrations and require preconcentration prior to analysis. Concentration techniques vary depending upon the concentration and form of nutrients. Those most commonly used include ion exchange, precipitation, and solvent extraction.

3. Derivation

In many cases, various forms of nutrients are converted or made to react with other reagents to form suitable derivatives, which can be quantified by an appropriate instrumental technique. For example, during the determination of orthophosphate by colorimetry, orthophosphate is made to react with ammonium molybdate to form the heteropoly molybdo-phosphoric acid. This is then reduced with stannous chloride in aqueous sulfuric acid medium to form molybdenum blue. The molybdenum blue color is measured in a colorimeter at 660-nm wavelength to determine the amount of orthophosphate in the sample.

There are a number of important factors that should be considered during the formation of suitable derivatives for quantitative analysis. These include pH, reagent concentration, order of addition of various chemicals, time, and stability. In routine analysis or methods development, these parameters must be closely adhered to or optimized in order to obtain precise results. For example, in many colorimetric determinations, pH and the type of buffer system must be closely controlled for the formation of a colored product according to its stoichiometry. This is particularly important when the color reaction involves metal complexes. The use of buffer systems containing anions which tend to form precipitates or strong metal complex, e.g., phosphate, citrate or pyridine, seems to affect the rate of formation of colored complexes. Therefore, in a routine analytical method, the recommended pH and buffer system should be closely followed. Any variation or modifcation should be checked and evaluated thoroughly to ensure that the precision and/or accuracy of a particular analytical method is not altered. Similarly, reagent concentration, time of reaction, etc. are important, and variations may affect the formation of the derivative.

4. Quantitative Estimation

The ultimate goal of an analyst is to obtain accurate quantitative data in a sample, i.e., a true level, which corresponds to the original concentration. The presence of foreign substances in a sample and contamination during sample pretreatment can adversely influence the final results. Therefore, the analyst must attempt to take appropriate precautions to minimize errors during the analysis.

5. Interferences

In environmental samples, the determinand is always present with a number of cations, anions, and other substances. The total concentration of other materials in a sample is generally much greater than the concentration of the determinand. Some substances that accompany the analyte may produce similar responses, while others may tend to produce quenching effects. This effect can be minimized or reduced by taking appropriate precautions. For example, if the interfering ions do not interact chemically with the determinand or the reagent, but interfere by their own absorbance at the wavelength of measurement, this interference may be reduced by measuring the absorbance of the solution before and after the complex formation and subtracting the two values.[15] An alternate method of removing interferences due to foreign ions or substaces is the use of masking agents. Masking agents preferentially form complexes with interfering substances or alter their chemical nature (oxidize or reduce) to a non-interfering state. It is important that masking agents do not react chemically with the substance to be determined or with the reagent used for the formation of colored products. In addition, these agents should not produce a similar response under the conditions of the determination. When suitable masking agent cannot be found, the interferences are usually removed by separation techniques.

6. Contamination

One of the most common sources of errors in analysis is contamination during sample preparation. Sources of error may include glassware, other apparatus, chemicals, and other reagents used during sample pretreatment, or contamination from dust and vapors (e.g., chloride, nitrate, ammonia, organic compounds) in the laboratory atmosphere.

Most errors can be minimized by the use of blank determinations, calibration curve, high-purity water and reagents, and standard addition of known quantites of analyte. The calibration curve and the slope of the curve obtained using standard additions should be compared with that in the published or standard method. The difference in slope between the curve generated in the laboratory and the published methods may reveal the possible error.

II. INSTRUMENTAL TECHNIQUES FOR MEASUREMENT OF NUTRIENTS

Numerous techniques are employed for quantitative analysis of nutrients. Comprehensive reviews and numerous books are available that deal solely with these many analytical techniques for trace constituents including nutrients.[14,18,21] The following section will give the reader a brief concept of the basic principles and scope of the instrumental techniques most widely used. These include colorimetry, fluorimetry, ion-selective electrodes, and infrared spectrometry. Descriptions of individual techniques will also include the factors which should be considered during optimization of methods in order to obtain a selective and sensitive response, and to avoid errors during analysis.

A. Colorimetry and Spectrophotometry

Colorimetric and spectrophotometric methods have been used extensively in the analysis of nutrients because of their speed, simplicity and accuracy. Nutrients such as nitrogen- or phosphorus-containing substances can be detected in microgram per liter range without any separation or pre-concentration.[24-26] These methods are based on absorption of visible or ultra-violet light by molecules, and the intensity of absorption is proportioned to the concentration of the absorbing species in the solution. When the absorption of light occurs in the visible range (approximately 400 to 700 nm) and can be monitored visually or by simple devices such as colorimeters or filters and photometers, the technique is referred to as colorimetry.

Spectrophotometry differs from colorimetry in that a narrower band of wavelength for absorption is used, and the absorption of a substance is measured over a wider wavelength range which includes the ultra-violet region of the spectrum (200 to 750 nm). In typical photometers, a monochromator is used for wavelength selection, and the degree of absorption of the substance is conveniently plotted as a function of wavelength. This plot is referred to as "absorption spectrum". The ultra-violet and visible absorption spectra provide a valuable tool in identification and quantification of substances.

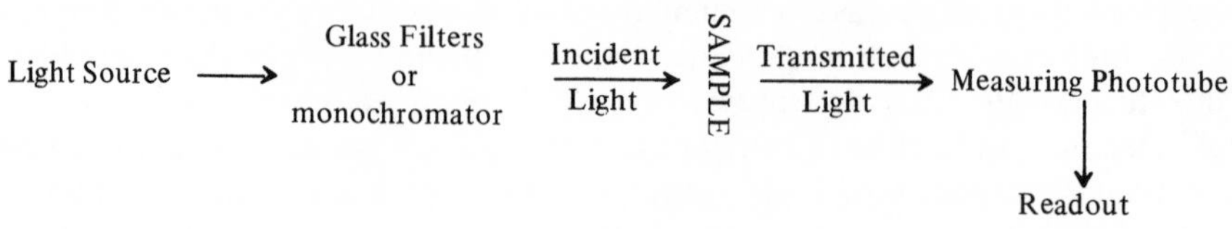

Light Source $\longrightarrow$ | Glass Filters or monochromator | $\xrightarrow[\text{Light}]{\text{Incident}}$ | SAMPLE | $\xrightarrow[\text{Light}]{\text{Transmitted}}$ Measuring Phototube

$\downarrow$

Readout

FIGURE 3.   Schematic representation of the basic components of a spectrophotometer.

In dilute solutions, the concentration of colored substance, the intensity of incident and transmitted light are interrelated. The relationship, known as Beer's Law, is the fundamental law of colorimetry and spectrophotometry.

$$\log \frac{I}{T} = A = abc$$

where A is the absorbance or optical density, I is the intensity of incident radiation, T is the intensity of transmitted light, a is the absorptivity or extinction coefficient, b is the length of light path, and c is the concentration.

It should be noted that Beer's Law is valid only when dilute solutions ($>10^{-3}$ μ) are used, and these solutions must be free of colloidal material. Other factors such as reflection, convergence, nonhomogeneity of radiation of sample, scattering, decomposition, etc. may also produce deviations from Beer's Law. Therefore, it is important that these factors be considered in order to obtain accurate results.

A typical spectrophotometer consists of five major constituents: (1) a source of radiant energy or light covering the spectrum range of interest for the measurement (tungsten and/or hydrogen lamps), (2) a spectral dispersing system to obtain the desired frequency or wavelength of absorption, (3) a sample compartment to hold a cuvette containing the solution of absorbing species, (4) a photometric system by which the absorbance of the solution may be determined, and (5) a read-out system to measure absorbance or changes in ratios of I:T. A large variety of spectrophotometers and colorimeters are commercially available. These instruments can be operated using standard cuvettes as well as in a continuous flow mode and are able to measure absorbances down to 0.001. A schematic representation showing the basic elements of a spectrophotometer or colorimeter is shown in Figure 3.

In the development of the methods using absorptiometry, the analyst may select a direct or derivative approach. In some instances, the substance itself absorbs light strong enough so that measurement can be made without chemical treatment or conversion of the parent ion or compound. For example, certain oxides of nitrogen can be determined directly by measuring their absorbance in UV-range. However, in practice, the direct methods tend to be less selective and are not used extensively. The other approach is to change the desired substance into an intensely colored compound which will absorb a specific wavelength different from other materials in the sample. A large number of such methods have been described in the literature to analyze various forms of nutrients in natural waters.[23,26]

Most of the absorptiometric methods for nutrients are based on the formation of colored products followed by the measurement of color intensity by colorimeter or spectrophotometer. Quantitative response of color reagent depends upon many factors. These include choice of reagent for the formation of colored product, spectral characteristics of the reagent and the resultant complex, reagent concentration, pH, order of addition, rate of color formation and stability of a complex, effect of temperature, and the presence of foreign substances which may interfere during the formation of the colored product or may produce interferences during the quantitative estimation. These factors must be controlled or optimized during routine analysis or methods development, in order to avoid errors.

Errors can also occur during color measurement due to instrumentation. These errors could be due to photocell response, type of cuvettes, temperature of solution during color measurement, light source slit, width of spectrophotometers, and the voltage/power fluctuation in the laboratories during the measurement of absorbance. The use of inappropriate calibration curves may also cause errors. For example, although colored systems generally follow Beer's Law, deviation may be encountered if the concentrations are too high or too low. Therefore, it is important to prepare a calibration curve for the desirable concentration range. The slope of the calibration curve may also change depending upon the concentration range and the chemical composition of the sample. For example, the calibration curve for determination of ammonia or phosphate may exhibit different slopes in fresh water and sea water. This is due mainly to high salinity differences between the two types of samples. Therefore, it is important to prepare calibration curves in synthetic solutions which closely correspond to the chemical composition of the sample. In many laboratories, it is a common practice to prepare the calibration curve in distilled/deionized water and employ it for various types of samples, ranging from lake water containing low concentrations of major ions to industrial wastes. This practice can produce significant errors in analytical results. Therefore, it is recommended that synthetic solutions be prepared to contain similar levels of major ions and use these solutions for the calibration curve. For example, if the analyst is analyzing lake water or sea water, he must prepare synthetic lake water and synthetic sea water with a similar matrix (main constituents of sample) for the calibration curve. An alternate method for avoiding errors of this type is to employ standard addition techniques. In this case, a known amount (X, 2X, 4X . . .) of standard solution of analyte is added to the sample, and the concentration is calculated by extrapolating the curve to zero.

B. Fluorimetry of Spectrofluorimetry

Methods based on fluorescence are becoming widely accepted analytical procedures because of their potentially high sensitivity and selectivity. Fluorescence is caused by the absorption of radiation energy and re-emission of this energy, usually at lower frequency/higher wavelength than that absorbed. For example, if ultraviolet light is absorbed, the emission may be in the longer wavelength of the spectrum, i.e., ultraviolet, blue, green, yellow or red. If blue light is absorbed, the material may fluoresce (or emit) green, yellow, or red, etc. The mechanism by which substances normally fluoresce can be described as follows: when a beam of light passes through a material, its energy will reappear in a variety of forms. Part of the light will be absorbed, part will be reflected, and part will be scattered in various ways. If part of the light is absorbed by molecules in a solution, then the transitions take place and the molecules are moved to an electronically excited state. Depending on the frequency of the light absorbed, the molecule is raised to a particular vibrational level — one of the upper electronically excited states, as shown in Figure 4. After excitation, almost all molecules rapidly drop back to the lowest vibrational level of the first excited state, and it is from this level that the molecule can return to ground state by emitting fluorescence. It is possible that the excited molecules may return to ground state by other mechanisms and hence no fluorescence.[27] The relationship between fluorescence intensity and concentration is given by the following equation:

$$F = [I_0 (1 - 10^{\epsilon cd})] [\phi]$$

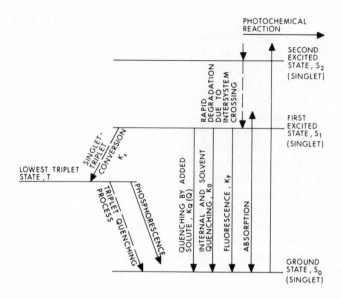

FIGURE 4.   Transitions giving rise to absorption and fluorescence.

F is the fluorescence intensity, I_o is the intensity of excited light, c is the concentration of fluorescing substance, d is the optical path, ε is the molar extinction coefficient, and ϕ is the quantum efficiency of fluorescence.

Ideally for dilute solutions:

$$F = I_o K_{\varepsilon cd} \quad \text{or} \quad F \alpha I_o, D, d$$

Any given fluorescent molecule is generally characterized by two spectra, the excitation spectrum, which causes the excitation of molecules, and the emission spectrum, which is used to determine the fluorescence efficiency of a molecule. Therefore, the instrument for fluorescence spectrometry must contain a light source, sample holder, detection system, and devices to select desired frequency for excitation and emission. A diagram of a typical, general-purpose spectrofluorometer is shown in Figure 5.[28]

For quantitative purposes, the desired wavelength of exciting light is selected by the filter or monochromator (called the primary filter or excitation monochromator) between the light source and the sample. The wavelength of fluoresnce is selected by a second optical filter or monochromator (called the secondary filter or emission monochromator) between sample and photo detector. Fluorescence is normally measured at right angles to eliminate any interference due to absorption or scattered light.

Methods based on fluorescence are normally more sensitive than absorptiometric methods. One reason for this difference is that the emitted light, the quantity related to concentration, is measured directly and can be increased by increasing the intensity of the light source used for excitation. In absorptiometry, however, the analogous quantity (absorbance or optical density) is measured indirectly as the ratio between the incident light and transmitted light (I_o/T). Since this ratio cannot be measured by the majority of laboratory instruments to an accuracy better than one part in thousand (0.001), the detection limits for colorimetric or spectrophotometric are normally higher than for fluorescence methods. In addition, the fluorescence intensity is directly proportional to the intensity of the excited source and, therefore, it is possible to obtain increased sensitivity by employing an intense source of excitation. On the other hand, in colorimetry absorbance, I_o/T is measured as a ratio of incident to transmitted light, and it is not possible to increase absorbance by changing the intensity of the source.

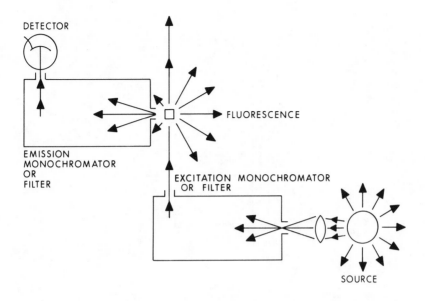

FIGURE 5. Schematic of a general-purpose spectrofluorometer.

Apart from increased sensitivity, the fluorescence methods are more specific than the absorptiometric methods. The most important factor affecting the selectivity of fluorometric methods is that since many compounds absorb light, only a few fluoresce. A potentially interfering substance which absorbs light but does not fluoresce is a positive interference in absorptiometric methods but is ignored during fluorescence measurement. A second important factor is that the simultaneous choice of two wavelengths in fluorescence, instead of one in absorptiometry, provides an extra degree of freedom — two compounds absorbing at the same wavelength need not emit at the same wavelength.

In order to develop a suitable method for quantitative analysis the following factors should be kept in mind:

1. Selection of optimum excitation wavelength
2. Selection of optimum emission wavelength
3. Choice of sample container
4. Effect of solvent
5. Effect of pH
6. Effect of temperature
7. Interference

Methods for quantitative analysis may be based on direct measurement of the fluorescence of the determinand or by converting it to a suitable fluorescent derivative prior to analysis. The earlier discussion on sources of error in colorimetry or spectrophotometry is also applicable to fluorescence. The conditions causing errors in color reaction may also affect fluorescent reactions.

C. Ion-Selective Electrodes

The instrumental setup used to make measurements with an ion-selective electrode involves the setting up of an electrochemical cell. The operation of the system is similar to a pH-meter, and the technique is concerned with the measurement of potentials, known as potentiometry. In a solution, an ion-selective electrode and a suitable refer-

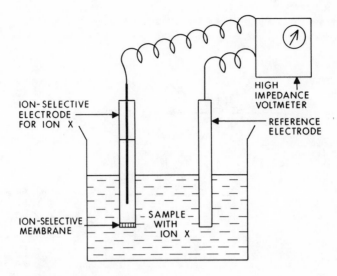

FIGURE 6.   Set-up for selective-ion electrode measurements.

ence electrode produce a net voltage which is measured with a high-impedance milli-voltmeter. The potential of the ion-selective electrode is a function of the activity or concentration of the substance or substances undergoing exchange of electrons respon-sible for the potential change at the surface of the electrode. Measurement of the elec-trode potential, therefore, allows the concentration of the substance to be found.

The measuring setup consists of the ion-selective electrode — sample — reference electrode. A configuration of the measuring system is shown in Figure 6. When the ion-selective electrode of ion X is bought into contact with a solution (X), there is a tendency for the electrode to establish electrical potential (E.M.F.) as a result of ex-change of electrons according to redox couple present:

$$Ox + ne^- \underset{\text{oxidation}}{\overset{\text{reduction}}{\rightleftharpoons}} Red$$

where Ox and Red are oxidized and reduced forms of an ion, respectively, and n is the number of electrons, e^-, involved in the reaction. In the reversible system, the electrical potential of the cell consisting of ion selective electrode X, reference elec-trode, and the activity of ion X in a solution can be expressed, when no current is flowing, by a classical Nernst equation or its expanded version.[29]

$$E = E^° \pm \frac{RT}{zF} \ln a_X$$

where $E^°$ is the standard potention of the cell, a_x is the activity of ion X, z is the number of charges on X, R is the gas constant, t is absolute temperature, and F is Faraday's constant.

Ion-selective electrodes measure activity of ions in solution. In analytical work it is convenient to measure concentration on molar or molal scale, or in equivalent units. In dilute solutions (below 10^{-4} *M*), the activity of the ion is virtually identical to molar concentration [X]; hence, a_x may be substituted by [X] without significant error in analytical results. It is also possible to simplify the Nernst's equation. For example, R and F are constant, most potentiometric measurements are made at 25°C, and most analysts use logarithms to the base 10; a more convenient form of Nernst's equation may be expressed as:

$$E = E° \pm \frac{59.12}{z} \log_{10} [X] \text{ at } 25°C$$

The use of the ion-selective electrode can be traced back to the early 1900s. The hydrogen electrode was one of the first of this type used to measure hydrogen ions in solution. It is only during the last ten to fifteen years that ion-selective electrodes and gas-sensing probes have achieved popularity, and the field has been subject to increasing research.

Ion-selective electrodes are increasingly being used in water quality. These electrodes offer a unique capability for measuring a wide variety of chemical and bio-chemical parameters that affect water quality. They find increasing use in analysis and monitoring of various forms of nutrients in natural waters, wastewaters, industrial effluents, and sediments. They are also used for continuous monitoring of rivers, streams, effluent, and municipal wastes because they are relatively inexpensive, simple, portable, and measurements can be made within minutes. In addition, the sample does not require extensive pretreatment. Opaque solutions and even slurries can be measured directly. Recently, Bailey[29] has published a comprehensive text covering all aspects of ion-selective electrodes, including analytical applications.

Many different types of selective ion electrodes have been reported in the literature. These can be classified into a few major categories:[29] glass electrodes, electrodes based on inorganic salts, electrodes based on organic ion exchangers, and neutral carriers and gas probes.[29]

Ion-selective electrodes based on inorganic salts and gas-sensing probes are most commonly used for analysis of ammonia, cyanide, and other inorganic ions in water and wastewaters. Inorganic salts employed for the preparation of these electrodes include silver halides,[30] silver sulfide,[31] lanthanum fluoride,[31] and mixtures of silversulfide and mercurous chloride.[33] Gas membrane probes are different from the typical ion-selective electrodes, because these probes are complete electrochemical cells and incorporate both an ion-selective electrode and reference electrode within the probe. These probes are used to measure gases and are very useful for determining ammonia, carbon dioxide, sulfur dioxide, cyanide, etc. There are two types of gas-sensing probes available: probes with membrane and probes without membrane (air-gap electrodes). The membrane probes contain an ion-selective electrode sensitive to determinand gas, reference electrode, and a thin film of internal electrolyte. The electrochemical cell is held by a special container with a gas permissible membrane at the bottom. When the probe is immersed into a solution, the determinand gas diffuses through the membrane and establishes an equilibrium partial pressure between the internal electrolyte of the probe and the sample solution. This changes the chemical composition of the thin film of internal electrolyte, resulting in a change of potential of the ion-selective electrode which is related to the concentration of the determinand gas. In gas membrane probes without the membranes, the membrane is replaced by a small air gap (few millimeters), and the probe is suspended over the sample solution in a specially designed vessel. The determinand gas from the sample is evolved by either stirring or chemical reaction. The gas thus evolved reacts with the thin film of electrolyte on the surface of the probe, resulting in a voltage change which gives a measure of the quantity of determinand gas in the sample. The thin film around the probe is rejuvenated after each determination. The theory and applications of these probes are extensively covered in the recent publication by Bailey.[29] These probes are used extensively for the analysis of ammonia in water, effluents, and sewage effluents.[33-35]

The practical procedures in the ion-selective electrode are based on measurement of potential changes and calculation of concentrations of analyte. The pH/millivoltmeter is used for measuring voltage changes. The meter should be able to read down to ±0.2 mV or better. The meter should possess at least one high impedance input (greater than 10^{12} Ω) where the ion-selective electrode is plugged. Modern millivoltmeters feature double high-impedance inputs, so that conventional reference electrodes may be replaced by other ion-selective electrodes as reference electrodes. This is advantageous when ion-selective electrodes are used to analyze very low concentrations of analyte. Conventional reference electrodes utilize an internal electrolyte which is either a saturated solution, such as potassium chloride, or contain high concentrations of chloride ion in order to maintain a stable reference potential. This, in most cases, is the major source of contamination when low concentrations are analyzed by this technique. For example, traditional pH/millivoltmeters have one low and one high impedance source and, therefore, these meters can only be operated using a standard mode, i.e., ion-selective electrode/ sample/ conventional reference electrode. In this configuration, the internal electrolyte can contaminate the sample and limit the sensitivity. However, using a modern pH/millivolt meter, containing two high-impedance inputs, it is possible to replace the standard reference electrode by other ion-selective electrodes and eliminate the possibility of contamination of the sample.

There are several methods whereby net potential changes can be converted to the concentration of the determinand. These include direct method, incremental (standard addition/subtraction) methods, Gran plot, and titrimetry. The choice of method usually depends upon such factors as composition of sample, the nature of the determinand, the concentration range, the time of analysis, and precision and accuracy. The direct method is the most simple and convenient technique and should be preferred whenever possible. In this method, the sample is pre-treated (adjustment of pH and ionic strength), and the electrochemical cell containing the ion-selective electrode and reference electrode is immersed in the sample to be analyzed. The equilibrium cell potential is measured, and the sample concentration is determined from the calibration curve prepared by using appropriate standardized solutions. The concentration can be computed using the calibrated pH/millivoltmeter. In direct methods, pre-treatment is used to eliminate matrix effects.

In incremental methods, a wide variety of spiking techniques are used, in which a known volume of standard solution is added to the sample, and changes in the initial and final potenial are recorded. The initial concentration of the determinand in the sample may then be calculated by employing appropriate equations.[29] This method is only applicable when the concentration range of the sample is known in advance in order to select appropriate increments.

In Gran's plot, electrode potentials are plotted on special graph paper, and the concentrations are determined graphically.[31] These plots are used to linearize data from multiple-point incremental methods. The theory associated with these plots and their application is discussed in more detail by Gran[36] and others.[29,37,38]

Potentiometric titrations allow unknown concentrations to be determined, when the reaction is known, by the additions of known quantities of solutions of known concentration. Titrimetry increases the precision and accuracy of measurement; however, it is very time-consuming and is not frequently used on a routine basis. The only advantage of this technique is in the determination of a species not directly sensed by ion-selective electrodes or gas sensing probes. For example, metals such as cadmium, nickel and zinc may be determined by titrating with EDTA using cupric ion-selective electrode as sensor.[39]

There are a number of factors which affect the performance and utility of a particular ion-selective electrode or gas-sensing probe. These are discussed in detail by Bailey.[29] They include response range and slope, selectivity, stability and reproducibility, response time, sensitivity to temperature/pressure/light, etc., frequency and ease of operation, mechanical design, availability, cost and lifetime. There are additional factors that must also be considered when ion-selective electrodes are used for analysis or methods development. These are also discussed in detail by Bailey.[29] It is recommended that newcomers to the field of ion-selective electrodes refer to this excellent book, which provides a comprehensive and up-to-date review of this comparatively new analytical technique.

D. Infra-Red Spectroscopy

Infra-red absorption has proven useful in characterizing and measuring total organic carbon in water, wastewater and other aquatic environmental samples. Organic carbon is determined by pyrolizing the sample at elevated temperatures to produce carbon dioxide. The resultant carbon dioxide is determined by the absorption band of carbon dioxide. Simple, non-dispersive filter instruments can be used to detect as low as 0.01 mg using commercially available instruments. The methods, based on measurement of infra-red absorption of carbon dioxide, have been used on a routine basis to determine total carbon in surface waters, wastewaters, and saline waters in the range 0.1 mg/ℓ to 50 mg/ℓ carbon.

Molecular absorption of infra-red occurs with a wide range of organic compounds, producing a characteristic absorption for various chemical groups (such as C = 0, −OH, − NH$_2$ and −COOH). The general principle of the technique is the same as molecular absorption. Infra-red radiation from a particular source is passed through the sample. The characteristic absorption spectrum is recorded and used as a fingerprint for a compound or mixture of compounds present in the sample. A wide variety of samples such as solutions and solids, as well as vapors, can be analyzed using infra-red spectroscopy. Its main application in water quality has been in the identification of oils in natural waters and industrial wastes. The reader may consult the pertinent text books[40,41] and reviews on the theory and application of this technique.[42,43]

E. Automation

When monitoring water quality for pollution, it is necessary to analyze numerous samples in order to obtain statistically significant data which can be related to the occurrence, distribution, and study of environmental effects of pollutants. Many wet-chemical manual methods have proven to be unsatisfactory and time consuming. Automation of manual methods has become a necessity in both research and monitoring of chemical constituents, especially in water pollution. In addition, automated methods have proven to provide improved precision and reliability of analytical results. Steps such as sample addition, reagent addition, pH adjustment, temperature, time, and readout systems are all carried out automatically. In a typical analytical system, automation is achieved by interconnecting series of modules which automate each step of the procedure. During the past fifteen to twenty years, various manufacturers have introduced a wide variety of systems where automation is achieved, using open systems or continuous-flowing air-segmented streams.

In an open system, each sample is treated separately. Discrete samples are transported in open tubes made of plastic or glass. The reagents are delivered by suitable dispensers. Other functions such as filtering, mixing, heating, digesting, and quantitation are achieved by interconnecting suitable devices. Syringes are usually employed for sampling, dilution and reagent addition.

In a continuously flowing air-segmented system, a proportioning pump is operated continuously. A wash solution of sample and a number of reagents are pumped through flexible tubes via a proportioning pump. The tubes, made of a suitable plastic or silicon, are placed between the roller and plateau of the pump. In a typical automated method, a multi-channel proportioning pump moves the samples, one after another, and a number of streams of reagents, and air into the system. Mixing coils are used to mix various reagents as well as provide the necessary delay time for the reaction to complete. Other parameters of a chemical reaction, such as heating, cooling, filtering, etc., are achieved by interconnecting various modules and passing the continuously flowing stream. The stream is air-segmented to keep the integrity of each sample, and minimize cross-contamination of the sample and blank segment within the system. Once the reaction is completed, the stream is passed through a suitable detector, via debubbler, and the response is measured using a recorder and/or printer. A brief description of the most important modules used to automate the analytical system is given below.

The sampler contains a tray where up to 40 cups containing samples or standards may be placed. These, together with a wash solution, are automatically introduced at a pre-selected rate. The pre-selected rate of sample, air, and wash solution is achieved by a suitable probe controlled via a cam in a sampler. A probe dips into each cup and introduces the contents for a set time. The probe then lifts out of the cup, aspirates air for a short time (approximately one second) and descends into a wash receptable where wash solution is aspirated into the system. Again, after a pre-set interval, the probe rises from the wash receptacle, aspirates air, and moves into the next sample cup. This sequence is repeated until all samples have entered the system.

A proportioning pump replaces the pipettes of a manual method. The pump continuously and precisely introduces various fluids necessary for chemical reaction into the system. The fluids are advanced on volume-per-time basis into the system. When the pump is operating, the roller moves across the tubes and squeezes the fluids ahead of them into the system. The volume of fluid in the stream is controlled by the internal diameter of the tubes employed in the manifold. A dialyzer is used to continuously separate interfering substances in the reaction mixture. A heating bath, containing glass coils, may be used to control the temperature and the time of the chemical reaction.

The most widely used continuous-flowing system is manufactured by Technicon® — "Auto-Analyzers". The open system is manufactured by Beckman® and others. The application of automated systems in analytical chemistry and water analysis has been discussed and reviewed by a number of authors.[44-48]

III. STANDARD METHODS FOR ANALYSIS OF NUTRIENTS

A number of agencies are involved in developing and adopting, and later standardizing, methodologies for a wide variety of physical, chemical and biological parameters. The American Society for Testing and Materials (ASTM), the American Public Health Association, and the American Water Works Association (APHS, AWWA) publish methods for analysis in water and waste water.[49,50] The majority of the methods published by these agencies are manual methods.[49,50]

In most water pollution laboratories, analysis is generally carried out using automated methodologies. Environment Canada and U.S. Environmental Protection Agency publish and update an Analytical Methods Manual. Most methods for inorganic constituents are automated methods. These methods are fully tested and standardized prior to inclusion in the methods manual. The following section provides details of the methods recommended by Environment Canada for use on a routine basis.[51] These methods include all forms of nutrients and are recommended for use in surface water, ground water, wastewater and sediments.

IV. ORGANIC CARBON

This method is applicable to the determination of organic carbon in surface waters, wastewaters, and saline waters in the range 1 mg/ℓ to 50 mg/ℓ C. Higher values may be determined by appropriate dilution of an aliquot.

The method involves determination by infrared spectrophotometry. Either a single or a double-channel instrument can be used.

Using a single-channel instrument a small, accurately measured volume of sample is injected into the combustion tube of the analyzer with a microliter syringe. The combustion tube is kept at 950°C and receives a continuous stream of oxygen. Cobalt oxide-impregnated asbestos packing is in the combustion tube and serves to disperse the sample and to catalyze oxidation. The carbonaceous material in the sample is oxidized, yielding carbon dioxide and steam. The carbon dioxide and steam are carried out of the furnace, where the steam is condensed and removed. The carbon dioxide is swept into an infrared analyzer, and a recorder then registers a signal that is proportional to the concentration of carbon dioxide in the oxygen stream and, therefore, to the carbon content of the sample.

Inorganic carbonates (carbonate and bicarbonate) should be removed prior to analysis, using the single-channel instrument in order to obtain the organic carbon content of a sample.

The dual-channel instrument is similar to the single-channel instrument except it utilizes a second operation. In the second operation, a sample of similar size is also injected by syringe into a flowing stream of air and swept into the second combustion tube, which contains quartz chips wetted with 85% phosphoric acid. This tube is enclosed in an electric heater set at 150°C, which is below the temperature at which organic matter would be oxidized. The acid-treated packing causes the release of carbon dioxide from inorganic carbonates, and the water is vaporized. The air flow carries the steam and carbon dioxide out of the furnace where the steam is condensed and removed. The carbon dioxide is measured in an infrared analyzer as in the single-channel instrument. Subtracting the results in the second operation from the results obtained in the first operation yields an analysis of organic carbon.

The samples must be homogeneous and suitable for reproducible injections into the apparatus by means of a microliter type syringe. Results of analyses on many samples for total organic carbon using a blended sample can be difficult to reproduce. This is thought to be due to the use of a small syringe for sample-taking and its inability to withdraw identical samples. Consequently, a filtered (0.45 μm) portion is used on such samples and the results termed "dissolved organic carbon".

Large particles in the sample that are not homogenized are serious interferences. Homogenizing permits the determination of carbon in insoluble carbonates and insoluble organic liquids and solids.

Waters containing greater than 5% concentration of salts, acids, or bases should be diluted. Samples higher in salt concentration may be analyzed, but a shorter catalyst life and poor precision are to be expected.

A. Equipment and Reagents

A Beckman® Carbonaceous single- or dual-channel infrared analyzer can be used. Hypodermic syringe used should be Hamilton® No. 705 NLT microliter (0.50 μℓ) syringe or equivalent. A Waring® blender or equivalent is employed. The distilled water used in the preparation of standards and dilution of samples should be of the highest quality in order to have a small blank.

Organic carbon, stock solution, 1000 mg/ℓ C: Dissolve 2.125 g anhydrous potassium biphthalate, $KHC_8H_4O_4$, in distilled water and dilute to 1 ℓ in a volumetric flask.

Organic carbon, standard solutions: Prepare standard solutions from the stock solution as required.

Inorganic carbon, stock solution, 1000 mg/ℓ C: Dissolve 3500 g sodium bicarbonate, NaHCO$_3$, and 4.418 g sodium carbonate, Na$_2$CO$_3$, in distilled water in a 1 ℓ volumetric flask and make up to the mark.

Inorganic carbon, standard solution: Prepare standards from the stock solution as required.

Packing for total carbon tube: Dissolve 20 g cobalt nitrate, Co(NO$_3$)$_2$·6H$_2$O, in 50 mℓ distilled water. Add this solution to 15 g long-fiber asbestos in a porcelain evaporating dish. Mix and evaporate to dryness on a steam bath. Place the dish in a cold muffle furnace and bring to 950°C. After 1 to 2 hr at this temperature, remove the dish and allow to cool. Break up any large lumps and mix adequately but not excessively. With the combustion tube held in a vertical position, taper joint up, first put about 12 mm of untreated asbestos into the tube, then transfer in small amounts approximately 1 g of catalyst into the tube with forceps or tweezers. As it is added, tap or push the material gently with a 6-mm glass rod. Do not force the packing. The weight of the rod itself is sufficient to compress the material. When completed, the length of the packing should be about 5 cm or 6 cm. Test the packed tube by measuring the flow rate of gas through it at room temperature, and then at 750°C. The rate should not drop more than 20%.

Packing for carbonate tube (dual-channel instrument): Place a small wad of quartz wool or asbestos near the exit end of the carbonate evolution tube. From the entrance end add 6-12 mesh quartz chips, permitting these to collect against the wad to a length of 10 cm. Pour an excess of 85% phosphoric acd, H$_3$PO$_4$, into the tube while holding it vertically, and allow the excess to drain out.

B. Procedure

The sample should be refrigerated for storage and kept tightly capped.

Total organic carbon: Blend a shaken sample for at least 5 min in a Waring® blender and acidify to a pH of 2 or less with concentrated hydrochloric acid (omit acidification if a dual-channel instrument is used). Purge sample with CO$_2$-free nitrogen gas for 5 to 10 min or let stand for several hours to remove inorganic carbonates.

Dissolved organic carbon: Filter (0.45 μm) a suitable sample aliquot and acidify to a pH of 2 or less with concentrated hydrochloric acid. Again remove inorganic carbonates as explained above.

Turn on the infrared analyzer, recorder and tube furnaces, setting the total carbon furnace at 950°C and the carbonate furnace at 175°C. Allow sufficient warm-up time for stable drift-free operation; about 2 hr is required. If used daily the analyzer can be left on continuously. Adjust the oxygen flow rate to 80 to 100 mℓ/min through the total carbon tube. Adjust the amplifier gain so that a 20 $\mu\ell$ sample of the 100 mg/ℓ organic carbon standard gives a peak height of approximately half the recorder scale. Other adjustments may be used such as setting a 50 mg/ℓ standard at half the recorder scale. At these settings the noise level should be less than 0.5% of full scale.

Immediately prior to carrying out calibrations or analyses, inject several portions of the appropriate 100 mg/ℓ or 50 mg/ℓ standard into the tube to be used, until constant readings are obtained.

1. Analysis, Dual-Channel Instrument

Successively introduce 20 $\mu\ell$ of each organic carbon standard (including a blank) and samples into the total carbon tube and record peak heights. Between injections allow the recorder pen to return to its base line. The actual injection technique is as follows.

Rinse the syringe several times with the solution to be analyzed; fill and adjust to 20 $\mu\ell$. Wipe off the excess with soft paper tissue, taking care that no lint adheres to the needle. Remove the plug from the syringe holder, insert the sample syringe, and inject the sample into the combustion tube with a single, rapid movement of the thumb. Leave the syringe in the holder until the flow rate returns to normal, then replace it with the plug.

In the same way, prepare a series of diluted carbonate standards containing 20 mg of inorganic carbon per liter. Turn the four-way valve of the apparatus to direct the gas flow through the low-temperature tube and to the analyzer. Adjust the flow rate to 80 to 100 mℓ/min, and allow the base line to become stabilized. Successively introduce 20 $\mu\ell$ of each standard and sample, and a water blank into the low-temperature tube, and record peak heights.

2. Analysis, Single-Channel Instrument

Using the same technique outlined for the dual-channel instrument, inject sample and standards, and record peak heights.

C. Calculations
1. Dual-Channel Instrument

Prepare calibration curves derived from the peak heights obtained with the standard total carbon and inorganic carbon solutions. Determine the concentration of the total carbon and inorganic carbon in the sample by comparing sample peak heights with the calibration curves. Determine the concentration of total organic carbon in the sample by subtracting the inorganic carbon value from the total carbon value.

2. Single-Channel Instrument

Prepare a calibration curve derived from the peak heights obtained with the standard total carbon solutions. Determine the concentration of organic carbon in the sample by comparing sample peak heights with the calibration curve.

D. Precision and Accuracy

In a single laboratory, the coefficients of variation at total organic carbon levels of 20 mg/ℓ and 30 mg/ℓ C were found to be ±3.08% and ±0.02%, respectively, using a single-channel instrument.

V. PARTICULATE ORGANIC CARBON

This method is applicable to the determination of particulate organic carbon in natural waters in the range 1 μg/ℓ to 100 mg/ℓ C. For best performance, about 0.6 mg to 0.8 mg of particulate matter should be analyzed.

The method[52,53] is based on thermal combustion (850°C) of the sample to form CO_2, H_2O and N_2 gases which are then separated chromatographically. Quantitation is made with a thermal conductivity detector, which is coupled with a recorder and an electronic integrator.

Several catalysts are employed to ensure complete conversion of C, H and N components to CO_2, H_2O and N_2. For instance, the combustion is performed in the presence of manganese dioxide to oxidize the carbonaceous material to CO_2 and H_2O. Any carbon monoxide formed is oxidized further to CO_2 by passing the gaseous products through a cupric oxide powder.

Particulate inorganic carbon, obtained by filtering the suspended solids, is dissolved by washing the solid residue with 0.3% (v/v) sulfuric acid. The dried residue remaining is particulate organic carbon and is ready for analysis.

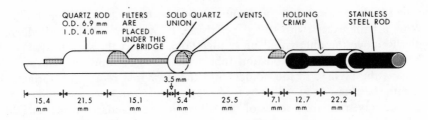

FIGURE 7.   Custom-made quartz ladles.

Incomplete removal of water-insoluble carbonates may produce high organic carbon values. There are no other interferences with this method.

A. Equipment and Reagents

A Hewlett-Packard® Model 185 CHN Analyzer equipped with recorder and integrator, or the equivalent can be used. Improved performance can be obtained by using a chromatographic column prepared from copper tubing, 2.4 m with 6.4 mm O.D., packed with 80 100 mesh Porapak Q.®

The custom-made quartz sample ladles which are used are shown in Figure 7.

Filteration is done through Whatman® GF/C filters.

Dilute sulfuric acid, H_2SO_4: Add 3.0 mℓ of concentrated sulfuric acid, H_2SO_4, to a 1 ℓ volumetric flask containing about 200 mℓ of distilled water. Stir briskly and dilute to 1 ℓ.

Standards to calibrate CHN Analyzer®: Weigh accurately 0.1 to 0.9 mg of microanalytical standard cyclohexanone 2,4-dinitrophenylhydrazone, $C_6H_{10}:NNHC_6H_3(NO_2)_2$ or microanalytical standard acetanilide, $C_6H_5NOCH_3$, in an aluminum weighing boat. Then add to this boat active manganese dioxide, MnO_2, and analyze in the same manner as the samples (refer to Section B below). The amount of C in these standards is obtained from the certificate of analysis.

Manganese dioxide catalyst: Pour 3 to 4 g of granular manganese dioxide, MnO_2, onto a clean watch glass. Heat at 250°C for 8 to 12 hr in a muffle furnace and then place in a desiccator for about 1 hr. It is convenient to fill a small airtight vial with just enough active MnO_2 for each day's analyses. Store unused MnO_2 in a desiccator at all times.

B. Procedure

Water samples collected by the usual procedures should be filtered as soon as possible. Prior to filteration no acids or preservatives should be used.

The dried sample (solid materials remaining on the GF/C filters) should be refrigerated in an airtight petri dish. The storage time may extend to several days or even several months.

The volume of water required for filtration in determining the particulate organic carbon content depends on the type of sample. The following guidelines may be useful. If the sample contains 10 mg/ℓ or more of suspended matter, use 500 mg/ℓ of water sample. Similarly, if there is 5 to 10 mg/ℓ, 2 to 5 mg/ℓ or less than 2 mg/ℓ of suspended matter, use 1 ℓ, 1 to 2 ℓ or more than 2 ℓ, respectively, of the water sample. These guidelines are only a rough approximation.

Clean the glass-fiber filters by placing them in a preheated muffle furnace (450°C) for 1 hr to remove organic matter. Do not use any filters that are bent, as they do not filter properly. Store filters in a sealed dry container.

Using a pair of tweezers, place a pretreated Whatman® GF/C filter on the filtration holder, taking care not to touch the filter with fingers. Place the filtration cup on the filter and clamp securely.

Filter an aliquot of the well-shaken water sample. The vacuum must be 5 to 7 psi, and the depth of water in the cup must be kept at about 2 to 4 cm. This is a rather slow filtration. Record the volume of water filtered.

After the water sample has passed through the filter, wash the inside of the filter cup with a few milliliters of carbon-distilled water to dislodge any solid matter. Maintain the vacuum at 5 to 7 psi in this washing step. Carefully add a few milliliters (ca. 2 to 5) of 0.1 N H_2SO_4 (0.3% v/v sulfuric acid) to the filter. Continue filtering. After the acid wash (removal of carbonates), rinse the filter with a few milliliters (approximately 2 to 5) of carbon-free distilled water.

Remove the filtration cup, and with the vacuum at 5 to 7 psi, continue filtering air to remove any excess water from the GF/C filter. When the filter paper is adequately drained, transfer it carefully to a labelled petri dish, and store in a suitable vacuum desiccator (20 psi) for 2 or 3 days. After the filter is completely dry, the petri dishes may be stored in an airtight container in a refrigerator (4°C) for several months before the analysis is performed.

Blank filters should be prepared occasionally to monitor the background by using the wash water and following the filteration steps above.

Instrument adjustments: The analysis of carbon content in particulate matter may be measured with various CHN Analyzer systems. The instructions below refer to the Hewlett-Packard® 185 Model CHN Analyzer.

Turn on main power, oven power, furnace poser, and power for gas purifier heater. Set the oxidation furnace and reduction furnace to 950°C and 560°C, respectively. Set the gas purifier heater to 400°C. Allow 4 to 5 hr for warm-up. Turn on the carrier gas, adjusting the flow rate to 100 mℓ/min and turn on the bridge current, which is set for 160 mA. Turn the recorder on, adjust the attenuation to ∞ and set the recorder to zero, using the zero control knob. Set the attenuation to eight, and adjust the recorder base line, using the fine balance control. The recorder is set on the normal mode. If used daily, the analyzer should be left on overnight.

The use of filter cups, with rectangular slotted bases, facilitates insertion of the rolled filter paper into the ladle. Cut out rectangular sample portion of the filter paper using precleaned scissors and tweezers; fold this portion lengthwise and sprinkle about 0.1 g of active MnO_2 along the fold. Fold the paper a second time, and slide it into the sample ladle. Guide the ladle into the sample port and leave it in the equilibration position. Quickly tighten large and small knurled nuts on the ladle. Adjust the attenuation to eight. When the recorder pen has returned to the base line, press the start button for the time bypass cycle, loosen the small knurled nut one quarter turn, push the ladle in as far as it will go, then retighten the small knurled nut. Approximately 50 sec after the bypass cycle has ended adjust the attenuation to two, and wait for the first peak. When the first peak is on its downslope, adjust the attenuation for the next peak (CO_2 peak). The attenuation used here is usually 8, 16, 32 or 64. Do not adjust the attenuation for the H_2O peak.

For running standards, the same technique is used as for samples, except that an accurately weighted amount of standard in a weighing boat is covered with active MnO_2 catalyst and placed in the ladle instead of a filter paper.

Filter paper blanks are analyzed in exactly the same manner as samples. MnO_2 blanks are handled in the same manner as standards.

C. Calculations

In recorders not equipped with an integrator, measure the CO_2 peak on its downslope for all samples, standards and blanks.

For each type of organic carbon standard analyzed, a calibration constant (K) is derived as follows:

$$K = \frac{A \times B}{C - D}$$

where A = theoretical fraction of element in the standard, B is the weight of standard used (μg), C is the peak height of carbon in standard, and D is the peak height of carbon in the blank. Calculate the K value of carbon for each standard run during the day and average these values.

Calculate the concentration of carbon for each sample as well as blanks by applying the following equation:

$$C = \frac{K_C \times (CO_2 \text{ peak height})}{V}$$

where C is the concentration of organic carbon (μg/ℓ), V is the volume (ℓ) of water sample analyzed, and K_C is the average value for the calibration constant (K) determined above. The true concentration of organic C is determined by subtracting the filter paper blanks from the sample values.

D. Precision and Accuracy

In a single laboratory, the coefficient of variation was 3.7% at 123 μg C/ℓ. This value refers to the overall method of analysis.

E. Notes

1. The step in the procedure where care should be exercised is in the filtration of the water sample.
2. The method outlined for suspended solids in waters may be modified to include the analysis of organic C in sediments.[54-59]
3. The analyst should be aware of additional thermal gravimetric analysis methods (i.e., DTA) in order to control carbonate interferences.[54-59]
4. Filter and reagent blanks must be monitored routinely on the CHN Analyzer. The appropriate correction factor should be made for each sample.
5. In applying the formulas to calculate the concentrations, one may substitute areas obtained from an area integrator rather than peak heights.

VI. AMMONIA

This automated method is applicable to the determination of ammonia in surface waters in the range 1 μg/ℓ to 150 μg/ℓ. Higher concentrations can be determined by appropriate dilution of an aliquot or by slight modification of the method.

Ammonia in a sample is chlorinated in the presence of phosphate buffer, excess hypochlorite is destroyed, and the chlorinated ammonia is determined with o-tolidine. There are no significant interferences. Ions that normally would interfere are removed by dialysis.

A. Equipment and Reagent

A Technicon Auto Analyzer® unit (Figure 8) or comparable equipment consisting of sampler, manifold, proportioning pump, dialyzer, heating bath, colorimeter (equipped with 50-mm flow cell and 420-mm filters), range expander, and recorder is used.

Alkaline complexing agent: Dissolve 52 g sodium hydroxide, NaOH, in 1 ℓ deionized

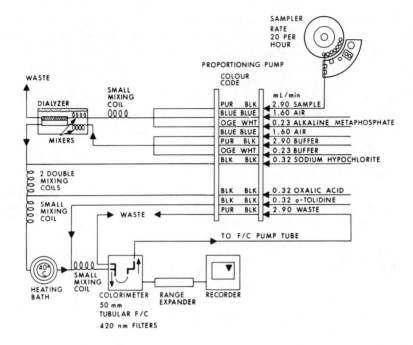

FIGURE 8.   Manifold for ammonia.

water. Dissolve 40 g sodium hexametaphosphate in 1 ℓ deionized water. Mix 100 mℓ of each of the solutions above daily.

Buffer: Dissolve 96 g disodium hydrogen phosphate hydrate and 10 g sodium dihydrogen phosphate in 5 ℓ deionized water (pH 7.5).

Sodium hypochlorite: Dilute sodium hypochlorite solution (Javex® is suitable) to approximately 0.004% available chlorine with deionized water.

Oxalic acid: Dissolve 20 g oxalic acid and 170 g monochloroacetic acid in deionized water and make up to 1 ℓ.

o-Tolidine: Prepare by heating 1.2 g o-tolidine dihydrochloride in 120 mℓ concentrated hydrochloric acid, HCl, at 60°C for 1 hr; then adjust to a volume of 1 ℓ with distilled water.

Stock ammonia solution: Dissolve 3,819 g anhydrous ammonium chloride, NH_4Cl, dried at 100°C, in distilled water and dilute to 1 ℓ. This solution contains 1000 mg/ℓ N.

Intermediate ammonia solution, 10 mg/ℓ: Dilute 10.00 mℓ stock ammonia solution to 1 ℓ with deionized water.

Standard ammonia solution, 1 mg/ℓ N: Dilute 100 mℓ intermediate ammonia solution to 1 ℓ with deionized water.

Working ammonia standards: Using standard ammonia solution prepare working ammonia standards in 100-mℓ volumetric flasks. Prepare fresh each week. The standards listed in Table 1 are suggested.

B. Procedure

The sample container should be tightly capped immediately after collection of the sample and analysis carried out the same day. The sample used for analyses should be obtained by decantation of the sample container. The sample used for analyses should be as free as possible from turbidity. Slight traces of turbidity will be removed by dialysis.

The sample tray (Sampler II®) is filled with samples and standards and covered with Saran® Wrap to prevent contamination from the atmosphere. The probe used with the sampler is a metallic needle-type that pricks the Saran® Wrap to withdraw samples. Samples and standards are run at a rate of 20 per hour, using the manifold shown in Figure 8. Samples high in ammonia may be determined by any one or a combination of two or more of the following:

Replace colorimeter (50 mm) with a colorimeter equipped with a 15- or 8-mm flow cell. Remove range expander. Replace sample line with a smaller line, and add a distilled water line to make up the volume difference. Dilution of the sample should be done prior to loading on the sample tray.

C. Calculations

Prepare a calibration curve derived from the peak heights obtained with the standard solutions. Determine the concentration of ammonia in the samples by comparing sample peak heights with the calibration curve.

D. Precision and Accuracy

In a single laboratory, the coefficients of variation at ammonia levels of 10 $\mu g/\ell$ and 50 $\mu g/\ell$ N were ±4.53% and ±1.69%, respectively.

VII. NITRATE AND NITRITE

This method is applicable to the determination of nitrate and nitrite, singly or combined in surface, ground and wastewaters in the range 0.005 mg/ℓ to 2.0 mg/ℓ N. The method is extremely sensitive and as little as 0.001 mg/ℓ-N can be determined with slight modification.

The sample is passed through a coil containing cadmium fillings that reduce the nitrates present to nitrites. The resulting nitrites plus those originally present are then reacted with sulfanilamide to form the diazo compound. The coupling reaction is carried out on the diazotized sample by the addition of N-(1-naphthyl) ethylenediamine dihydrochloride to form the azo dye. The azo dye intensity, which is proportional to the nitrate concentration, is then measured. Separate nitrate and nitrite values can be obtained by carrying out the procedure without the cadmium reduction step. Sample color that absorbs in the photometric range used for analysis will interfere.

A. Equipment and Reagent

A Technicon® AutoAnalyzer unit (Figure 9) or comparable equipment consisting of the following components: Sampler II®, manifold (see Figure 9), colorimeter equipped with a 50-mm flow cell (or a 15-mm cell for high range) and 550-nm filters, range expander, and recorder is used.

EDTA reagent (2.25% w/v): Dissolve 10.5 g disodium salt of ethylenediaminetetraacetic acid, $C_{10}H_{14}O_8N_2Na_2$, in 450 mℓ distilled water. Adjust pH to 6.5 to 7.0 with dilute NaOH. Bring to final volume of 500 mℓ with distilled water.

Sulfanilamide reagent: Dissolve 1.0 g sulfanilamide, $C_6H_8N_2O_2S$, in a mixture of 90 mℓ distilled water and 10 mℓ concentrated hydrochloric acid.

Naphthylethylenediamine reagent: Prepare a 0.1% w/v solution of N-(1-naphthyl) ethylenediamine dihydrochloride, $C_{12}H_{14}N_2 \cdot 2HCl$, and store in an amber glass bottle. This reagent is stable for a few weeks, after which it should be discarded.

Cadmium reduction column: A short mixing coil is packed with cadmium fillings (mesh 20). Metal fillings can be obtained by using an open-type rasp on cadmium metal sticks. Suitably prepared cadmium may be obtained commercially from Technicon Instruments.

Stock nitrate solution, 1000 mg/ℓ NO$_3$ as N: Dissolve 7.218 g potassium nitrate, KNO$_3$, in distilled water and dilute to 1 ℓ.

Table 1
WORKING AMMONIA
STANDARDS

Ammonia — N (mg/ℓ)	Milliliters of standard solution per 100 mℓ
0.002	0.2
0.01	1.0
0.02	2.0
0.05	5.0
0.10	10.0
0.15	15.0
0.20	20.0

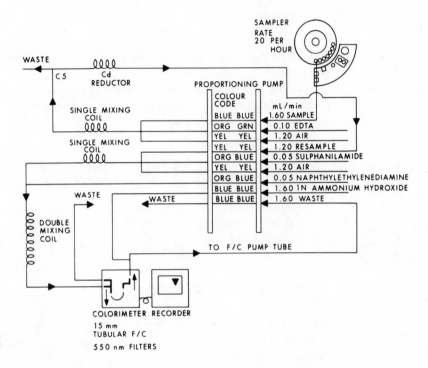

FIGURE 9.   Manifold for nitrate and nitrite.

Stock nitrite solution, 1000 mg/ℓ NO_2 as N: Dissolve 6.072 g potassium nitrite, KNO_2, in distilled water and dilute to 1 ℓ. This solution is unstable, and should be prepared as required.

Standard nitrate solution, 10 mg/ℓ NO_3 as N: Dilute 10.0 mℓ stock nitrate solution to 1 ℓ with distilled water.

Standard nitrite solution, 10 mg/ℓ NO_2 as N: Dilute 10.0 mℓ stock nitrite solution to 1 ℓ with distilled water.

Working nitrate and nitrite solution: Using the standard nitrate or nitrite solutions prepare working standards in 100-mℓ volumetric flasks as shown in Table 2.

B. Procedure

Analyses should be carried out as soon as possible after collection of the sample, preferably the same day. The sample aliquot used for the analysis either should be free from turbidity or should be filtered through a 0.45 μm membrane filter.

Table 2

WORKING NITRATE AND NITRITE
SOLUTIONS

Concentration mg NO₃ N or NO₂ N/ℓ	Volume of standard (10 mg/ℓ) diluted to 100 mℓ
0.00	0.0
0.02	0.2
0.05	0.5
0.10	1.0
0.20	2.0
0.50	5.0
1.0	10.0
2.0	20.0
4.0	40.0
5.0	50.0

Procedure for high range, 0.01 mg/ℓ to 5 mg/ℓ-N: Run the samples and standards at 20/hr, using the manifold as shown in Figure 9.

Procedure for low range, 0.001 mg/ℓ to 0.20 mg/ℓ-N: Run the samples and standards as shown in Figure 9, except replace the 15-mm colorimeter with a 50-mm colorimeter, and install a range expander for use at X 4 or X 10.

C. Calculations

Prepare a calibration curve derived from the peak heights obtained with the standard solutions. Determine the concentration of nitrate + nitrite in the samples by comparing sample peak heights with the calibration curve.

D. Precision and Accuracy

In a single laboratory, the coefficient of variation at a concentration of 50 μg/ℓ NO₃-N was ± 1.6%.

VIII. ORGANIC NITROGEN

This method pertains to the determination of organic nitrogen in unpolluted or slightly polluted natural waters. The applicable range is from 25 μg/ℓ TO 1000 μg/ℓ organic nitrogen-N. The detection limit for this method is 25 μg/ℓ organic N.

In this automated procedure organic nitrogen compounds, including ammonia in a sample, are oxidized by ultraviolet irradiation to a mixture of nitrate and nitrite with the measurement of the resultant mixture.[60] Differentiation of different forms of nitrogen is accomplished by determining, independently, the nitrate and nitrite and ammonia content without irradiation; organic nitrogen is then determined by subtracting these values obtained from the nitrate and nitrite value obtained on the same sample after irradiation.

There are no significant interferences found in normal waters. Because the concentration of organic nitrogen is calculated by subtracting the concentrations of ammonia, and nitrate and nitrite from the total nitrogen content, large concentrations of inorganic nitrogen (ammonia, nitrate or nitrite) originally present in a sample make it difficult to measure low concentrations or organic nitrogen.

A. Equipment and Reagents

A Technical AutoAnalyzer® unit or comparable unit (e.g., Figure 10) consisting of Sampler II®, two proportioning pumps, three colorimeters (two 15-mm flow cells and one 50-mm flow cell and two sets of 550- and one set of 420-nm filters), recorders

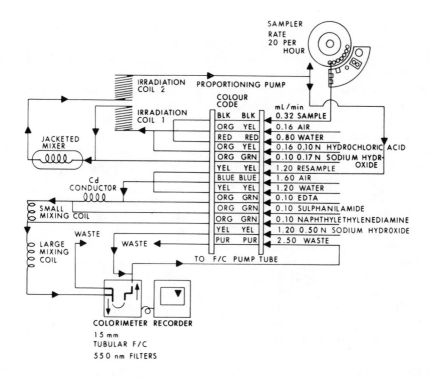

FIGURE 10.   Manifold for organic nitrogen.

(one two-pen and one single-pen), three manifolds (one for ammonia and two for nitrate and nitrite), dialyzer, heating bath, 40°C, and UV reacter are used.

The UV reactor, 32 cm by 29.5 cm by 37 cm, is constructed using a sheet metal box to house components. A 550-W photo-chemical lamp is placed inside a fused quartz lamp protection jacket and mounted axially in the center of the box. Two quartz coils are placed on top of each other so that the same lamp can be used to irradiate both coils. Irradiating coils in the UV reactor are cooled by a high-speed fan fitted on the side of the reactor box; an exhaust is fitted on the top of the reactor. The quartz coils are made of Purcil® 453 quality fused silica tubing of 3 mm I.D. and 0.6-mm wall thickness and with a coil diameter of approximately 12.7 cm. Prior to entering the manifold, the sample is segmented by purified air, which is passed through a scrubber containing 10% sulfuric acid.

1. Reagents for Total Nitrogen, and Nitrate plus Nitrite

Hydrochloric acid solution: dilute 38.5 mℓ concentrated hydrochloric acid, HCl, to 5 ℓ with distilled-deionized water.

Sodium hydroxide solution: dissolve 34.0 g sodium hydroxide (NaOH), 29.8 g disodium hydrogen phosphate (Na_2HPO_4), and 5.6 g potassium dihydrogen phosphate (KH_2PO_4) in distilled-deionized water, and dilute to 5 ℓ.

EDTA solution: dissolve 50.0 g disodium salt of ethylenediaminetetraacetic acid, $C_{10}H_{14}O_8N_2Na_2$, in 700 mℓ distilled-deionized water, and adjust the pH of the solution to 6.5 to 7.0 with dilute sodium hydroxide, and dilute to 1 ℓ.

Sulfanilamide solution: dissolve 5.0 g sulfanilamide, $C_6H_8N_2O_2S$, in distilled-deionized water; add 50 mℓ concentrated hydrochloric acid, and dilute to 1 ℓ with distilled-deionized water.

Naphthylethylenediamine solution: dissolve 0.5 g naphthylethylenediamine dihydrochloride in distilled-deionized water, and dilute to 1 ℓ.

Stock nitrate solution, 1000 mg/ℓ NO_3 as N: dissolve 7.218 g potassium nitrate, KNO_3, in distilled-deionized water, and dilute to 1 ℓ.

Intermediate nitrate solution, 10 mg/ℓ NO_3 as N: dilute 10 mℓ stock nitrate solution to 1 ℓ with distilled-deionized water.

Standard nitrate solutions: from the intermediate nitrate solution prepare standard solutions of nitrate to cover the expected range of samples.

2. Reagents for Ammonia

Alkaline hexametaphosphate solution: dissolve 40.0 g sodium hexametaphosphate, $(NaPO_3)_6$, in 1 ℓ distilled-deionized water and 52.0 g sodium hydroxide in 1 ℓ distilled-deionized water. Mix 100 mℓ of each solution daily.

Buffer solution: dissolve 89.7 g disodium hydrogen phosphate hydrate and 10.0 g disodium hydrogen phosphate in distilled-deionized water, and dilute to 5 ℓ.

Hypochlorite stock solution: dilute 100 mℓ commercial chlorine bleach to 500 mℓ with distilled-deionized water, and filter through a 0.45 μm filter.

Hypochlorite reagent solution: transfer 5 mℓ hypochlorite stock solution to a 1 ℓ volumetric flask, and dilute to 1 ℓ with distilled-deionized water.

Reducing agent: dissolve 40.0 g oxalic acid, HO_2CCO_2H, and 340.0 g monochloroacetic acid, $C_2H_3O_2Cl$, in distilled-deionized water, and dilute to 2 ℓ.

o-Tolidine: prepare by heating 1.2 g o-tolidine dihydrochloride, $C_{14}H_{16}O_2$, in 120 mℓ concentrated hydrochloric acid, HCl, at 60°C for 1 hr; then adjust to a volume of 1 ℓ with distilled water.

Standard ammonia solutions: see "Nitrogen: Ammonia" (automated) method described elsewhere in this manual.

B. Procedure

The sample should be collected in a polyethylene bottle, and analyzed as soon as possible. If the analysis cannot be carried out soon after collection (within two days), it should be frozen for storage.

If the sample is relatively free from turbidity, such as a sample taken in open waters of any of the Great Lakes, a shaken sample should be placed in the sample cup for analysis and the result called total organic nitrogen. If the sample contains large quantities of suspended solids it should be filtered through a 0.45 μm filter and the result, after analysis, called soluble organic nitrogen. To obtain a total organic nitrogen value on such a sample, the method outlined elsewhere in this manual for "Total Organic Nitrogen" should be followed.

Fill the sample tray (Sampler II) with samples and standards and cover with Saran® Wrap to prevent contamination from the atmosphere. The probe used with the sampler is a metallic needle type that pricks the Saran® Wrap to withdraw samples. Run samples and standards at a rate of 20 per hour. Run samples and standards using the manifolds shown in Figures 10 to 12, as follows.

Figure 10 shows the flow diagram for total soluble nitrogen, which involves the oxidation of organic nitrogenous compounds by UV irradiation to a mixture of nitrate and nitrite and the measurement of the resultant mixture formed. The flow diagrams for nitrate and nitrite (Figure 11) without irradiation and ammonia (Figure 12) are essentially the same as Figures 8 and 9, except that two additional lines containing acid and alkali have been added to simulate the addition of these chemicals used in the nitrate-nitrite manifold with irradiation for the determination of total organic nitrogen.

C. Calculations

Prepare calibration curves for nitrate + nitrite and ammonia without irradiation

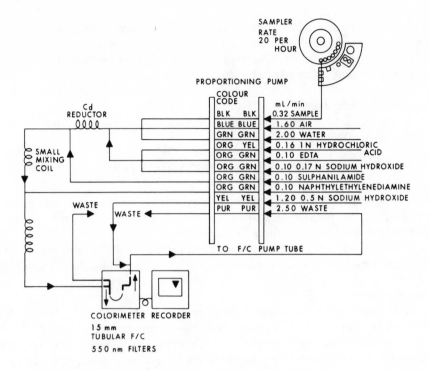

FIGURE 11.   Manifold for nitrate and nitrite (similar to Figure 8).

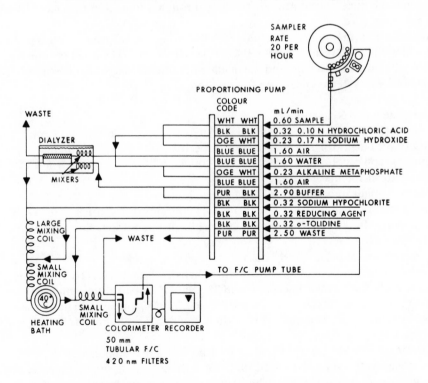

FIGURE 12.   Manifold for ammonia (similar to Figure 9).

and for total nitrogen after irradiation derived from the peak heights obtained with the standard solutions. Determine the concentrations in the samples of nitrate + nitrite and ammonia without irradiation by comparing sample peak heights with the calibration curves. Determine the concentration of organic nitrogen by subtracting the concentration of nitrate + nitrite and ammonia obtained without irradiation from the total nitrogen content obtained after irradiation.

D. Precision and Accuracy

In a single laboratory the coefficient of variation at a concentration of 200 $\mu g/\ell$ organic N was $\pm 2.02\%$.

IX. PARTICULATE ORGANIC NITROGEN

This method is applicable to the determination of particulate organic nitrogen in natural waters in the range 1 $\mu g/\ell$ to 100 mg/ℓ N. For best performance about 0.6 mg to 0.8 mg of particulate matter should be analyzed.

The method[52,53] is based on thermal combustion (850°C) of the sample to form CO_2, H_2O and N_2 gases, which are then separated chromatographically. Quantitation is made with a thermal conductivity detector, which is coupled with a recorder and an electronic integrator.

Several catalysts are employed to ensure complete conversion of C, H, and N components to CO_2, H_2O, and N_2. For instance, the combustion is performed in the presence of manganese dioxide to oxidize the carbonaceous material to CO_2, H_2O and oxides of nitrogen. Any carbon monoxide formed is oxidized further to CO_2 by passing the gaseous products through a cupric oxide powder. In the same manner, oxides of nitrogen are reduced by passing these gases through a reduction furnace (560°C) containing copper wire.

Particulate inorganic carbon obtained by filtering the suspended solids is dissolved by washing the solid residue with 0.3% (v/v) sulfuric acid. The dried residue remaining is particulate organic carbon and particulate organic nitrogen, and is ready for analysis. There are no interferences with this method.

A. Equipment and Reagents

A Hewlett-Packard® Model 185 CHN Analyzer or comparable unit, equipped with recorder and integrator is used. Improved performance[59] can be obtained by using a chromotographic column prepared from copper tubing, 2.4 m with 6.4 mm O.D., packed with 80-100 mesh Porapak Q®.

A custom-made quartz sample ladle (Figure 7) is used.

Dilute sulfuric acid, H_2SO_4: add 3.0 mℓ of concentrated sulfuric acid, H_2SO_4, to a 1 ℓ volumetric flask containing about 200 mℓ of distilled water. Stir briskly and dilute to 1 ℓ.

Standards to calibrate CHN Analyzer: weigh accurately 0.1 to 0.9 mg of microanalytical standard cyclohesanone 2,4-dinitrophenylhydrazone, $C_6H_{10}:NNHC_6H_3(NO_2)_2$ or microanalytical standard acetanilide, $C_6H_5NOCH_3$, in an aluminum weighing-boat. Then to this boat add active manganese dioxide, MnO_2, and analyze in the same manner as the samples. The amount of N in these standards is obtained from the certificate of analysis.

Manganese dioxide catalyst: pour 3 to 4 g of granular manganese dioxide, MnO_2, onto a clean watch glass. Heat at 250°C for 8 to 12 hr in a muffle furnace, and then place in a desiccator for about 1 hr. It will be convenient to fill a small airtight vial with just enough active MnO_2 in a desiccator at all times.

B. Procedure

Water samples should be filtered as soon as possible. Prior to filteration no acid or preservative should be used.

The dried sample (solid materials remaining on the GF/C filters) should be refrigerated in an airtight petri dish. The storage time may extend to several days or even several months.

The volume of water that will be required for filtration in determining the particulate organic carbon and organic nitrogen content depends on the type of sample. The following guidelines may be useful. If the sample contains 10 mg/ℓ or more of suspended matter, use 500 mℓ of water sample. Similarly, if there is 5 to 10 mg/ℓ, 2 to 5 mg/ℓ, or less than 2 mg/ℓ suspended matter, use 1 ℓ, 1 to 2 ℓ, or more than 2 ℓ, respectively, of the water sample. These guidelines are only a rough approximation.

Clean the glass-fiber filters (Whatman® GF/C) by placing them in a preheated muffle furnace at 450°C for 1 hr to remove organic matter. Do not use any filters that are bent, as they will not filter properly. Store filters in a sealed dry container.

Using a pair of tweezers, place a pretreated Whatman® GF/C filter on the filtration holder, taking care not to touch the filter with fingers. Place the filtration cup on the filter, and clamp securely.

Filter an aliquot of the well-shaken water sample. The vacuum must be 5 to 7 psi, and the depth of water in the cup must be kept at about 2 to 4 cm. This is a rather slow filtration. Record the volume of water filtered.

After the water sample has passed through the filter, wash the inside of the filter cup with a few milliliters of nitrogen-free distilled water to dislodge any solid matter. Maintain the vacuum at 5 to 7 psi in this wash step.

Remove the filtration cup, and with the vacuum at 5 to 7 psi, continue filtering air to remove any excess water from the GF/C filter. When the filter paper is adequately drained, transfer it carefully to a labeled petri dish, and store in a suitable vacuum desiccator (20 psi) for two or three days. After the filter is completely dry, the petri dishes may be stored in an airtight container in a refrigerator (4°C) for several months before the analysis is performed.

Using the wash water and following the filtration steps above, blank filters should be prepared occasionally to monitor the background.

Instrument adjustments: the analysis of carbon and nitrogen content in particulate matter may be measured with various CHN Analyzer systems. The instructions below refer to the Hewlett-Packard® 185 Model CHN Analyzer.

Turn on main power, oven power, furnace power, and power for gas purifier heater. Set the oxidation furnace, and reduction furnace to ~950°C and ~560°C, respectively. Set the gas purfier heater to 400°C. Allow 4 to 5 hr for warm-up. Turn on the carrier gas, adjusting the flow rate to 100 mℓ/min, and turn on the bridge current, which is set for 160 mA. Turn the recorder on, adjust the attenuation to ∞, and set the recorder to zero, using the zero control knob. Set the attenuation to eight, and adjust the recorder base line, using the fine balance control. The recorder is set on the normal mode. If used daily, the analyzer should be left on overnight.

The use of filter cups, with rectangular slotted bases, facilitates insertion of the rolled filter paper into the ladle. Cut out rectangular sample portion of the filter paper using precleaned scissors and tweezers; fold this portion lengthwise, and sprinkle about 0.1 g of active MnO_2 along the fold. Fold the paper a second time, and slide it into the sample ladle. Guide the ladle into the sample port, and leave it in the equilibration position. Quickly tighten large and small knurled nuts on the ladle. Adjust the attenuation to eight. When the recorder pen has returned to the base line, press the start button for the time bypass cycle, loosen the small knurled nut one quarter turn, push the ladle in as far as it will go, then retighten the small knurled nut. Approximately 50 sec after the bypass cycle has ended, adjust the attenuation to two, and wait for the first peak (N_2 peak). When the first peak is on its downslope, adjust the attenuation for the next peak (CO_2 peak). The attenuation used here is usually 8, 16, 32, or 64. Do not adjust the attenuation for the H_2O peak.

For running standards, the same technique is used as for samples, except that an accurately weighted amount of standard in a weighing-boat is covered with active MnO_2 catalyst, and placed in the ladle instead of a filter paper. Filter paper blanks are analyzed in exactly the same manner as samples. MnO_2 blanks are handled in the same manner as standards.

C. Calculations

In recorders not equipped with an integrator, measure the N_2 peak height on its upslope for all samples, standards and blanks. Measure all N_2 peak heights in reference to the attenuation factor of eight. For each type of organic nitrogen standard analyzed, a calibration constant (K) is derived as follows:

$$K = \frac{A \times B}{C - D}$$

where A = theoretical fraction of element in the standard, B = weight of standard used (μg), C = peak height of nitrogen in standard, and D = peak height of nitrogen in the blank. Calculate the K value for each element of each standard run during the day. Average these values for nitrogen (K_N).

Calculate the concentration of nitrogen for each sample as well as blanks by applying the following equation:

$$N = \frac{K_N \times (N_2 \text{ peak height})}{V}$$

where N = concentration ($\mu g/\ell$) of organic nitrogen, V = volume (ℓ) of water sample analyzed, and K_N = average value for the calibration constant determined above. The true concentration of organic N is determined by subtracting the filter paper blanks from the sample values.

D. Precision and Accuracy

In a single laboratory, the coefficient of variation was 4.2% at 21 $\mu g/\ell$-N. This value refers to the overall method of analysis.

E. Notes

1.   The step in the procedure where care should be exercised is in the filtration of the water sample.
2.   The method outlined for suspended solids in waters may be modified to include the analysis of organic N in sediments.[54-59]
3.   Filter and reagent blanks must be monitored routinely on the CHN Analyzer. The appropriate correction factor should be made for each sample.
4.   In applying the formulas to calculate the concentrations, one may substitute areas obtained from an area integrator rather than peak heights.

X. TOTAL KJELDAHL NITROGEN

This method is applicable to the determination of total Kjeldahl nitrogen in surface waters, wastewaters, and saline waters. Concentrations of Kjeldahl nitrogen greater than 0.5 mg/ℓ-N may be accurately determined, whereas values which fall below this value are questionable.

Total Kjeldahl nitrogen is the sum of the free ammonia and organic nitrogen compounds which are converted to ammonium bisulfate under the conditions of digestion described below.

Organic nitrogen is the difference obtained by subtracting the free ammonia value from the total Kjeldahl nitrogen value. Organic nitrogen may be determined directly by removal of the ammonia before the digestion step.

In the presence of sulfuric acid, potassium sulfate, and mercuric sulfate (the latter as a catalyst), nitrogen of many organic materials is converted to ammonium bisulfate. The ammonia is distilled from an alkaline medium, and absorbed in boric acid. The ammonia is then determined by titration with standard sulfuric acid, using a mixed indicator. Nitrogen compounds from outside sources will cause high results.

A. Equipment and Reagents

A Kjeldahl apparatus, a combination digestion and distillation unit available from Labconco® Corporation, is used. Kjeldahl flasks, 800-ml capacity, are required.

Phosphate buffer solution, 0.5 M: dissolve 14.3 g anhydrous potassium dihydrogen phosphate, KH_2PO_4, and 68.8 g anhydrous dipotassium hydrogen phosphate, K_2HPO_4, in distilled water, and dilute to 1 l.

Mercuric sulfate solution: dissolve 8 g red mercuric oxide, HgO, in 50 ml 1 + 5 H_2SO_4, and dilute to 100 ml with distilled water.

Sulfuric acid-mercuric sulfate-potassium sulfate solution: dissolve 267 g potassium sulfate, K_2SO_4, in 1300 ml distilled water and add 400 ml concentrated H_2SO_4. Add 50 ml mercuric sulfate solution, and dilute to 2 l. This reagent crystallizes at temperatures lower than 14°C.

Sodium hydroxide-sodium thiosulfate solution: dissolve 500 g sodium hydroxide, NaOH, and 25 g sodium thiosulfate, $Na_2S_2O_3 \cdot 5H_2O$, in distilled water, and dilute to 1 l.

Phenolphthalein indicator solution: dissolve 5 g phenolphthalein disodium salt in distilled water, and dilute to 1 l. If necessary, add 0.02 N NaOH in drops until a faint pink color appears.

End-Point indicator solution: especially made for Kjeldahl test, available from Matheson Coleman and Bell Manufacturing Chemists®, Norwood, Ohio.

Boric acid solution: dissolve 20 g boric acid, H_3BO_3, in distilled water, and dilute to 1 l. Make fresh every 20 days.

Standard sulfuric acid titrant, 0.02 N: In this strength, 1.0 ml = 0.28 mg N.

B. Procedure

Samples should be stored at 4°C and may be preserved by the addition of 2 ml of concentrated H_2SO_4 per liter. Samples should be analyzed within 24 hr of sampling, because conversion of organic nitrogen to ammonia may occur. Shaken samples are used for this test.

Place measured sample in an 800-ml Kjeldahl flask. The sample size can be determined from Table 3. Dilute the sample to 500 ml if necessary.

Add 25 ml phosphate buffer solution, and boil off the free ammonia (boil to about one third the original volume and let cool).

Add 50 ml acid-sulfate solution, and evaporate the mixture in the Kjeldahl apparatus until SO_3 fumes are given off and the solution turns colorless. Continue boiling for an additional 20 min. Cool the residue and add 300 ml distilled water.

Make alkaline by careful addition (approximately 50 ml) of sodium hydroxide-thiosulfate solution, without mixing, by slowly pouring the caustic solution down the neck of the digestion flask while holding the flask in a tilted position. Do not mix until

Table 3
SAMPLE SIZES

Organic nitrogen in sample (mg/l N)	Sample size (ml)
0—5	500
5—10	250
10—20	100
20—50	50.0
50—100	25.0

the digestion flask has been connected to the distillation apparatus, as ammonia may be lost.

Connect the Kjeldahl flask to the distillation apparatus with the tip of the condenser below the level of the boric acid solution in the receiving flask. Distill into 50 ml boric acid solution until about 200 ml distillate has been collected.

Titrate the distillate with 0.02 $N\,H_2SO_4$, using 0.5 ml of indicator until the indicator turns pinkish brown. Run a blank using distilled water instead of sample and carry through the procedure.

C. Calculations

Depending on the procedure used, calculate either the total Kjeldahl nitrogen or organic nitrogen in the sample, as follows:

$$mg/\ell\text{-N} = \frac{(A\text{-}B) \times N \times 14000}{C}$$

where A = H_2SO_4 (ml) for sample, B = H_2SO_4 (ml) for blank, C = sample (ml), and N = normality of sulfuric acid solution. If the normality of the sulfuric acid solution is exactly 0.02 N, the formula is shortened to:

$$mg/\ell\text{-N} = \frac{(A\text{-}B) \times 280}{C}$$

D. Precision and Accuracy

In a single laboratory, the coefficient of variation at a total Kjeldahl nitrogen concentration of 25 mg/l-N was ±4.58%.

XI. INORGANIC PHOSPHORUS

This colorimetric method is applicable to surface and groundwaters with inorganic phosphorus levels in the range 1 to 500 µg/l-P. Samples having higher concentrations than this can be measured by appropriate dilution of an aliquot. The detection limit for this method is 0.001 mg/l-P.

The polyphosphates in a sample are hydrolyzed to orthophosphate by heating the sample in the presence of sulfuric acid. The resulting orthophosphate plus the orthophosphate originally present is then reacted with ammonium molybdate to form heteropoly molybdophosphoric acid, $H_3(PMo_3O_{10})_4$. This is then reduced with stannous chloride in an aqueous sulfuric acid medium to form molybdenum blue. The molybdenum blue color is measured in a colorimeter at 660-nm wavelength.

The only known interferences with the method are mercury and arsenic. Mercury at levels above 1 mg/l Hg gives a precipitate of mercurous chloride and mercury in the reduction step. This is not a problem with natural waters unless mercuric chloride has

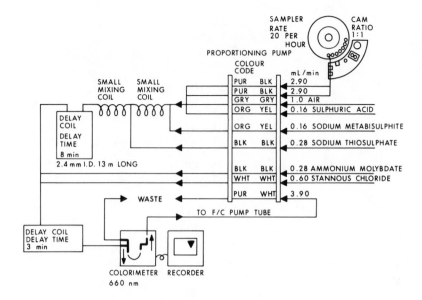

FIGURE 13. Manifold for inorganic phosphorus (presence of As).

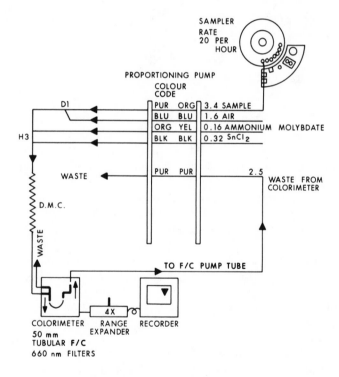

FIGURE 14. Manifold for inorganic phosphorus.

been used to "preserve" the sample. If it is suspected that arsenic is present in a sample, the interference can be eliminated by employing the manifold shown in Figure 13.[61] At the concentration of sulfuric acid used in the method, silica does not interfere.

A. Equipment and Reagents

A Technicon® AutoAnalyzer unit (e.g. Figure 14) or comparable equipment consisting of sampler, heating bath at 30°C, manifold, two proportioning pumps, two

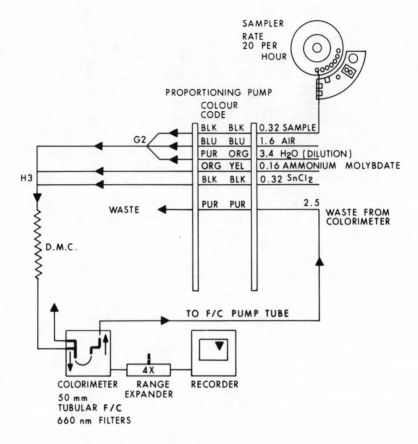

FIGURE 15.   Manifold for inorganic phosphorus (50 to 500 $\mu g/\ell$)

colorimeters with 50-mm flow cells and two sets of 660-nm filters, recorder (two-pen), two range expanders, is used. A Castle® No. 999-C electrically heated autoclave or equivalent is employed.

Ammonium molybdate solution: dissolve 25 g ammonium molybdate, $(NH_4)_6Mo_7O_{24}\cdot4H_2O$, in 175 m$\ell$ distilled water; to 400 mℓ distilled water add 280 mℓ concentrated H_2SO_4. Add the molybdate solution to the acid solution, and dilute to 1 ℓ.

Stock stannous chloride solution: dissolve 5 g stannous chloride, $SnCl_2\cdot2H_2O$, in 25 mℓ conc. HCl; dilute to 500 mℓ with distilled water. This stock is stable for 2 weeks at 5°C storage.

Stannous chloride working solution: to 30-mℓ stock solution add 25 mℓ concentrated HCl, and dilute to 500 mℓ with distilled water. This solution is stable for 12 hr and generally is sufficient for 1 day.

Strong acid solution: to 600 mℓ distilled water add 300 mℓ concentrated H_2SO_4. Cool, then add 4 mℓ concentrated HNO_3, and dilute to 1 ℓ.

Standard phosphorus solutions: see method Orthophosphate (Section XII) for preparation.

"Blank" reagents: prepare "blank" ammonium molybdate and stannous chloride reagents as above, but do not add the ammonium molybdate and stannous chloride, respectively.

Sodium thiosulfate solution, 3% w/v: dissolve 30.0 g of sodium thiosulfate, $Na_2S_2O_3$, in 1 ℓ of deionized water.

Sodium metabisulfite solution, 15% w/v: dissolve 150 g of sodium metabisulfite in 1 ℓ of deionized water.

B. Procedure

A measured 100-mℓ aliquot of sample should be transferred to a 125-mℓ glass Erlenmeyer flask and acidified with 1.0 mℓ of 30% sulfuric acid. The sample should be cooled to about 4°C as soon as it has been taken.

If dissolved inorganic phosphorus is to be determined, the sample aliquot used for the analysis should be free from turbidity or filtered through a 0.45-μm membrane filter. If total inorganic phosphorus is to be determined, the sample should be well mixed before the aliquot is taken for analysis.

Add 50 mℓ of the standard solutions to each 100-mℓ beaker; to each beaker add 0.5 mℓ strong acid solution. Add 50 mℓ of water samples to each 100-mℓ beaker; to each beaker add 0.5 mℓ strong acid solution. Autoclave the beakers by the following procedure.

Place beakers in the autoclave. Turn handle to "Autoclave" position. When chamber pressure reaches 10 psig (121°C), time the digestion for 30 min. After digestion, turn the handle to "Liquids Cool" position, and permit chamber pressure to drop to zero. Turn handle to "Standby and Vent" position. Remove beakers and permit to cool. If there is any turbidity present after the autoclaving, filter the solutions through a 0.45-μm membrane filter.

Run the standards and samples after autoclaving, at 20 samples per hour using the manifolds as shown in Figures 13 to 15. The manifold in Figure 14 is used to measure concentrations in the range from 1 μg/ℓ to 50 μg/ℓ-P. The manifold in Figure 15 is used to measure concentrations in the range from 50 μg/ℓ to 500 μg/ℓ. The manifold in Figure 13 is employed if arsenic is suspected of being present in the sample. Wash water should be acidified with sulfuric acid at the rate of 1 mℓ 30% H_2SO_4/100 mℓ distilled water.

Every time a new batch of stannous chloride reagent is used, a set of standard solutions must be run. Standard solutions should be run periodically to check the validity of the calibration curve.

If any of the samples after autoclaving and filtering contain a significant level of color, a correction should be made by rerunning these samples through the system with the blank reagents. (A decision about what constitutes a significant level of color can only be made on experience gained by rerunning samples of "doubtful" levels of color.)

C. Calculations

Prepare a calibration curve derived from the peak heights obtained with the standard solutions. Determine the concentration of inorganic phosphorus in the samples by comparing sample peak heights with the calibration cuve. If a sample has been rerun with "blank" reagents, the inorganic phosphorus level in the original sample is determined by subtracting the "apparent" phosphorus level obtained with the blank reagents from the "phosphorus" level obtained with the sample and "regular" reagents.

D. Precision and Accuracy

In a single laboratory, the coefficients of variation at phosphorus levels of 7.9 μg/ℓ and 19.6 μg/ℓ P were ±9.98% and ±4.71%, respectively.

XII. ORTHOPHOSPHATE

This colorimetric method is applicable to surface and groundwaters with orthophosphate levels in the range from 1 μg/ℓ to 500 μg/ℓ-P. Samples having higher concentrations than this can be measured by appropriate dilution of an aliquot. The detection limit for this method is 0.001 mg/ℓ.

The orthophsophate in a sample is reacted with ammonium molybdate to form the heteropoly molybdophosphoric acid, $H_3(PMo_3O_{10})_4$. This is then reduced with stannous chloride in aqueous sulfuric acid medium to form molybdenum blue. The molybdenum blue color is measured in a colorimeter at 660-nm wavelength.

The only known interferences with the method are mercury and arsenic. Mercury at levels above 1 mg/ℓ Hg gives a precipitate of mercurous chloride and mercury in the reduction step. This is not a problem with natural waters unless mercuric chloride has been used to "preserve" the sample. If it is suspected that arsenic is present in a sample, the interference can be eliminated by employing the manifold shown in Figure 13. At the concentration of sulfuric acid used in the method, silica does not interfere.

A. Equipment and Reagents

An autosampler similar to that used in the inorganic phosphorus procedure is used. Ammonium molybdate solution: dissolve 25 g ammonium molybdate, $(NH_4)_6Mo_7O_{24} \cdot 4H_2O$, in 175 m$\ell$ distilled water; to 400 mℓ distilled water add 280 mℓ concentrated H_2SO_4. Add the molybdate solution to the acid solution, and dilute to 1 ℓ.

Stock stannous chloride solution: dissolve 5 g stannous chloride, $SnCl_2 \cdot 2H_2O$, in 25 mℓ concentrated HCl and dilute to 500 mℓ with distilled water; this stock solution is stable for 2 weeks at 5°C storage.

Stannous chloride working solution: to 30 mℓ stock solution add 25 mℓ concentrated HCl, and dilute to 500 mℓ with distilled water; this solution is stable for 12 hr and generally is sufficient for 1 day.

Stock phosphorus solution, 1000 mg/ℓ-P: dissolve 4.393 g anhydrous potassium dihydrogen phosphate, KH_2PO_4 (oven-dried at 105°C), in distilled water, and dilute to 1 ℓ. Store in amber or dark-colored bottle.

Intermediate phosphorus solution, 10 mg/ℓ-P: pipette 10 mℓ stock solution into a 1-ℓ volumetric flask, and dilute to the mark.

Standard phosphorus solution, 1000 μg/ℓ-P: pipette 50 mℓ intermediate solution into a 500-mℓ volumetric flask, and dilute to the mark; prepare daily.

Working phosphorus solutions (see Table 4): using the standard solution, prepare the following working solutions in 100-mℓ volumetric flasks:

Table 4

Milliliters of standard solution (1000 μg/ℓ P)	Concentration (μg/ℓ)
0	0
0.5	5
1.0	10
3.0	30
5.0	50
8.0	80
10.0	100
15.0	150
20.0	200
40.0	400
50.0	500

"Blank" reagents: prepare "blank" ammonium molybdate and "blank" stannous chloride reagents as above, but do not add the ammonium molybdate and stannous chloride, respectively.

Sodium thiosulfate solution, 3% w/v: dissolve 30.0 g of sodium thiosulfate, $Na_2S_2O_3$, in 1 ℓ of deionized water.

Sodium metabisulfite solution:psolution, 15% w/v. dissolve 150 g of sodium metabisulfite in 1 ℓ of deionized water.

B. Procedure

The sample should be cooled to about 4°C as soon as it has been taken. Analysis should be carried out the same day.

The sample aliquot used for the analysis should either be free from turbidity or should be filtered through a 0.45 μm membrane filter.

Run the standards and samples at 20 per hour using the appropriate manifold as shown in Figures 13, 14, and 15. System A is used to measure concentrations in the range from 1 μg/ℓ to 50 μg/ℓ-P; system B is used to measure concentrations in the range from 50 μg/ℓ to 500 μg/ℓ. System C is used if arsenic is suspected of being present in the sample.

Every time a new batch of stannous chloride reagent is used, a set of standard solutions must be run. Standard solutions should be run periodically to check the validity of the calibration curve.

If any samples contain a significant level of color, a correction should be made by rerunning these samples through the system with the "blank" reagents. (A decision about what constitutes a significant level of color can only be made on experience gained by rerunning samples of "doubtful" levels of color.)

C. Calculations

Prepare a calibration curve derived from the peak heights obtained with the standard solutions. Determine the concentration of ortho-phosphorus in the samples by comparing sample peak heights with the calibration curve. If a sample has been rerun with "blank" reagents, the ortho-phosphorus level in the original sample is determined by subtracting the "apparent" ortho-phosphorus level obtained with the blank reagents from the "ortho-phosphorus" level obtained with the sample and regular reagents.

D. Precision and Accuracy

In a single laboratory, the coefficients of variation at ortho-phosphorus levels of 4.0 μg/ℓ and 10.0 μg/ℓ-P were ±4.65% and ±3.98%, respectively.

XIII. TOTAL PHOSPHORUS

This colorimetric method is applicable to surface and groundwaters with total phosphorus levels in the range 1 to 500 μg/ℓ-P. Samples having higher concentrations can be measured by appropriate dilution of an aliquot. The detection limit for this method is 0.001 mg/ℓ.

The organic material in a sample is destroyed by digestion with a sulfuric acid-persulfate mixture. This releases the organically bound phosphorus as phosphate. The digestion with acid also hydrolyzes polyphosphates to orthophosphate. This orthophosphate is then reacted with ammonium molybdate to form heteropoly molybdophosphoric acid. $H_3(PMo_3O_{10})_4$. This is reduced with stannous chloride in an aqueous sulfuric acid medium to form molybdenum blue. The molybdenum blue color is measured in a colorimeter at 660-nm wavelength.

The only known interferences with the method are mercury and arsenic. Mercury at levels above 1 mg/ℓ Hg gives a precipitate of mercurous chloride and mercury in the reduction step. This is not a problem with natural waters unless mercuric chloride has been used to "preserve" the samples. If it is suspected that arsenic is present in a sample, the interference can be eliminated by employing the manifold shown in Figure 13.[63] At the concentration of sulfuric acid used in the method, silica does not interfere.

A. Equipment and Reagents

An autoanalyzer similar to that used in the inorganic phosphorus procedure is used. A Castle® No. 999-C electrically heated autoclave or equivalent is employed.

Ammonium molybdate solution: dissolve 25 g ammonium molybdate, $(NH_4)_6Mo_6O_{24} \cdot 4H_2O$, in 175 m$\ell$ distilled water. To 400 mℓ distilled water add 280 mℓ concentrated H_2SO_4. Add the molybdate solution to the acid solution, and dilute to 1 ℓ.

Stock stannous chloride solution: dissolve 5 g stannous chloride, $SnCl_2 \cdot 2H_2O$, in 25 mℓ concentrated HCl, and dilute to 500 mℓ with distilled water. This stock is stable for 2 weeks at 5°C storage.

Stannous chloride working solution: to 30 mℓ of stock solution add 25 mℓ concentrated HCl, and dilute to 500 mℓ with distilled water. This solution is stable for 12 hr and generally is sufficient for 1 day.

Strong acid solution: to 600 mℓ distilled water add 300 mℓ of concentrated H_2SO_4. Cool; add 4 mℓ concentrated HNO_3, and dilute to 1 ℓ.

Standard phosphorus solutions: see method Orthophosphate (Section XII) for preparation.

"Blank" reagents: prepare "blank" ammonium molybdate and stannous chloride reagents but do not add the ammonium molybdate and stannous chloride, respectively.

Potassium persulfate: solid form, ACS grade.

Sodium thiosulfate solution, 3% w/v: dissolve 30.0 g of sodium thiosulfate, $Na_2S_2O_3$, in 1 ℓ of deionized water.

Sodium metabisulfite solution, 15% w/v: dissolve 150 g sodium metabisulfite in 1 ℓ of deionized water.

B. Procedure

A measured 100-mℓ aliquot of sample should be transferred to a 125-mℓ glass Erlenmeyer flask and acidified with 1.0 mℓ of 30% sulfuric acid. The sample should be cooled to about 4°C as soon as it has been taken. The sample should be well mixed before the aliquot is taken for analysis.

Add 50 mℓ of the standard solutions to each 100-mℓ beaker. Add 0.5 mℓ strong acid solution. Add 0.2 g potassium persulfate. (N.B. It is acceptable to add the potassium persulfate by a volumetric measure, once such a measure has been calibrated.) Add 50 mℓ of each water sample to a 100-mℓ beaker; treat as above. If the samples have been stored in 125-mℓ Erlenmeyer flasks, add the strong acid and potassium persulfate directly to these flasks. Autoclave the samples or standard solutions in beakers (or flasks) by the following procedure.

Place beakers in the autoclave. Turn handle to "Autoclave" position. When chamber pressure reaches 15 psig (121°C), time the digestion for 30 min. After digestion, turn the handle to "Liquids Cool" position, and permit chamber pressure to drop to zero. Turn handle to "Standby and Vent" position. Remove beakers and permit to cool. If there is any turbidity present after autoclaving, filter the solutions through a 0.45-μm membrane filter.

Run the standards and samples after autoclaving at 20 samples per hour using the appropriate manifold as shown in Section XI. System A is used to measure concentrations in the range from 1 μg/ℓ to 50 μg/ℓ-P; system B is used to measure concentrations in the range from 50 μg/ℓ to 500 μg/ℓ. System C is used if arsenic is suspected of being present in the sample. Wash water should be acidified with sulfuric acid at the rate of 1 mℓ H_2SO_4/100 mℓ distilled water.

Every time a new batch of stannous chloride reagent is used, a set of standard solutions must be run. Standard solutions should be run periodically to check the validity of the calibration curve.

If any of the samples after autoclaving and filtering contain a significant level of color, a correction should be made by rerunning these samples through the system with the blank reagents. (A decision about what constitutes a significant level of color can only be made on experience gained by rerunning samples of "doubtful" levels of color.)

C. Calculations

Prepare a calibration curve derived from the peak heights obtained with the standard solutions. Determine the concentrations of phosphorus in the calibration curve. If a sample has been rerun with "blank" reagents, the phosphorus level in the original sample is determined by subtracting the "apparent" phosphorus level obtained with the blank reagents from the "phosphorus" level obtained with the sample and regular reagents.

D. Precision and Accuracy

In a single laboratory, the coefficients of variation at phosphorus levels of 8.0 $\mu g/\ell$ and 20.0 $\mu g/\ell$-P were $\pm 7.30\%$ and $\pm 4.11\%$, respectively.

XIV. PHOSPHORUS, ALL FORMS

This colorimetric method is applicable to surface waters, domestic and industrial wastewaters, and saline samples with orthophosphate levels in the range of 3 $\mu g/\ell$ to 100 $\mu g/\ell$-P. Samples having higher concentrations than this can be measured by appropriate dilution of an aliquot.

Total and inorganic forms of phosphorus can be measured by this method, with the appropriate preliminary pretreatment, which includes persulfate digestion for total P and acid hydrolysis with sulfuric acid for inorganic phosphorus, which changes the phosphorus to the ortho state, which then can be measured directly. The detection limit for this method is 0.003 mg/ℓ-P.

Ammonium molybdate and potassium antimonyl tartrate react in an acid medium with dilute solutions of phosphorus to form an antimony-phospho-molybdate complex. This complex is reduced to an intensely blue-colored complex by ascorbic acid. The color is proportional to the phosphorus concentration. The blue color is measured in a colorimeter at 880-nm wavelength.

Interference from copper, iron, and silicate is insignificant at concentrations many times greater than their reported concentrations in sea water. High iron concentrations, however, can cause precipitation of phosphorus through the formation of clumps in the bottom of the sample.[63] The salt error for samples ranging from 5% to 20% salt content was found to be less than 1%.[63] Although arsenate produces a color similar to phosphate, sea water rarely contains arsenate in concentrations high enough to interfere.[63] Mercury chloride, if used as a preservative, interferes. This interference is overcome by substitution of a sodium chloride (2.5 g/ℓ) in place of the distilled water (line ORN-WHT, 0.23 mℓ/min, in Figure 16).[64]

A. Equipment and Reagents

A Technicon AutoAnalyzer II® (Figure 16) or equivalent unit consisting of: sampler IV, manifold, proportioning pump III, heating bath of 37.5°C, colorimeter (equipped with 50-mm tubular flow cell and 880-nm filters), recorder, and hot plate or autoclave is used.

All glassware used in the determination should be washed with hot 1:1 HCl and rinsed with distilled water. The acid-washed glassware should be filled with distilled water and treated with all the reagents to remove the last traces of phosphorus that

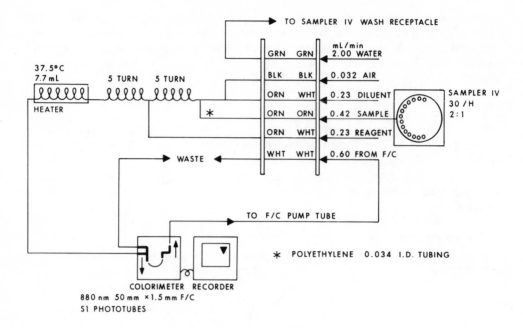

FIGURE 16.   Manifold for phosphorus (all forms).

might be absorbed on the glassware. Preferably, this glassware should be used only for the determination of phosphorus, and after use, it should be rinsed with distilled water and kept covered until needed again. If this is followed, the washing with 1:1 HCl is only required occasionally. Commercial detergents should never be used.

Sulfuric acid, 4.9 N: dilute 136 mℓ of concentrated H_2SO_4 with deionized-distilled water to 1000 mℓ.

Potassium antimonyl tartrate solution: weigh 1.5 g $K(SbO)C_4H_4O_6 \cdot 1/2H_2O$, and dissolve in 300 m$\ell$ deionized-distilled H_2O in a 500-mℓ volumetric flask; dilute to volume. Store at 4°C in a dark glass-stoppered bottle.

Ammonium molybdate solution: dissolve 20 g $(NH_4)_6Mo_7O_{24} \cdot 4H_2O)$ in 500 mℓ deionized-distilled water. Store in plastic bottle at 4°C.

Ascorbic acid, 0.1 M: dissolve 9.0 g of ascorbic acid in 500 mℓ of deionized-distilled water. The solution is stable for one to two months if kept at 4°C. The container, however, must be kept tightly stoppered.

Combined working reagent: mix the reagents above in the following proportions for 100 mℓ of the mixed reagent; 50 mℓ of 4.9 N H_2SO_4, 5 mℓ of potassium antimonyl tartrate solution, 15 mℓ of ammonium molybdate solution, and 30 mℓ of ascorbic acid solution. Mix after addition of each reagent. All reagents must reach room temperature before they are mixed and must be mixed in the order given. If turbidity occurs in the combined reagent, shake and let it stand for a few minutes until the turbidity disappears before proceeding.

Water diluent: to deionized-distilled water add 3.0 mℓ Levor® per liter.

Strong acid solution: slowly add 310 mℓ concentrated H_2SO_4 to 600 mℓ deionized-distilled water. When cool, dilute to 1 ℓ.

Wash water: add 40 mℓ of strong acid solution to 1 ℓ of deionized-distilled H_2O, and dilute to 2 ℓ (not to be used when only orthophosphate is being determined).

Stock solution (50 mg/ℓ): dissolve 0.2197 g of predried KH_2PO_4 in deionized-distilled H_2O, and dilute to 1 ℓ. Add mℓ chloroform as a preservative.

1 mℓ = 0.05 mg P

Standard solution A (5 mg/ℓ): dilute 100 mℓ of stock solution to 1 ℓ.

$$1 \text{ m}\ell = 0.005 \text{ mg P}$$

Standard solution B (0.5 mg/ℓ): dilute 10 mℓ of standard solution A to 100 mℓ.

$$1 \text{ m}\ell = 0.5 \text{ } \mu\text{g P}$$

Prepare a series of standards by diluting suitable volumes of standard solution B to 100.0 mℓ with deionized-distilled H_2O (see Table 5).

Standard solution B must be prepared fresh daily. When saline samples are being analyzed, substitute ocean water (SOW) should be used for preparing the standards; otherwise distilled H_2O is used. A tabulation of SOW composition is given in Table 6.

B. Procedure

The sample should be cooled to 4°C as soon as it has been taken. Analysis should be carried out the same day. The sample aliquot used for the analysis should either be free from turbidity or should be filtered through a 0.45-μm membrane filter.

1. Total Phosphorus

Add 0.5 mℓ strong acid solution to a 50-mℓ sample in a 125-mℓ Erlenmeyer flask. Add 0.2 g of potassium persulfate. Autoclave flasks for 30 min at 121°C (15 to 20 psi). Cool; if samples are not clear at this point, filter through a 0.45-μm membrane filter. Determine phosphorus as outlined under "Orthophosphate" (Section XII).

2. Inorganic Phosphorus

Add 0.5 mℓ of strong acid solution to a 50-mℓ sample in a 125-mℓ Erleymeyer flask. Autoclave flasks for 30 min at 121°C (15 to 20 psi). Cool; if sample is not clear at this point, filter through a 0.45-μm membrane filter. Determine phosphorus in sample as outlined under "Orthophosphate" (Section XII).

3. Orthophosphate

Run samples and standards at 30 per hour using manifold as shown in Figure 16. Place standards in sampler in order of decreasing concentration. Complete filling of sampler with unknowns.

C. Calculations

Prepare a calibration curve by plotting peak heights of standard solutions. Determine the concentrations of phosphorus in the sample by comparing sample peak heights with the calibration curve. Any sample whose peak height is less than 5% of its predecessor must be rerun.

D. Precision and Accuracy

In a single laboratory, using surface water samples at concentrations of 0.005 mg/ℓ, 0.026 mg/ℓ, 0.051 mg/ℓ and 0.090 mg/ℓ-P, standard deviations were ±0.0005, ±0.0004, ±0.0008, and ±0.0007, respectively. In a single laboratory using surface water samples at concentrations of 0.010 mg/ℓ and 0.049 mg/ℓ-P, recoveries were 95% and 96%, respectively.

XV. REACTIVE SILICA

This automated method is applicable to the determination of silica in surface,

Table 5
RECOMMENDED DILUTIONS

Milliliters of standard B	Concentration ($\mu g/l$ P)
0.6	3
1.0	5
2.0	10
6.0	30
10.0	50
15.0	75
20.0	100

Table 6
COMPOSITION OF
SUBSTITUTE OCEAN
WATER

Compound	Concentration (g/l)
NaCl	24.53
CaCl$_2$	1.16
KBr	0.10
NaF	0.003
MgCl$_2$	5.20
KCl	0.70
H$_3$BO$_3$	0.03
N$_2$SO$_4$	4.09
NaHCO$_3$	0.20
SrCl$_2$	0.03

ground and sea waters. The applicable range is 0.5 mg/l to 20 mg/l SiO$_2$; the range, however, can be extended down to 0.005 mg/l by slight modification and upward by dilution of the original sample.

Ammonium molybdate at pH of approximately 1.2 reacts with silica and also with any phosphate present to produce heteropoly acids. Oxalic acid is added to destroy the molybdophosphoric acid but not the molybdosilicic acid. The yellow color produced is reduced by means of aminonaphthosulfonic acid to heteropoly blue. The blue color is more intense than the yellow color and is proportional to the "molybdate-reactive" silica. The color is measured in a colorimeter at 660 nm. There are no significant interferences.

A. Equipment and Reagents

A Technicon® AutoAnalyzer unit (Figure 17) or comparable equipment consisting of: sampler, manifold, proportioning pump, colorimeter, (equipped with a 50-mm flow cell and 660-nm filters), recorder and range expander, is used.

Stock silica solution, 1000 mg/l SiO$_2$: dissolve 4.73 g reagent grade sodium metasilicate, Na$_2$SiO$_3 \cdot$9H$_2$O, in distilled water, and dilute to 1 l. This solution should be checked using a gravimetric procedure.[65] Adjust the remainder of the solution to contain exactly 1000 mg/l SiO$_2$. Alternatively, an accurate stock silica solution may be purchased from Hartman-Leddon® Company (Canlab).

Standard silica solutions: prepare a series of standard silica solutions by diluting suitable aliquots of stock silica solution to 1 l with distilled water.

Ammonium molybdate solution: dissolve 40 g ammonium molybdate, (NH$_4$)$_6$Mo$_7$O$_{24} \cdot$4H$_2$O, in 400 ml distilled water and dilute to 2 l. Filter.

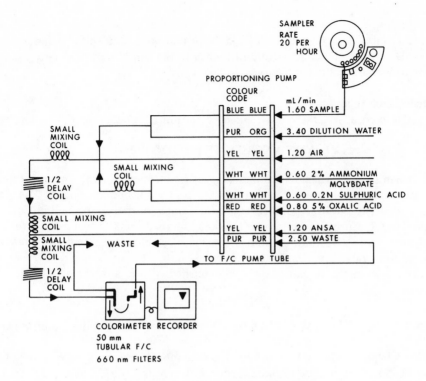

FIGURE 17. Manifold for reactive silica.

Sulfuric acid solution, 0.2 *N*: To 500 m*l* distilled water, add 11.1 m*l* concentrated sulfuric acid, and dilute to 2 *l*.

Oxalic acid solution: dissolve 100 g oxalic acid, $H_2C_2O_4\cdot 2H_2O$, in 800 m*l* distilled water, and dilute to 2 *l*.

Stock ANSA solution: (A) Dissolve 150 g sodium bisulfite, $NaHSO_3$, in about 1 *l* of distilled water. (B) Dissolve 5.0 g anhydrous sodium sulfite, $NaSO_3$, in about 40 m*l* of distilled water, and then dissolve 2.5 g 1-amino-2-naphthol-4-sulfonic acid (ANSA) in the sodium sulfite solution with gentle warming if necessary. Mix solutions A and B. Dilute to 1250 m*l* and filter. Store in a dark bottle.

Working ANSA solution: dilute 200 m*l* stock ANSA solution to 2 *l* with distilled water.

Synthetic sea water: dissolve 25 g sodium chloride, NaCl, and 8 g magnesium sulfate heptahydrate, $MgSO_4\cdot 7H_2O$, in distilled water, and dilute to 1 *l*.

B. Procedure

Sample should be colleted in a polyethylene bottle. Glass containers should not be used. For low-level silica analysis such as found in waters of the Greak Lakes, the test should be carried out within 8 hr after collection.

The sample aliquot either should be free from turbidity or should be filtered through a 0.45-μm membrane filter.

If sea water samples are to be determined, it is necessary to run the base line using synthetic sea water solution instead of distilled water. Also, the standard silica solutions must be made up using the synthetic sea water solution as dilution water. Run the samples and standards at 20 per hour using the manifold as shown in Figure 17. If lower limits of detection are required, replace the dilution line (see Figure 17) for a sample line, and install a range expander (X 10) in the system. With these changes, it is possible to determine accurately as low as 5 μg/l SiO_2.

C. Calculations

Prepare a calibration curve derived from the peak heights obtained with the standard solutions. Determine the concentration of silica in the samples by comparing sample peak heights with the calibration curve.

D. Precision and Accuracy

In a single laboratory, the coefficient of variation at the 200 $\mu g/\ell$ SiO_2 level was ±0.7%.

XVI. ADVANCED METHODOLOGIES

Improvements in analytical techniques to measure increasingly lower concentrations and analyze samples with complex matrices are continuously taking place. These advances are regularly published and reviewed in various journals published in Canada, the United States, Europe and other countries. The reader may consult these reviews for analyzing water, wastewater, and sediment in publications such as *Analytical Chemistry, Analytical Chimica Acta, Water Research, The Analyst,* etc. A few examples of the recent methods that have been published for the analysis of nutrients are given below.

A. Determination of Organic Carbon by UV-Irradiation and Conductometric Determination

PHOTO Chem® Organic Carbon Analyzer, manufactured by Sybron/Barnstead[65] has been proposed to analyze organic carbon in water and wastewater. It is based on oxidation of organic matter by UV-irradiation to carbon dioxide and the subsequent determination of the resultant carbon dioxide conductometry. It is claimed that photochemical oxidation offers a number of advantages over the conventional high temperature oxidation — the technique enables the introduction of significantly larger samples in order to achieve increased sensitivity and to simplify the operation. Sensitivity in low parts per billion is obtained for a wide variety of organic compounds.

B. Determination of Ammonia by Automated Direct Potentiometry

Sekerka and Lechner[35] have proposed an automated method based on direct potentiometry. The method can be operated at the sampling rate of 20 samples per hour. A wide variety of samples have been analyzed by this method, and the direct potentiometric method compares favorably with the standard methods of analysis for ammonium ion. However, the proposed method[35] is simpler, faster and cost effective. The reproducibility of the method ranges from 2.1 to 11% for a concentration range of 10 to 0.1 mg/ℓ. The accuracy of the method was compared with flame photometric and colorimetric methods. The relative error in the range of ±2% was obtained using automated direct potentiometry.

C. Determination of Nitrate in Sediment

The application of methods used for analysis of nitrate in water may not give precise and accurate results when applied to sediment samples or samples containing high concentrations of humic acid, fulvic acid and sulfide.[51] Afghan and Ryan[66] have described a modification to the most commonly used colorimetric method,[51] whereby such samples can be analyzed for nitrate content with a higher degree of precision and accuracy.

The method essentially consists of two steps. In the first step, the interfering substances are removed by precipitation with copper. The resultant solution, after filtration, is then analyzed by a colorimetric approach, using cadmium reduction and sulfanilamide plus naphthylene diamine.[51] The method was tested on a wide variety of natural waters, sediments, clays, and sands and, in each case, accurate results were obtained.

D. Selective Method for the Direct Determination of Nitrate

A method developed by Afghan and Ryan[66] utilizes 2,2′-dihydroxy-4,4′-dimethoxy-benzophenone as a sensitive fluorometric reagent for the determination of nitrate. The procedure to eliminate possible interferences from high concentrations of chloride, sulfide and humic acid substances is also incorporated. The method has been automated to analyze 20 samples per hour, and the rate of analysis can be increased to 40 samples per hour, depending upon the concentration and type of sample. The method can be used to detect nitrate as low as 5 $\mu g/\ell$ nitrate-nitrogen. In this method, other forms of nitrogen such as nitrite, ammonia or organic nitrogen compounds do not react; hence, this method can be used to determine nitrate directly. Most methods published in the literature and used on a routine basis determine nitrate indirectly, i.e., converting nitrate to nitrite and determining the resultant nitrite. This indirect approach may produce errors if there is a significant amount of nitrite in the sample or if the sample contains substances which inhibit or prevent the initial conversion of nitrate to nitrite. The direct method was also critically evaluated against the routine method presently used on a universal basis.[51] It was found that the direct method proposed by Afghan and Ryan[66] gave accurate results when the sample contained large amounts of calcium, sulfide and fulvic acid. The performance of the method was also superior when analyzing clay sediments.

E. Method for Extraction and Determination of "Available" Phosphorus from Lake Sediments

Wentz and Lee[67] have described a method for determining the amount of phorphorus that is probably most often implicated as a possible limiting factor in the biological production of a lake, i.e., the amount of phosphorus which is biologically usable to produce algae or plant growth in an aquatic environment. This method utilizes the extractant of dilute hydrochloric acid-sulfuric acid at pH 1.1. The resultant phosphate is analyzed using vanadomolybdophosphoric yellow procedure instead of the more commonly used molybdophosphoric blue procedure. The main reason for this was that pH adjustment was not required, and the quantitation was achieved without incorporating a reduction step such as stannous chloride or ascorbic acid. In addition, the vanadomolybdophosphoric yellow was found to be stable for at least two weeks, and there was no interference from any ions present in a sample. The above procedure was found to extract relatively loosely bound forms of phosphorus present as apatite and certain organic phosphorus compounds.

ACKNOWLEDGMENT

The author thanks Mr. J. P. Lively, Chief, Laboratory Operations Division, Water Quality Branch, Ottawa, Ontario for giving permission to reproduce the details of the methods published in the Analytical Methods Manual.[51]

REFERENCES

1. Vallentyne, J. R., The Algal bowl — lakes and man, Miscellaneous Special Publication 22, Canada Fisheries and Marine Service, Environment Canada, Ottawa, 1974, 17.

2. Stumm, W. and Morgan, J. J., *Aquatic Chemistry — An Introduction Emphasizing Chemical Equilibria in Natural Waters,* Wiley-Interscience, N.Y., 1979, 429.
3. Vollenweider, R. A., Scientific fundamentals of eutrophication of lakes and flowing waters with particular reference to nitrogen and phosphorus as factors in eutrophication, Technical Report Organization for Economic Cooperation and Development (OECD), 1968.
4. Prince, A. T. and Bruce, J. P., *Nutrient Control Policies in Canada, Nutrients in Natural Waters,* Allen, H. E. and Kramer, J. R., Eds., Wiley-Interscience, 1972, 433.
5. Walter, W. G. and McBee, R. H., *General Microbiology,* D. Van Nostrand, 1964, 210.
6. Kerr, P. C., Brockway, D. L., Paris, D. F., and Sanders, W. W., The carbon cycle in aquatic ecosystem, in *Nutrients in Natural Waters,* Allen, H. E. and Kramer, J. R., Eds., Wiley-Interscience, N.Y., 1972, 101.
7. *Methods for Microbiological Analysis of Waters, Wastewaters and Sediments,* Canada Centre for Inland Waters, Environment Canada, Burlington, Ontario, 1978, 3, 1.
8. Christensen, M. H. and Harremoes, P., Biological Nitrification in Wastewaters, Report 2-72, Department of Sanitary Engineering, Technical University of Denmark, 1972.
9. Vollenweider, R. A., unpublished data, Canada Centre for Inland Waters, Environment Canada, Burlington, Ontario, 1970.
10. Thomas, E. A., Epirische and Experimetelle Untersuchungen Zur Konntnis der Minemumstoffe, in 46 Seen der Schweiz und angrenzender Gebiete, Schweiz., Ver, Ges. und Wasserfachem, 1969, cited in *Chemical Abstracts,* 73: 91075.
11. Report to the International Joint Commission on the Pollution of Lake Erie, Lake Ontario and the International Section of the St. Lawrence River, Vol. I to III, International Joint Commission, Windsor, Ontario, Canada, 1969.
12. Special Report — Chemistry and the Environment, American Chemical Society, 1967, 33A.
13. Elder, F. C., Kramer, J. R., and Kuntz, K. W., *Mass Contribution of the Atmosphere to Lake Huron and Lake Superior, Environmental Analysis,* Academic Press, N.Y., 1978, 1.
14. Morrison, G. H. and Skogerboe, R., General aspects of trace analysis, in Trace Analysis; Physical Methods, Morrison, G. H., Ed., Interscience, N.Y., 1965, 2.
15. Kirkbright, G. F., Development and Publication of New Spectrophotometric Methods of Analysis, *Talanta,* 13(1), 1, 1966.
16. Bauer, E. L., *A Statistical Manual for Chemists,* Academic Press, N.Y., 1971, 6.
17. Youden, W. J., *Statistical Techniques for Collaborative Tests,* Association of Official Analytical Chemists, Washington, D.C., 1969, 21.
18. Wilson, A. L., *The Chemical Analysis of Water,* Society for Analytical Chemistry, London, England, 1974, 45.
19. Afghan, B. K., Analysis of Water Containaints with Special Reference to Heavy Metals. Technical Bulletin No. 52, Inland Waters Branch, Department of Environment, Ottawa, 1971.
20. Mancy, K. H., and Weber, W. J., in Kolthoff, I. M., Elving, P. J., and Stross, F. H., Eds., *Treatise on Analytical Chemistry,* Vol. 2, Part III, Wiley-Interscience, N.Y., 1971, p. 413.
21. Klein, L., *River Pollution I, Chemical Analysis,* Butterworths, Mass., 1962.
22. Jenkins, D., *Determination of Forms of Nitrogen and Phosphorus in Natural Waters,* American Chemical Society, Division Water Waste Chemistry, Reprints 7(1), 121, 1967.
23. Armstrong, F. A. J., Analysis of phosphorus compounds in natural waters, in Halmann, M., *Analytical Chemistry of Phosphorus and Its Compounds,* John Wiley & Sons, N.Y., 1972, chap. 20.
24. Goulden, P. D., The chemical analysis of nutrients, paper presented before the Division of Water, Air and Waste Chemistry, American Chemical Society, Los Angeles, Calif., March 28-April 12, 1971.
25. Grasshoff, K., A Simultaneous Multichannel System for Nutrient Analysis in Sea Water with Analog and Digital Data Record, Advances in Automated Analysis, Technicon International Congress, Vol. 2, Futura Publishing, Mount Kisco, N.Y., 1969, 133.
26. Harwood, J. E., Van Steenderen, R. A., and Kuhn, A. L., A comparison of some methods for total phosphate analyses, *Water Res.,* 3, 425, 1969.
27. Parker, C. A., *Photoluminescence of Solution,* Elsevier Publishing Company, London, 1968.
28. Parker, C. A. and Rees, W. T., Fluorescence spectrophotometry — a review, *Analyst,* 87, 83, 1962.
29. Bailey, P. L., Analysis with Ion-Selective Electrodes, Hyden and Sons Ltd., London, 1976.
30. Pungor, E. and Totch, K., Ion-selective membrane electrodes, *Analyst,* 95, 625, 1979.
31. *Analytical Methods Guide,* 9th ed., Orion Research Inc., Cambridge, Mass., December, 1978.
32. Frant, M. S. and Ross, J. W., Fluoride-sensitive electrode, *Science,* 154, 3756, 1966.
33. Beckett, M. J. and Wilson, A. L., Manual determination of ammonia in fresh waters using an ammonia-sensitive membrane electrode, *Water Res.,* 8, 333, 1974.
34. Evans, W. H. and Partridge, B. F., Determination of ammonia levels in water and wastewaters with an ammonia probe, *Analyst,* 99, 367, 1974.
35. Sekerka, I. and Lechner, J., Simultaneous determination of sodium, potassium and ammonium ions by automated direct potentiometry, *Anal. Lett.,* 7, 463, 1974.

36. Gran, G., Determination of equivalent point in potentiometric titration, *Analyst,* 77, 661, 1952.
37. Orion Research Inc., Newsletter 2, 1979, p. 49.
38. Orion Research Inc., Newsletter 3, 1971m p. 1.
39. Ross, J. W. and Frant, M. W., Cupric-selective electrode with copper (II) — EDTA for end point detection in chelometric titration of ions, *Anal. Chem.,* 41, 2072, 1969.
40. Cross, A. D. and Jones, R. A., *Practical Infra-red Spectroscopy,* 3rd ed., Butterworths, London, 1969.
41. Stewart, J. E., *Infra-red Spectroscopy: Experimental Methods and Techniques,* Dekker, N.Y., 1970.
42. McDonald, R. S., Infra-red spectroscopy, *Anal. Chem.,* 50(5), 283, 1978.
43. Fisherman, M. J. and Erdmann, D. E., Water analysis, *Anal. Chem.,* 51(5), 317, 1979.
44. Koch, W., Automation of analytical processes, *Pure Appl. Chem.,* 18, 1, 1969.
45. Dijkstra, A., General aspects of automation, *Chem.* Week B 1., 64(40), 19, 1968.
46. Babcock, R. H., Automation of water analysis and analytical processes instrumentation, *J. Am. Water Works Assoc.,* 62(13), 145, 1970.
47. Zaleiko, N. W., Automating water and wastewater analysis, *Ind. Water Eng.,* 1, 32, October, 1967.
48. Technicon, Internation Conferences on Advances in Automated Analysis, 1966—1972, Futura Publishing, Mt. Kisco, N.Y.
49. American Society for Testing Materials, Annual Book of ASTM Standards, Part 31, Water, 1979.
50. Standard Methods for the Examination of Water and Wastewater, 14th ed., American Public Health Association, N.Y., 1975.
51. Environment Canada, *Analytical Methods Manual,* Water Quality Branch, Inland Waters Directorate, Ottawa, 1979.
52. Erhardt, M., The particulate organic carbon and nitrogen and the dissolved organic carbon in the Gotland Deep, May 1968, *Kiel Meeresforsch.* XXV(1) 1969.
53. Hewlett-Packard, Instruction Manual for the Model 185 CHN analyzer.
54. Konrad, J. G., Chestens, G., and Keeney, D. R., *J. Therm. Anal.,* 2, 199, 1970.
55. Scesser, W. G. and Rodgers, J., *Deep Sea Res.,* 18, 135, 1971.
56. Dean, W. E., Jr., *J. Sediment Petrol.,* 44, 242, 1974.
57. Roberts, A. A., Palacus, J. G., and Frost, I. C., *J. Sediment Petrol.,* 43, 1157, 1973.
58. Pella, E. and Colombo, B., *Mikrochim. Acta,* 697, 1973.
59. Philbert, F. J. and Traversy, W. J., Proc. 16th Conf. on Great Lakes Research, 294, 1973.
60. Afghan, B. K., Goulden, P. D., and Ryan, J. F., *Adv. in Automated Analysis, Technicon International Congress, 1970,* Vol. 11, Futura Publishing, N.Y., 1971, 291.
61. Goulden, P. D. and Brooksbank, P., *Limnol. Oceanogr.,* 19, 705, 1974.
62. Murphy, J. and Riley, J., *Anal. Chim. Acta,* 27, 31, 1962.
63. U.S. Environmental Protection Agency, Methods for Chemical Analysis of Waters and Wastes, 1971.
64. Standard Methods for the Examination of Water and Wastewater, 12th ed., American Public Health Association, N.Y., 1965.
65. Sybron/Barnstead, *Organic Carbon Determinations with Accuracy, Simplicity and Economy,* Brochure published by Barnstead Company, Boston, 1978.
66. Afghan, B. K. and Ryan, J. F., Substituted Benzophenone as Fluorometric Reagent in Automated Determination of Nitrate, *Anal. Chem.,* 47(14), 2347, 1975.
67. Wentz, D. A. and Lee, G. V., Sedimentary Phosphorus in Lake Cores — Analytical Procedure, *Environ. Sci. Tech.,* 3(8), 750, 1969.

Chapter 4

DISSOLVED OXYGEN AND MINOR IONS

B. K. Afghan

TABLE OF CONTENTS

I. INTRODUCTION*

In this chapter the term "minor ion" refers to those chemical constituents that are present at low concentrations, i.e., mg/ℓ to μg/ℓ, such as dissolved oxygen, cyanide, sulfide, boron, fluoride and iodide. The discussion will be mainly restricted to the sources of these constituents in surface waters and the possible forms in which they may be present. Details of the most commonly used methods for analysis of individual constituents will be presented. Finally, the analytical potential of recently developed advanced methodologies will also be discussed.

A. Occurrence and Forms of Minor Ions
1. Oxygen
The most frequently measured gas in water is oxygen, because it is essential in most biological processes in an aquatic environment. Dissolved oxygen levels in a basin may be indicative of the type of physical, chemical and biological activities occurring in a particular body of water.

Oxygen occurs mainly in molecular form (O_2) in the atmosphere, and as ozone (O_3) at high altitudes. It is also present as oxides and silicates of metals in minerals. In the

---

* Methods in this chapter are reproduced by permission of the Ministry of Supply and Services, Canada.[13]

atmosphere, the constant concentration of ~20% O_2 by volume of oxygen is maintained by the photosynthesis-respiration cycle.

There are two main sources of oxygen in surface water — from the atmosphere and from the photosynthetic process. The most significant source of oxygen in water is atmospheric. The extent of reoxygenation of water bodies depends upon water temperature, atmospheric pressure of oxygen, and salt content of the water body.[3] Aquatic organisms need an appropriate amount of oxygen for survival.

In many cases, fish kill is a result of lack of sufficient dissolved oxygen in a reservoir rather than from the direct toxicity of pollutants. For example, if an effluent containing large amounts of degradable organic compounds (manure, etc.) suddenly enters a reservoir, dissolved oxygen will be consumed during degradation of the organic matter. This may result in depletion of dissolved oxygen to such a low level that fish in the reservoir die due to suffocation.

The Great Lakes Water Quality Board of the International Joint Commission has recommended specific water quality objectives for dissolved oxygen in the connecting channes in the upper waters of the lakes to be not less than 6.0 mg/ℓ at any time. In hypolimnetic waters, it should not be less than the amount necessary for the support of fishlife, particularly cold water species.[4]

2. Cyanide

Cyanide and its compounds enter the aquatic environment via industrial effluents, but are also found in the environment because of natural sources. Many plants contain cyanogenic glycosides, and they release hydrogen cyanide upon hydrolysis. The major source of cyanide, however, is from industrial effluents such as metal and ore processing plants.

Cyanide exists in water in several forms such as molecular cyanide, cyanide ion, simple cyanide, and complex cyanide. The term "simple cyanide" refers to those cyanide compounds which readily dissociate in water to produce cyanide ion. The term "complex cyanide" is used for metal cyanides such as nickel, copper, and iron cyanides.

The persistence and occurrence of a specific form of cyanide that is present in the aquatic environment depends upon many physical and chemical factors. The most important factor that governs the formation of a predominant species is the pH of the aquatic environment.

The most toxic form of cyanide is hydrogen cyanide, which shows a very specific type of toxic action, related to inhibition of oxygen metabolism. A number of studies are reported dealing with the acute toxicity of cyanides to fish. The minimum lethal (threshold) concentration of free cyanide ranges from 12 min to 10 days, depending upon the species of fish and the concentration of cyanide. A review of many studies has shown that free cyanide concentrations in the range of 50 to 100 μg/ℓ as cyanide have proven eventually fatal to many sensitive fish, and levels much above 200 μg/ℓ probably are rapidly fatal to most species.

Based on chronic effects on fish growth and reproduction, an objective of 5 μg/ℓ free cyanide as HCN is recommended by the Water Quality Board of the International Joint Commission as one of the water quality objectives.[5]

An excellent overview of the literature on cyanide chemistry, fate, toxicity and detection in surface waters has been prepared by Ecological Analysts, Inc.[6]

3. Sulfide

Sulfides are discharged into water bodies through industrial sources. Sulfides are also produced in the aquatic environment by microorganisms through the sulfur cycle process. Sulfides react in water to produce various forms, depending upon pH, temperature, and chemical composition of the water body.

In the aquatic environment, sulfides react with the available hydrogen ions to form HS^- or H_2S, depending upon the pH. At pH 9, about 99% of sulfide occurs in HS^- form.

Toxicity of hydrogen sulfide is dependent on temperature, pH, and dissolved oxygen. The presence of dissolved oxygen promotes the conversion of hydrogen sulfide to sulfate resulting in reduced toxicity. A number of authors have reported that concentrations as low as 0.02 mg/ℓ can produce undesirable effects. For example, Colby and Smith[7] report that walleye eggs held in trays in zones where hydrogen sulfide levels were commonly 0.1 to 0.02 mg/ℓ did not hatch. The same study reports a 96-bour LD_{50} of 0.05 mg/ℓ for walleye fry in a laboratory bioassay.

Based on the protection of a balanced population of complete aquatic life cycles, it is recommended that the maximum undissociated hydrogen sulfide concentration should not exceed 0.002 mg/ℓ.[5]

4. Boron

Boron occurs naturally in most unpolluted waters in trace amounts. Additional amounts may be produced by cleaning material or industrial effluents. The most common form of boron in water is undissociated orthoboric acid, H_3BO_3. Sodium tetraborate, $Na_2B_4O_7$, is also used in detergent formulations. In 1970, Waggot[9] published an excellent article discussing the investigation of the potential problem of increasing boron concentrations in rivers and water courses. In this article, the sources of pollution by boron are summarized, the chemical properties of borax or boric acid are given, and toxicity in relation to water pollution is discussed.

The toxicity of boron in water is not well established. Trace quantities of boron are considered essential for algae and other plants; however, boron can be toxic to some plants at higher concentrations ($\simeq$ mg/ℓ or over).[7]

5. Fluoride

Fluoride enters the aquatic environment through the weathering process of igneous and sedimentary rocks. The main source of fluoride is discharges from phosphate and aluminum industries. It is estimated that these industries discharge between 10,000 to 35,000 tons of fluoride into the United States waters annually. Fluoridation of municipal water supplies also adds another 20,000 tons each year.[5]

Fluoride concentration in most fresh water streams is less than 0.2 mg/ℓ. Fluoride ion is considered to be the main ion responsible for solubilizing berylium, aluminum, iron, tin and other metals.[7]

The effects of fluoride on vegetation and aquatic animals have been studied.[7] These studies have concluded that fluoride is relatively non-toxic.

6. Iodide

Iodide ion in fresh water may enter the aquatic environment from industrial waste waters, natural brines, and sea water. Iodide is not a common element in fresh waters. Anderson[10] has reported on the water quality of surface waters of the Ohio River basin and the St. Lawrence River basin. His report covers composite samples collected during 1964-65 at 267 stations on numerous streams for studying the chemical and physical characteristics of surface water. This study included iodide levels. The results show that iodide was present in the nanogram range. This paper also provides an explanation of the presence, problems and significance of various chemicals and properties in water. Many other studies have also been reported to determine the levels of iodide and other microelements in surface waters around the world. All studies have indicated that iodide is present at trace levels and is not considered a significant element.[11,12] The toxicity and environmental effects of iodide have also not been investigated in detail.

B. Analytical Methods

Dissolved oxygen and other minor ions described above are measured on a regular basis in many laboratories involved in water quality studies and waste water effluent characterization and control. The techniques most commonly used for the determination of these elements include absorptiometry, fluorescence and ion-selective electrodes. These techniques are already described in detail in a previous chapter. Therefore, this section will only include details concerning the analytical methods for routine analysis. The majority of the methods described hereafter are taken from the Analytical Methods Manual of the Water Quality Branch, Environment Canada.[13] Sulfide is not included in the above manual, because it is not determined on a regular basis and it is present at very low concentrations.

II. OXYGEN

This titration method is applicable to the determination of dissolved oxygen in surface waters and many wastewaters.

The sample is treated with manganous sulfate and a strongly alkaline iodide reagent. The manganous hydroxide formed reacts with the dissolved oxygen in the sample to form a brown precipitate, manganic hydroxide, $MnO(OH)_2$. Upon acidification, in the presence of iodide, iodine is liberated to an amount equivalent to the dissolved oxygen originally present. The iodine is then titrated with standard sodium thiosulfate. The method is not applicable under the following conditions:

1. Samples containing more than 1 mg/ℓ of ferrous iron
2. Samples containing sulfite, thiosulfite, polythionate, free chlorine, or hypochlorite
3. Samples high in suspended solids
4. Samples containing other oxidizing or reducing materials

If 1 mℓ fluoride solution is added before acidifying the sample and there is no delay in titration, the method is also applicable in the presence of 100 to 200 mg/ℓ ferrous iron. In instances where the azide modification is not applicable, the DO probe should be used.

A. Equipment and Reagents

Three hundred mℓ BOD bottles are used. A 100-mℓ automatic pipette with a two-way stopcock is employed. A small laboratory pump such as Cole-Parmer Instrument and Equipment Company Roll-Flex® Pump 7022 is sufficient for the vacuum system. A 10-mℓ burette with graduations every 0.05 mℓ and with automatic zero adjust is used.

Two hundred and fifty mℓ Erlenmeyer flasks painted white on the outside over the base and two thirds of the way around the sides are employed. During the titrations, the flask is illuminated through the unpainted portion of the side by means of an ordinary light bulb; stir the contents of the flask with a white magnetic stirring bar. A lamp and stand suitable for titrations is needed.

Manganous sulfate solution: dissolve 365 g manganous sulfate monohydrate,

$MnSO_4 \cdot H_2O$, or 400 g manganous sulfate dihydrate, $MnSO_4 \cdot 2H_2O$, or 480 g manganous sulfate tetrahydrate, $MnSO_4 \cdot 4H_2O$, in freshly boiled distilled water and dilute to 1 ℓ. A magnetic stirrer will speed up dissolution. Store in a glass bottle.

Alkaline-iodide-azide solution: dissolve 400 g sodium hydroxide pellets, NaOH, in 500 mℓ of freshly boiled distilled water. (Add the sodium hydroxide in small increments, and dissolve each addition before proceeding). Then add 900 g sodium iodide, NaI, while the solution is still hot. Dissolve 10 g sodium azide, NaN_3, in 40 mℓ distilled water. Add the latter to the former and dilute, if necessary to 1 ℓ.

Sodium thiosulfate solution, 0.0187 N, 1 mℓ = 1.5 mg DO (the thiosulfate concentration is such that 10 mℓ is required for an oxygen concentration of 15.5 mg/ℓ, near the maximum value to be found in waters such as the Great Lakes): Dissolve 4.8 g sodium thiosulfate pentahydrate, $Na_2S_2O_3 \cdot 5H_2O$, and 0.1 g sodium carbonate, Na_2CO_3, in 1 ℓ of distilled water. Add a few drops of chloroform as a preservative.

Standardization procedure: fill a BOD bottle with water (not chlorinated tap water). Add 1.0 mℓ alkaline-iodide-azide reagent, replace the stopper, and mix. Add 1.0 mℓ concentrated sulfuric acid, and mix. Add 1.0 mℓ manganous sulfate and mix again. Transfer 100 mℓ of solution from the BOD bottle to an Erlenmeyer flask, and add 10.00 mℓ potassium bi-iodate. Let the solution stand for at least 2 min in the dark, then titrate the liberated iodine with thiosulfate. (See Section B, Procedure, for details of titration.) Normality *(N)* of thiosulfate is obtained from the expression

$$N_1 \times V_1 = N_2 \times V_2$$

where N_1 = normality of sodium thiosulfate, V_1 = volume of thiosulfate, N_2 = normality of the standard bi-iodate solution, and V_2 = volume of potassium bi-iodate solution. (This is always 10.0.)

Starch indicator solution: add 30 g soluble starch to 1 ℓ of glycerol. Heat the mixture to 180°C, and maintain that temperature until the solution is transparent.

Potassium bi-iodate, 0.01 N: dry a suitable amount of primary standard grade bi-iodate, $KH(IO_3)$, at 105°C for an hour. After cooling in desiccator, weigh out 0.3249 g and dissolve in distilled water. Dilute to 1 ℓ, and store this solution in a tightly stoppered dark glass bottle.

Potassium fluoride solution: dissolve 40 g $KF \cdot 2H_2O$ in distilled water and dilute to 100 mℓ.

B. Procedure

Collect the sample, if possible, in a 300-mℓ BOD (biochemical oxygen demand) bottle, taking precaution to avoid entrainment or dissolution of atmospheric oxygen. Where samples must be collected using other types of samplers, such as Knudsen bottles, the BOD bottle should be filled from the sampler within 15 min of sampling.

To fill the BOD bottle, a length of flexible tubing should be attached to the outlet of the sampler and the tubing be extended to the bottom of the BOD bottle. Care must be taken to prevent turbulence and air entrainment, especially when just starting to fill the bottle. Keep the end of the tubing in the BOD bottle under the surface of the water while the bottle is filling. Allow the water to overflow from the top of the BOD bottle, and replace the stopper at once. The sample is ready for oxygen determination.

Remove stopper from the BOD bottle and add, by means of pipettes placed just below the water in the BOD bottle, 1.0 mℓ of manganous sulfate reagent followed immediately by 1.0 mℓ of alkaline-iodide-azide solution. Restopper the bottle at once, and mix the contents by shaking vigorously for at least 20 sec, or until the precipitated manganous and manganic hydroxide is evenly dispersed. No air bubbles should be trapped in the bottle. After 2 to 3 min, shake the bottle again. Allow the precipitate in the sample to settle at least one third of the way down the bottle — about 1 hr is required. The sample may be allowed to stand indefinitely at this stage, as long as air is not drawn into the bottle. Add 1.0 mℓ concentrated H_2SO_4 by placing the pipette just beneath the surface. Restopper the bottle and shake until dissolution of precipitate.

By means of the two-way pipette and vacuum system, transfer 100 mℓ of solution from the BOD bottle to a specially painted Erlenmeyer flask containing a magnetic stirring bar. Titrate at once with thiosulfate solution until the solution is a very pale straw color. Then add four drops of starch solution. Continue the titration until the blue color just disappears.

C. Calculations
Calculate the dissolved oxygen content of the sample as follows:

$$Mg/\ell\ DO\ =\ \frac{N \times V \times 8 \times 1000}{m\ell\ \text{sample titrated}}$$

where N = normality of sodium thiosulfate solution and V = volume of sodium thiosulfate for sample. The volume of treated sample is 100 mℓ, but the volume of the water sample titrated is

$$100 \times \frac{(300-2)}{300}\text{, i.e., 99.3}$$

D. Precision and Accuracy
In a single laboratory, the coefficient of variation at a dissolved oxygen concentration of 4.1 mg/ℓ was ±1.8%.

III. CYANIDE (COLORIMETRIC - CHLORAMINE-T, PYRIDINE-PYRAZOLONE)

This method is applicable to the determination of cyanide in surface, waste and sea water. When the distillation assembly described is used, the method is sensitive to about 2 μg/ℓ.

Distillation of the sample in the presence of sulfuric acid converts complex cyanides into hydrocyanic acid, HCN. The HCN gas is absorbed in a solution of sodium hydroxide, and the cyanide is determined colorimetrically.

Equations:

$$Fe(CN)_6{}^{4-} + 6H^+ \longrightarrow 6HCN + Fe^{2+}$$

$$HCN + NaOH \longrightarrow NaCN + H_2O$$

In the colorimetric measurement, the cyanide in the sodium hydroxide solution after distillation is converted to cyanogen chloride, CNCl, by reaction with chloramine-T, $CH_3C_6H_4SO_2NClNa \cdot 3H_2O$. The CNCl then forms a blue dye on the addition of pyridine-pyrazolone ($C_{10}H_{10}N_2O$) reagent, and the absorbance is measured at 620 nm.

The reaction may be expressed by the following equations:[14]

$$NaCN + [\text{ring, } SO_2N(Na)Cl, CH_3] \longrightarrow CNCl + [\text{ring, } SO_2N(Na)_2, CH_3]$$

$$[\text{piperidine ring, } N] + CNCl \longrightarrow [\text{piperidinium ring, } N^+\text{-CN}] + Cl^-$$

$$[\text{piperidinium ring, } N^+\text{-CN}]\ Cl^- + H_2O \longrightarrow O{=}C{-}CH{=}C{-}\overset{H}{\underset{H}{C}}{-}C{=}O$$

Sulfides interfere and should be removed prior to distillation. Other interfering substances are removed by distillation.

A. Equipment and Reagents

A 1000-mℓ three-necked flask fitted with ground-glass joints is used. To the three necks are fitted a thistle funnel which extends below the liquid level, an air sparging tube drawn to a fine tip, and a Friedrichs® condenser (Figure 1).

A gas absorption column is made from Pyrex® tubing, 22 mm I.D. and 450 mm long, and is filled with glass helices (3 mm I.D.).

A Technicon® proportioning pump fitted with 8 pumping tubes, 2.8 mm I.D. is used.

A 100-mℓ graduate cylinder is mounted with a clamp on a retort stand so that its position can be adjusted.

A spectrophotometer for use at 620 nm is used together with 1 cm cells.

Sodium hydroxide solution, 1 N: dissolve 50 g sodium hydroxide, NaOH, in 1 ℓ distilled water.

Mercuric chloride solution: dissolve 34 g mercuric chloride, HgCL$_2$, in 500 mℓ distilled water. (Caution: toxic; take care to avoid ingestion.)

Magnesium chloride solution: dissolve 51 g magnesium chloride, MgCl$_2\cdot6H_2O$, in 100 mℓ distilled water.

Sodium hydroxide solution, 0.2 N: dissolve 8 g sodium hydroxide, NaOH, in 1 ℓ distilled water.

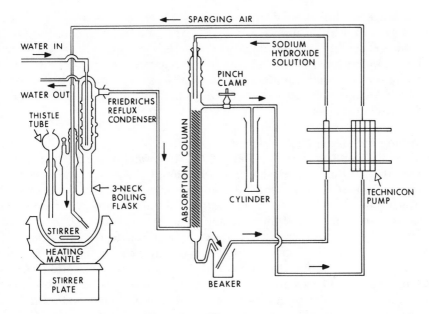

FIGURE 1.   Cyanide distillation apparatus. Friedrichs® and Technicon® are registered trademarks of the Friedrichs Company and Technicon, Inc., respectively.

Acetic acid: Dilute 1:4 with water.

Stock cyanide solution: dissolve 2.51 g potassium cyanide, KCN, in 1 l water. Standardize with 0.0192 N silver nitrate, $AgNO_3$. The solution loses strength gradually and must be rechecked every week. Approximate strength, 1 ml = 1 mg CN. (Caution: toxic; take care to avoid ingestion.)

Standard cyanide solution: dilute 10 ml stock cyanide solution to 1 l with distilled water; mix and make a second dilution of 10 ml to 100 ml, 1.00 ml = 1.0 μg. CN. This solution must be prepared daily. Caution: Toxic; take care to avoid ingestion.)

Chloramine-T solution: dissolve 1 g of chloramine-T, $CH_3C_6H_4SO_2NClNa \cdot 3H_2O$, in 100 m$l$ water. Prepare daily.

1-Phenyl-3-methyl-5-pyrazolone solution: prepare a saturated aqueous solution (approximately 0.5 g/100 ml) by adding the pyrazolone, $C_{10}H_{10}N_2O$, to water at approximately 75°C. Agitate occasionally as the solution cools to room temperature. If necessary, the pyrazolone (melting point 127°C to 128°C) can be purified by recrystallization from ethyl alcohol. Usually this is not required.

Mixed pyridine-pyrazolone reagent: Mix 125 ml of the filtered, saturated aqueous solution of pyrazolone with a filtered solution containing 0.025 g bis-pyrazolone dissolved in 25 ml pyridine. Several minutes of mixing is usually necessary to dissolve the bis-pyrazolone in pyridine. The mixed reagent develops a pink color on standing, but if used within 24 hr, this does not affect the color production with cyanide. Prepare daily.

Standard silver nitrate solution: dissolve 3.27 g of silver nitrate, $AgNO_3$, in 1 l of distilled water. Store in a dark bottle. One ml of this solution is equivalent to 1.00 mg CN.

B. Procedure

The sample should be collected in a 2-l plastic bottle and analyzed as soon as possible after collection. Samples should be preserved by addition of sufficient NaOH to raise the pH to 11.0 or above, and be stored in a cool place.

Decant a sample aliquot for distillation. Sulfide is removed by treating the alkaline sample at pH 11.0 with small increments of powdered lead carbonate, $PbCO_3$. Black lead sulfide precipitates in samples containing sulfide. Repeat this operation until no more lead sulfide forms. Filter and rinse the precipitate, add the rinse water to the filtrate, and use an aliquot for analysis. Avoid a large excess of lead carbonate and a long period of contact in order to minimize complexing or occlusion of the cyanide with the precipitated material.

1. Procedure for Distillation

Close pinch clamp (Figure 1). Add about 80 mℓ of water to the graduated cylinder, and adjust the position of the cylinder so that the tube dips near the bottom of the cylinder. Add 750 mℓ sample to the flask; insert stirring bar, and start the stirrer. Add 50 mℓ 1 N NaOH solution to the 250-mℓ beaker and start the pump. When the absorp-tion column is wet with NaOH solution, add 20 mℓ mercuric chloride solution and 10 mℓ magnesium chloride solution, described above, through the thistle tube. Slowly add 37 mℓ concentrated sulfuric acid to the flask through the thistle tube. Turn on the heating mantle and allow the contents of the flask to boil vigorously. (Excess air as the flask heats up will excape from the equipment through the dip tube in the grad-uated cylinder.) After 1 hr of boiling, turn off the heating mantle and open the pinch clamp. Remove the tube pumping NaOH from the NaOH solution; rinse the contents of the column into the NaOH beaker by pumping approximately 150 mℓ distilled water through the absorption column, and collect in the NaOH beaker. Transfer quantita-tively to 1 250-mℓ volumetric flask, and make up to the mark.

2. Procedure for Colorimetric Measurement

To prepare standard solutions for the calibration curve, use cyanide standard 1 mℓ = 1 μg CN. Pipette 0 (blank), 0.2 mℓ, 0.5 mℓ, 0.8 mℓ, and 1.0 mℓ into 50-mℓ beakers, and make up to 15 mℓ with NaOH, 0.2 N. Proceed as below, treating samples and standards in the same manner.

Adjust pH to 6 to 7 with 1:4 acetic acid. Transfer to 25-mℓ volumetric flask. Add 0.2 mℓ chloramine-T solution and mix. Allow 1 to 2 min for the reaction. Add 5.0 mℓ mixed pyridine-pyrazolone reagent and make up to the mark; mix. Allow 20 min for color development. Read absorbance at 620 nm in a 1-cm cell.

As a check on the distillation step, periodically process cyanide standard solutions through the complete procedure.

C. Calculations

Prepare a calibration curve derived from the peak heights obtained with the standard solutions. Determine the micrograms of cyanide in the samples by comparing sample peak heights with the calibration graph. Calculate the cyanide concentration as fol-lows:

$$\text{mg}/\ell \text{ CN} = \frac{F \times D}{C \times E}$$

where E = cyanide determined (μg), D = diluted absorbing solution (mℓ), C = origi-nal sample (mℓ), and E = aliquot used (mℓ).

IV. CYANIDE (AUTOMATED PYRIDINE-PYRAZOLONE)

This method is applicable to the determination of cyanide in water and wastewaters in the range from 0.001 mg/ℓ to 0.100 mg/ℓ HCN. The detection limit for this method is 0.001 mg/ℓ CN^-.

Complex cyanides are converted to hydrocyanic acid, HCN, by irradiation with ultraviolet light. After irradiation the sample is distilled from phosphoric acid solution into a sodium acetate absorbing solution.

In the colorimetric measurement, the cyanide in sodium acetate solution is converted to cyanogen chloride, CNCl, by reaction with chloramine-T. The CNCl then forms a blue dye on the addition of pyridine-pyrazolone reagent, and the absorbance is measured at 620 nm.

The reaction may be expressed by the following equations:[14]

$$NaCN + \underset{\underset{CH_3}{\big|}}{\underset{}{\bigcirc}}\text{—}SO_2N\overset{Na}{\underset{Cl}{\diagdown}} \longrightarrow CNCl + \underset{\underset{CH_3}{\big|}}{\underset{}{\bigcirc}}\text{—}SO_2N\overset{Na}{\underset{Na}{\diagdown}}$$

$$\underset{N}{\bigcirc} + CNCl \longrightarrow \underset{\overset{+}{N}\text{—}CN}{\bigcirc} + Cl^-$$

$$\underset{\underset{CN}{\big|}}{\underset{N^+}{\bigcirc}}\ Cl^- + H_2O \longrightarrow O{=}C\text{—}CH{=}C\underset{H}{\overset{H}{\underset{|}{\overset{|}{C}}}}\text{—}C{=}O$$

with the hydrogen substituents: $O{=}C\text{—}CH{=}C\text{—}C\text{—}C{=}O$ (bearing H atoms: on first C an H below, and H H H across the chain).

[pyrazolone condensation intermediate with pyridine-derived polymethine chain:]

$$\bigcirc\text{—}N\text{—}C{=}O \; ,\; N{=}C\text{—}CH_3 \; ... \; CH_2 O{=}C\text{—}C{=}C\text{—}C\text{—}C{=}OH_2 \; C ... \; C{-}N{-}\bigcirc,\; C{=}N,\; CH_3$$

$$\downarrow$$

$$\bigcirc\text{—}N\text{—}C{=}O,\; N{=}C\text{—}CH_3 \; \diagup C{=}C\text{—}C{=}C\text{—}C\text{—}C{=}C \diagdown C{-}N{-}\bigcirc,\; C{=}N,\; CH_3 \quad (H\ H\ H\ H\ \text{and}\ H)$$

Sulfides interfere and should be removed prior to analysis (see Procedure below). Thiocyanate gives an apparent total cyanide on an equimolar basis. Some organic cyanide compounds such as nitriles are decomposed under test conditions. Other interfering substances are removed in the distillation step.

A. Equipment and Reagents

A Technicon Auto-Analyzer® unit or the equivalent as shown in Figure 2 is used. The colorimeter is equipped with a 50-mm flow cell and a 620-nm filter.

UV reactor: the coil consists of quartz tubing, 3 mm I.D. and 12 m long, wound to a diameter of 13 cm. A 400-W mercury lamp is used. For description of the reactor, see the "Nitrogen: Organic" (automated ultraviolet digestion) method.

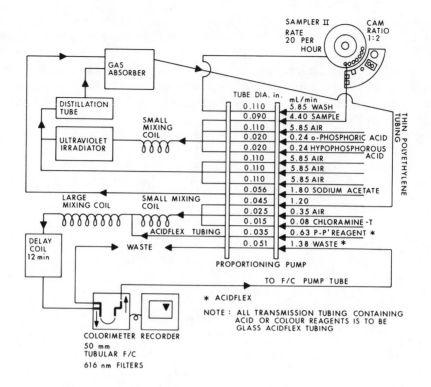

FIGURE 2.   Cyanide manifold. Sampler II® and Acidflex® are registered trade-marks of Technicon, Inc.

Distillation equipment (Figure 3): the distillation column consists of borosilicate glass tubing, 12 mm I.D. and 350 mm long, wrapped with 22-gauge heating wire to give a resistance of about 24 ohms. Approximately 65 V AC is applied to the heating wire, using a variable transformer.

Gas absorption column: for details of construction, see Figure 4.

Stock cyanide solution, 1000 mg/ℓ as CN: dissolve 2.51 g potassium cyanide, KCN, in distilled water. Add 10 mℓ of 10% sodium hydroxide solution, and make up to 1 ℓ in a volumetric flask. Standardize with 0.0192 N silver nitrate solution using indicator solution. The solution gradually loses strength and must be rechecked every 2 weeks. (Caution: toxic; take care to avoid ingestion.)

Sodium hydroxide solution, 10%: dissolve 100 g sodium hydroxide, NaOH, in approximately 750 mℓ distilled water and make up to 1 ℓ.

Standard silver nitrate solution, 0.0192 N: dissolve 3.270 g silver nitrate, AgNO₃, in distilled water, and dilute to 1 ℓ in a volumetric flask (1 mℓ of this solution is equivalent to 1 mg CN).

Indicator solution: dissolve 0.02 g rhodanine (p-dimethylamino-benzylidene) in 100 mℓ acetone.

Intermediate cyanide solution, 10 mg/ℓ as CN: dilute 10 mℓ stock cyanide solution and 10 mℓ stock cyanide solution and 10 mℓ sodium hydroxide solution to 1 ℓ with distilled water.

Standard cyanide solution, 0.1 mg/ℓ as CN: dilute 10 mℓ intermediate cyanide solution and 10 mℓ sodium hydroxide solution to 1 ℓ with distilled water.

Working cyanide standards: prepare a series of standards to cover the required concentration range by appropriate dilutions of the standard cyanide solution. All standards should be preserved with sodium hydroxide solution at the rate of 1 mℓ of 10% NaOH per 100 mℓ solution.

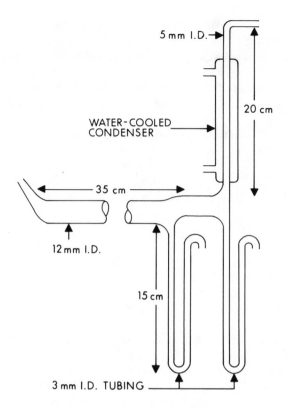

FIGURE 3.   Continuous distillation apparatus.

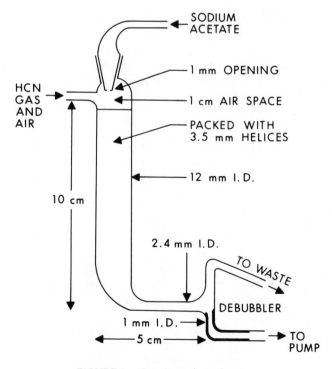

FIGURE 4.   Gas absorption column.

Phosphoric acid solution: dilute reagent grade of phosphoric acid, H_3PO_4, with an equal volume of distilled water.

Hypophosphorus acid solution: dilute 50% hypophosphorous acid, $H(H_2PO_2)$, with an equal volume of distilled water.

Sodium acetate solution, 2%: dissolve 20 g sodium acetate CH_3COONa, in distilled water and make up to 1 ℓ. Adjust the pH of this solution to 6.5 with acetic acid (glacial).

Wash solution, 0.1% sodium hydroxide solution: dilute 20 mℓ of 10% sodium hydroxide solution to 2 ℓ with distilled water.

Mixed pyridine-pyrazolone solution: dissolve 0.75 g 3-methyl-1-phenyl-2-pyrazolin-5-one and 0.05 g, 3,3'-dimethyl-1, 1'-diphenyl-(4,4'-bi-2-pyrazoline)-5,5'-dione in 50 mℓ pyridine. Add 250 mℓ distilled water and mix. Prepare daily.

Chloramine-T solution: dissolve 1 g chloramine-T, $CH_3C_6H_4SO_2NCIN\cdot3H_2O$, in 100 m$\ell$ distilled water. Prepare daily.

B. Procedure

The sample should be collected in a 125-mℓ plastic bottle and analyzed as soon as possible after collection. Samples should be preserved by addition of sufficient NaOH to raise the pH to 11. Usually 1 mℓ of 10% sodium hydroxide solution per 100-mℓ samples is adequate.

If sulfide is present, decant a 100-mℓ aliquot of sample and add 0.5 g citric acid and 0.5 g sodium bisulfite, $NaHSO_3$. Stir the solution and allow to stand for 15 min. Decant the supernatant from the precipitated sulfur, and neutralize the excess sulfite with hydrogen peroxide (30%), using starch-iodide paper as an indicator. Then immediately treat the sample aliquot with sodium hydroxide solution to raise the pH to above 10. Treat samples containing free chlorine with sodium bisulfite using starch-iodide paper as an indicator.

Set the system in operation and allow to warm up. Determine the rate of distillation by measuring the volume of liquid that overflows from each of the two U-tube traps. Adjust the power to the distillation column heater so that approximately 20% of the liquid is being vaporized in the tube. Check to see that the conical joint at the top of the gas absorber column is filled with sodium acetate solution. Run sodium hydroxide-treated standards and samples at 20/hr.

C. Calculations

Prepare a calibration curve from the peak heights obtained with the standard solution. Determine the concentration of cyanide in the samples by comparing sample peak heights with the calibration curve.

D. Precision and Accuracy

In a single laboratory, a natural water sample was analyzed, and the coefficient of variation at a level of 0.023 mg/ℓ was ± 3.8%. The sample was spiked with 0.025 mg/ℓ cyanide, and the mean concentration found was 0.046 mg/ℓ with a coefficient of variation of ±2.2%. The average percent recovery was 94%.

The recoveries of standard solutions of complex cyanides were compared with the recoveries of simple cyanide standard solutions (KCN).

Complex cyanide	Recovery (%)
Potassium hexacyanocobaltate	78—80
Potassium tetracyanonickelate	75—80
Potassium ferrocyanide	95—100
Potassium ferricyanide	95—100

V. BORON (CURCUMIN)

This method is applicable to surface and groundwaters in the range from 0.10 mg/ℓ to 1.0 mg/ℓ B. Higher concentrations may be determined using an appropriate dilution of the original sample.

The red-colored product called rosocyanine, obtained when a sample of water containing boron is acidified and evaporated in the presence of curcumin, is taken up in ethyl alcohol and measured in a spectrophotometer.

Nitrate concentrations above 20 mg/ℓ NO_3^- interfere with this method. There are no other serious interferences.

A. Equipment and Reagents

A spectrophotometer for use at 540 with a minimum light path of 1 cm is used. A water bath, operated at 55° ±2°C is required.

Glass-stoppered volumetric flasks, 25-mℓ capacity must be used.

Stock boron solution, 100 mg/ℓ B: dissolve 0.5716 g anhydrous boric acid, H_3BO_3, in distilled water and dilute to 1 ℓ. Because anhydrous H_3BO_3 loses weight on drying at 150°C, use a reagent meeting ACS specifications and keep the bottle tightly stoppered to prevent entrance of air moisture.

Standard boron solution, 1.0 mg/ℓ B(1.0 mℓ = 1.0 μg B): dilute 10.00 mℓ stock boron solution to 1 ℓ with distilled water.

Curcumin reagent: dissolve 0.040 g finely ground curcumin, $C_{21}H_2O_4 \cdot 2H_2O$, in 80 mℓ of 95% ethyl alcohol. Add 4.2 mℓ concentrated HCl, and make the solution up to 100 mℓ with ethyl alcohol in a 100-mℓ volumetric flask. (In place of ethyl alcohol, 95% isopropyl alcohol may be used.) This reagent will be stable for several days if stored in a refrigerator.

Ethyl alcohol, C_2H_5OH, 95%, is also needed.

B. Procedure

Samples should be stored in polyethylene bottles.

The sample aliquot used for analysis either should be free from turbidity or should be filtered through a 0.45-μm membrane filter.

To prepare calibration curve, pipette O (blank), 0.25 μg, 0.50 μg, 0.75 μg and 1.00 μg standard boron solution into evaporating dishes (dishes should be of the same size and shape). Pipette 1.00 mℓ sample or dilution into an evaporating dish. Add 4.0 mℓ curcumin reagent to each, and swirl dishes gently to mix. Float the dishes on a water bath set at 55° + 2°C, and evaporate the contents to complete dryness.

Remove each dish from the water bath as soon as the contents appear dry and the odor of HCl has gone. After the dishes have cooled to room temperature, add 10.0 mℓ 95% ethyl alcohol to each dish, stirring gently with a polyethylene rod to ensure complete dissolution of the red-colored product. Wash the contents of each dish into separate 25-mℓ volumetric flasks using 95% ethyl alcohol. Make up to the mark on each flask with 95% ethyl alcohol, and mix by inverting the flasks.

Read absorbancies at 540 nm after setting the reagent blank at zero absorbance. Readings should be taken within 1 hr after samples have dried.

C. Calculations

Prepare a calibration curve derived from the readings obtained with the standard solutions. Determine the concentration of boron in the sample by comparing sample reading with the calibration curve.

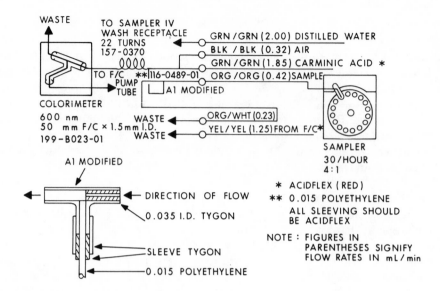

FIGURE 5.   Boron manifold—carminic acid method. Sampler IV® and Acidflex® are registered trademarks of Technicon, Inc.

D. Precision and Accuracy

The American Public Health Association[15] reports a standard deviation of ±0.06 mg/ℓ at a boron level of 0.18 mg/ℓ B as determined by nine laboratories.

VI. BORON (AUTOMATED CARMINIC ACID)

This method[16] is applicable to surface and groundwaters in the range 0.05 mg/ℓ to 1.0 mg/ℓ. Higher concentrations may be determined using an appropriate dilution of the original sample. The detection limit for this method is 0.02 mg/ℓ B.

Boron is reacted with carminic acid in concentrated sulfuric acid. The color of carminic acid changes from bright red to bluish red to blue, depending on the concentration of boron.

A phenol concentration of 0.05% (w/v) in the carminic acid solution suppresses nitrate interference[17,18] when the concentration of nitrogen is less than 40 mg/ℓ.

A. Equipment and Reagents

A Technicon AutoAnalyzer II® system, or the equivalent comparable to that shown in Figure 5, is used.

A colorimeter equipped with flow cell, 50 mm by 1.5 mm, and 600-nm filters is employed.

Stock boron solution, 100 mg/ℓ boron: dissolve 0.5716 g anhydrous boric acid, H_3BO_3, in distilled water and dilute to 1 ℓ. Because anhydrous H_3BO_3 loses weight on drying at 150°C, use a reagent meeting ACS specifications and keep the bottle tightly stoppered to prevent the entrance of air moisture.

Standard boron solution, 1.0 mg/ℓ boron: dilute 10.00 mℓ stock boron solution to 1 ℓ. Working standards in the range 1.0 mg/ℓ to 0.01 mg/ℓ are prepared from this solution.

Carminic acid (stock): dissolve 0.5 g of carminic acid in 500 mℓ concentrated H_2SO_4. This may take as long as 1 hr to dissolve. Store in a polyethylene bottle.

Carminic acid (working solution): dissolve 5.0 g phenol in 100 mℓ of stock carminic acid solution. Add 900 mℓ of concentrated H_2SO_4. Store in polyethylene bottle.

B. Procedure

Samples should be stored in polyethylene bottles. The sample used for analysis should be free from turbidity or should be filtered through a 0.45-μm membrane filter. Run the samples and standards up to 30/hr, as shown in Figure 5. If any samples contain a significant level of color, a correction should be made by rerunning these samples with the blank reagents.

C. Calculations

Prepare a calibration curve derived from the peak heights obtained with the standard solutions. Determine the concentration of boron in the samples by comparing sample peak heights with the calibration curve. If a sample has been rerun with "blank" reagents, the boron level in the sample is determined by subtracting the "apparent" boron level obtained with the blank reagents (as determined by the peak height and calibration curve) from the "boron" level obtained with the sample and regular reagents.

D. Precision and Accuracy

In a single laboratory, the average recovery was 101% for six spiked samples ranging from 0.05 mg/ℓ to 0.38 mg/ℓ B. The average error for these six samples was 6.1%.

E. Notes

1. The possibility that reagents and standards may take up boron from borosilicate glassware should be noted.
2. The working solution of carminic acid must be made up fresh before use (daily) in order to avoid an unstable base line. A noise level of one chart division (1%) is not uncommon in routine operation.
3. The use of the modified T joint shown in Figure 5 is essential for pulse suppression.

VII. BORON (AUTOMATED FLUOROMETRIC)

This automated method[19] is applicable to the determination of boron in natural waters, detergents and sewage effluents in the range of 0.005 mg/ℓ to 0.100 mg/ℓ B. Higher concentrations of boron can be determined using an appropriate dilution of the sample. The detection limit for this method is 0.005 mg/ℓ B.

The method is based upon the reaction of boron with 4'-chloro-2-hydroxy-4-methoxybenzophenone (CHMB) to form a fluorescent complex. The product is then quantitated by fluorescence spectroscopy.

There are no significant interferences in normal waters. Vanadium interferes[19] at concentrations greater than 0.1 mg/ℓ. Interference resulting from background fluorescence of organics has been eliminated by UV oxidation of the sample prior to the addition of the reagent.

A. Equipment and Reagents

A Technicon AutoAnalyzer® unit or comparable equipment as shown in Figure 6 is used.

A heating bath run at 52°C is employed.

A Fluorometer II® complete with flow cell and a 355-nm narrow pass primary filter and a 460-nm narrow pass secondary filter is required.

A Haake FK® constant temperature circulator, 20°C, is used.

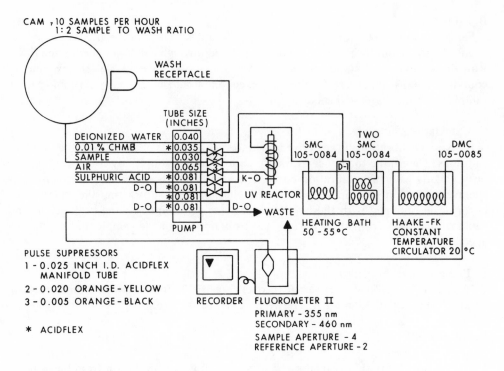

FIGURE 6.   Boron manifold—fluorometric method. Haake-Fk® is a registered trademark of Haake, Berlin. Fluorometer II® is a registered trademark of Farrand Optical Company.

A UV reactor is constructed using a sheet metal cylinder, 43 cm high and with a 19-cm diameter, to house components. A 550-W photochemical lamp is placed in a fused quartz lamp protection jacket and mounted axially in the center of the cylinder. A quartz coil made of G.E. clear fused quartz, 3 mm I.D. with 0.6-mm wall thickness and a coil diameter of 12 cm, is mounted coaxially around the lamp. The irradiating coil in the UV reactor is cooled by a high-speed fan fitted on the bottom of the reactor; a tapered exhaust chimney is fitted on top of the reactor (Figure 7).

CHMB solution (0.01%): dissolve 0.050 g 4'-chloro-2-hydroxy-4-methoxybenzophenone (CHMB) in sulfuric acid, H_2SO_4, and dilute to 500 mℓ with sulfuric acid, H_2SO_4.*

Stock boron solution, 100 mg/ℓ B: dissolve 0.5716 g of anhydrous boric acid, H_3BO_3, in distilled-deionized water.

Standard boron solution, 1 mg/ℓ B: dilute 10 mℓ of stock boron solution to 1 ℓ with distilled-deionized water.

Working boron solutions: prepare a series of calibration standards containing 0.005 mg/ℓ, 0.025 mg/ℓ, 0.050 mg/ℓ, 0.075 mg/ℓ, and 0.100 mg/ℓ boron by appropriate dilution of the standard boron solution.

B. Procedure

Samples should be collected and stored in polyethylene bottles.

The sample aliquot should be free from turbidity or filtered through a 0.45-μm membrane filter. Hydrogen peroxide (30%) should be added to the filtered aliquot at the rate of 2 drops (50 μℓ) per 50 mℓ of sample.

---

*   CHMB is available from Aldrich Chemical Company Ltd.

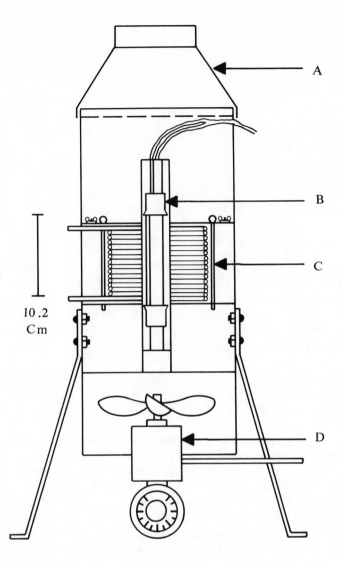

FIGURE 7. UV reactor; (a) chimney, (b) ultraviolet lamp,
(c) quartz coil, (d) cooling fan.

Set up the automated system as shown in Figure 6, and allow the system to warm
up for at least 1 hr to reach equilibrium. Do not flush Acidflex® pump tubes with
water at the end of each day; just pump dry with air. (Pumping the system clear with
water extends the time required each morning for the system to reach equilibrium.)

With a sample aperture of 4 and a reference aperture of 2, set the base line at 10%
T and set the highest standard, 0.100 mg/ℓ boron, at 90% T using the blank control
dial and the full-scale record dial, respectively. Run standards and peroxide-treated
samples at 10/hr. Rise time and descent time for sample response are rather slow. The
noise level for blanks and samples will be about 1 to 2% full-scale deflection.

C. Calculations

Prepare a calibration curve derived from the peak heights obtained with the standard
solutions. Determine the concentration of boron in the samples by comparing sample
peak heights with the calibration curve.

D. Precision and Accuracy

In a single laboratory, the coefficient of variation at a level of 0.100 mg/ℓ B was ±1.87%.

E. Notes

1. The possibility that reagents and standards may take up boron from borosilicate glassware should be noted.
2. The identity of the fluorescent product formed in the reaction of boric acid with CHMB is not clearly defined.[20]
3. For samples containing greater than 0.1 mg/ℓ of boron, it is possible to place a dilution line on the manifold as well as to remove the UV reactor.

VIII. FLUORIDE (SELECTIVE-ION ELECTRODE)

This method is applicable to the determination of fluoride in natural waters and wastewaters. All fluoride concentrations greater than 0.05 mg/ℓ F can be determined.

Fluoride is determined potentiometrically using a specific ion fluoride electrode in conjunction with a standard reference electrode and a pH meter that has an expanded scale capability. Specific ion meters are available that have a direct concentration scale for fluoride. There are no known significant interferences.

A. Equipment and Reagents

A pH meter with expanded scale or fluoride-specific ion meter, e.g., Corning® Model 12 pH meter or Orion® Model 801 specific ion meter, or equivalent of either instrument, is employed.

A fluoride-specific ion electrode (e.g., Orion® No. 94-09) together with a reference electrode (saturated calomel-sleeve type) are used.

Total ionic strength adjustment buffer (TISAB): to 57 mℓ glacial acetic acid, CH_3COOH, add 58 g sodium chloride, NaCl, and 1 g cyclohexanediaminetetraacetic acid (CDTA); dilute to 500 mℓ with distilled water. Adjust the solution to pH 5.8 with 5 M sodium hydroxide (NaOH) solution and dilute to 1 ℓ in a volumetric flask.

Stock fluoride solution, 100 mg/ℓ F: dissolve 0.2210 g anhydrous sodium fluoride, NaF, in a distilled water and dilute to 1 ℓ in a volumetric flask.

Standard fluoride solutions: prepare calibration standards to cover the desired range by appropriate dilution of the stock solution.

B. Procedure

The plastic sample container should be tightly closed immediately after collection, and the concentration of fluoride ions should be determined as soon as possible thereafter.

Pipette 10.0 mℓ of water sample and standards (0.5 mg/ℓ, 1.0 mg/ℓ and 2.0 mg/ℓ F) into 50-mℓ beakers. Pipette 10.0 mℓ of the TISAB solution into each beaker. Stir.

1. Procedure Using an Expanded-Scale pH Meter

Calibrate electrodes and instrument. Set pH meter to the expanded millivolt position. Set the Temperature Compensator to the temperature of the sample being measured. All samples, standards and electrodes must be at the same constant room temperature. Stir all solutions for about 2 min before measurement; continue constant stirring (use Teflon®-coated stirrer) during measurement.

Place electrodes successively in each standardizing solution; wait until the reading becomes stable (2 min); record millivolt readings. Blot electrodes dry between measurement.

Place electrodes in the sample. Record millivolt reading (wait for stabilization). Check the value of the 1 mg/ℓ standardizing solution after each determination, and recalibrate electrodes and instrument if necessary.

2. Procedure Using a Specific-Ion Meter

Place the electrodes in the 1 mg/ℓ fluoride solution, and set the specific-ion meter to the "F" position. Adjust the Calibration Control knob to obtain a mid-scale reading on the uppermost meter scale. All readings are divided by 100, so that the mid-scale value is read as 1 mg/ℓ.

Wipe both electrodes and repeat the procedure using the 0.5 mg/ℓ standard. Turn the Temperature Compensator knob so that the upper meter scale reads 50 (0.5 mg/ℓ). The specific ion meter is now calibrated to read fluoride unknowns. Recheck the calibration three or four times a day. Read the fluoride concentration of the unknown directly from the meter scale.

C. Calculations

1. Expanded-Scale pH Meter

Prepare a calibration curve by plotting on semi-logarithm graph paper electrode potentials developed in the solutions (linear axis) vs. fluoride concentration (log axis). Determine concentration of fluoride in sample by comparing reading with the calibration curve.

2. Specific-Ion Meter

The specific-ion meters read directly in milligrams per liter fluoride.

D. Precision and Accuracy

In a single laboratory, the coefficients of variation at fluoride levels of 0.2 mg/ℓ and 1.0 mg/ℓ F were $\pm 0.79\%$ and $\pm 0.93\%$, respectively, using an expanded-scale pH meter.

IX. IODIDE (CERIC)

This colorimetric method is applicable to the determination of free iodide ion in water supplies from both surface and ground sources in the range 0.5 μg/ℓ to 10 μg/ℓ I^-.

Iodide ions catalyze the reduction of ceric ions by arsenious acid, H_3AsO_3; the effect is proportional, but nonlinearly to the amount of iodide present. Owing to the rapidly fading color, the loss of ceric ions cannot easily be followed photometrically. In this method, however, the reaction is stopped after a given time interval by the addition of ferrous ammonium sulfate, $Fe(NH_4)_2(SO_4)_2 \cdot 6H_2O$; the resulting ferric ions, which are directly proportional to the remaining ceric ions, develop a color complex with potassium thiocyanate, KSCN, that is relatively stable.

$$2Ce^{4+} + H_3AsO_3 + H_2O \underset{}{\overset{I^-}{\rightleftharpoons}} 2Ce^{3+} + H_3AsO_4 + 2H^+$$

$$Ce^{4+} + Fe^{2+} \rightleftharpoons Fe^{3+} + Ce^{3+}$$

$$Fe^{3+} + SCN^- \rightleftharpoons Fe(SCN)^{2+}$$

Chloride ion interference in the original sample is eliminated by adding excess sodium chloride to give a maximum stable chloride ion concentration to sensitize the reaction.

A. Equipment and Reagents

A constant temperature water bath operating at 30°C ±0.5°C is used.

A spectrophotometer, measuring absorbance at 510 nm or 525 nm using 1-cm path length cells is required.

All stock solutions should be stored in tightly stoppered containers in a dark place.

Sodium chloride solution: dissolve 200.0 g sodium chloride, NaCl, in distilled water and dilute to 1 ℓ. If an interfering amount of iodine is present, recrystallize the NaCl using water-ethanol mixture.

Arsenious acid, 0.1 N: dissolve 4.946 g arsenious oxide, As_2O_3, in distilled water; add 0.2 mℓ concentrated H_2SO_4, and dilute to 500 mℓ in a volumetric flask.

Ceric ammonium sulfate solution, 0.02 N: dissolve 13.38 g ceric ammonium sulfate, $Ce(NH_4)_4(SO_4)_4 \cdot 4H_2O$, in distilled water; add 44 mℓ concentration H_2SO_4, and dilute to 1 ℓ in a volumetric flask.

Ferrous ammonium sulfate reagent: dissolve 1.50 g ferrous ammonium sulfate, $Fe(NH_4)_2(SO_4)_2 \cdot 6H_2O$, in 100 m$\ell$ distilled water containing 0.6 mℓ concentrated H_2SO_4. This requires daily preparation.

Potassium thiocyanate solution: dissolve 4.0 g potassium thiocyanate, KSCN, in 100 mℓ distilled water.

Stock iodide solution, 1.00 mℓ = 200 μg I$^-$: dissolve 0.2616 g anhydrous potassium iodide, KI, in distilled water, and dilute to 1 ℓ in a volumetric flask.

Intermediate iodide solution, 1.00 mℓ = 4.00 μg I$^-$: dilute 20.00 mℓ stock solution to 1 ℓ with distilled water.

Standard iodide solution, 1.00 mℓ = 0.100 μg I$^-$: dilute 25.00 mℓ intermediate iodide solution to 1 ℓ with distilled water.

Working iodide solution, 1.00 mℓ = 0.010 μg I$^-$: dilute 100 mℓ standard iodide solution to 1 ℓ with distilled water.

B. Procedure

The sample container should be tightly capped immediately after sample collection. No special storage precautions are necessary prior to analysis.

The sample aliquot either should be free from turbidity or should be filtered through a 0.45-μm membrane filter.

Add 10.00 mℓ water sample or aliquot made up to 10.00 mℓ with distilled water to a test tube, 2 cm by 15 cm. Add reagents to the sample in the following order: 1.00 mℓ NaCl solution, 0.50 mℓ arsenious acid solution, and 0.50 mℓ concentrated H_2SO_4. Place the reaction mixture and the ceric ammonium sulfate solution in the 30°C water bath and allow to come to temperature equilibrium. Add 1.00 mℓ ceric ammonium sulfate solution; mix the contents of the test tube with a Vortex Genie®, and start the stop watch to time the reaction.

After 15 min ± 0.1 min, remove the sample from the water bath and immediately add 1.00 mℓ ferrous ammonium sulfate reagent while mixing, whereupon the yellow ceric ion color should disappear. Then, while mixing, add 1.00 mℓ potassium thiocyanate solution. Replace the sample in the water bath.

Within 1 hr after the thiocyanate addition, read the red color in the spectrophotometer. As the method depends on a reduction in color, set the highest standard at 0 absorbance to adjust spectrophotometer.

Treat standards containing 0, 0.02 μg, 0.04 μg, 0.06 μg, 0.08 μg and 0.10 μg I$^-$ per 10.00 mℓ of solution. Run the standards with each set of samples to establish a calibration curve. Although it is possible to run higher standards, most samples fall within the range of the standards listed.

C. Calculations
Calculate the concentration of iodide in the sample by the following equation:

$$mg/\ell = \frac{\mu g \; I^-}{m\ell \; sample}$$

D. Precision and Accuracy
The method is reproted to be accurate to ±0.3 $\mu g/\ell$ I^-.[21]

X. OTHER METHODS

A. Cyanide
Measurement of cyanide in waste effluents is very difficult. Several authors have discussed the difficulties and problems associated with the analysis of various forms of cyanides in waste effluents.[6,22] Many analysts have concluded that the current acceptable methods for cyanide in natural waters do not produce accurate results.

Recently, Knechtel and Conn[23] have developed on accurate and reliable method for the determination of total cyanide in waste waters. The method incorporates distillation with hydrochloric acid and hydroxylamine hydrochloride, and the measurement of cyanide by modified colorimetric finish. This method is capable of quantitative determination of cyanide in both coking plant and gold mill samples containing sulfide, thiocyanate, copper and iron. The detection limit of the method, based on 100 mℓ sample aliquot, is 10 $\mu g/\ell$ of cyanide as CN.

Potentiometric determination of low levels of simple and total cyanide is also reported by Sekerka and Lechner.[24] Cyanide ion, down to 2 $\mu g/\ell$, can be determined by direct potentiometry using cyanide ion-selective electrodes and the known addition-known dilution techniques in a wide variety of water samples. The method can be operated either manually or automatically. The results obtained by the potentiometric method compare well with those obtained from conventional techniques.

B. Sulfide
Sekerka and Lechner[25] have developed a sensitive method for the determination of sulfide, iodide and cyanide, using solid-state sulfide electrodes. The method is capable of detecting concentrations down to ca. 1 × 10^{-8} M solution of the above ions, including sulfide. Sekerka and Lechner[26] have also developed a new technique entitled "Automated Zero-Current Chronopotentiometry". This technique can determine sulfide, cyanide and sulfite down to a low $\mu g/\ell$ range. It consists of an automated sensor, a rotating sampler and a gas pump. All signals necessary for timing and controlling the assembly are generated by a control module. The technique can be operated at the rate of 15 samples per hr. It offers an improved and selective analytical tool for determining sulfide and other volatile constitutents in environmental samples.

ACKNOWLEDGMENT

The author thanks Mr. J. P. Lively, Chief, Laboratory Operations Division, Water Quality Branch, Ottawa, Ontario, for giving permission to reproduce the details of the methods published in the Analytical Methods Manual.[13]

REFERENCES

1. Ardon, M., *Oxygen — Elementary Forms and Hydrogen Peroxide,* W. A. Benjamin, N.Y., 1965, 2.
2. The Chemical Society, *Environmental Chemistry,* Vol. 1, Burlington House, London, 1975, 19.
3. Mancy, K. H. and Jaffe, T., Analysis of dissolved oxygen in natural waters, Environmental Health Series, Department of Health, Education, and Welfare, Cincinnati, Ohio, 1966.
4. International Joint Commission, United States and Canada, New and Revised Specific Water Quality Objectives, 1978, p. 14.
5. International Joint Commission, United States and Canada, New and Revised Specific Water Quality Objectives, January 1978, Windsor, Ontario, Canada.
6. Ecological Analysts, Inc., Cyanide — An Overview and Analysis of the Literature on Chemistry, Fate, Toxicity and Detection in Surface Waters, June 1979, Townson, Md.
7. Colby, P. J. and Smith, L. L., Survival of walleye eggs and fry on paper sludge deposits in Rainy River, Minnesota, *Trans. Am. Fish. Soc.,* 96, 1967, 278.
8. Manahan, S. E., Environmental Chemistry, Willard Grant Press, Boston, 1972, 178.
9. Waggot, A., An Investigation of the Potential Problem of Increasing Boron Concentrations in Rivers and Water Courses, *Water Research,* 3, 1969, 749.
10. Anderson, B. A., Quality of Surface Water of United States — Ohio River Basin and St. Lawrence River Basin, *Chem. Abstr.,* 73, 1970, 59152m.
11. Sarkovic, M., Should We Consider Geochemistry an Important Exploratory Technique?, *Chem. Abstr.,* 84, 1976, 59364t.
12. Collins, A. G., Zelinski, W. P., and Pearson, C., Bromide and Iodide in Oilfield Brines in some Tertiary and Cretaceous Formations in Mississippi and Alabama, *Chem. Abstr.,* 67, 1967, 56210u.
13. Environment Canada, *Analytical Methods Manual,* Water Quality Branch, Inland Waters Directorate, Environment Canada, Ottawa, 1979.
14. Epstien, J., *Anal. Chem.,* Vol. 19, 1947, 272.
15. Standard Methods for Examination of Water and Wastewater, 12th ed., American Public Health Association, N.Y., 1965.
16. Technicon Instruments Corporation, Auto Analyser Methodology, Method B, Boron, Tarrytown, N.Y.
17. Lionnel, L. J., *Analyst,* 95, 1970, 194.
18. Standard Methods for Examination of Water and Wastewater, 13th ed., American Public Health Association, Washington, 1971.
19. Afghan, B. K., Goulden, P. D., and Ryan, J. F., *Water Res.,* 6, 1972, 1475.
20. Monnier, D., and Marcantomatos, M., *Anal. Chim. Acta,* 36, 1966, 360.
21. Standard Methods for Examination of Water and Wastewater, 12th ed., American Public Health Association, N.Y., 1965.
22. McKee, G. D., Status of Analytical Methods for Cyanide, Methods for Chemical Analysis of Water and Wastes, Environmental Monitoring and Support Laboratory, Environmental Protection Agency, Cincinnati, 1979, p. 131.
23. Knetchel, J. R. and Conn, K., Accurate and Reliable Determination of Total Cyanide in Wastewaters, unpublished report, Wastewater Technology Centre, E.P.S., Burlington, Canada, 1979. (Submitted for publication.)
24. Sekerka, I. and Lechner, J. F., Determination of sulfite and sulfur oxide by zero-current chronopotentiometry, *Analytical Chimica Acta,* 99, 1978, 99.
25. Sekerka, I. and Lechner, J. F., The response of the sulfide-selective electrode to sulfide, iodide and cyanide, *Analytica Chimica Acta,* 93, 1977, 139.
26. Sekerka, I. and Lechner, J. F., Automation of Zero-Current Chronopotentiometry, submitted to *Analytica Chimica Acta,* 1979.

Chapter 5

TRACE METALS

Jon C. Van Loon

TABLE OF CONTENTS

I. INTRODUCTION

The category "Trace Metals" in this chapter refers to not only the transition metals but those elements in groups IV A, V A, and VI A which form covalent hydrides. Of the many elements present in these sections of the periodic table, only those generally analyzed in waters are covered.

The analysis of water for trace metals would seem, at first glance, to be the simplest of all chemical analyses of metals. This common assumption has lead to a shocking amount of erroneous trace metal data.

Water samples span a wide variety of matrices. Sea and brackish waters and some effluents contain high levels of dissolved salts. In the case of effluents the main matrix constituents can vary widely from sample to sample. Waters with high dissolved salt content must, in most cases, be treated by solvent extraction or ion exchange to remove the desired trace constituents from interfering matrix constituents prior to analysis.

Gibbs[1] identified five mechanisms by which metal can be transported in waters; in solution and organic complexes, adsorbed, precipitated and coprecipitated in organic solids, and in crystalline sediment material. Whether one agrees with such a classification or not, it is obvious that metals in some forms are more available for reaction with biological components of the environment than others. It would be impossible to give a meaningful discussion of metal availability in a book of this nature, mainly because of the wide range of opinion that exists. In spite of this it is probably useful to distinguish between "easily extractable" metal and "total" metal in waters. "Extractable" metal in this context is that metal, both bound and unbound, which can be extracted by the organic reagents used according to a proposed procedure. "Total" metal refers to that which is released by the acid digestion of the sample. This metal is termed "total" because the present author's experience indicates that over 90% of the metal in the water is released by acid digestion.

No general accord exists on the method of water pretreatment prior to analysis. Most authors agree on an acidification following filtration. The filtration should be done at the time of sample collection or within a few hours thereafter. It is common to filter waters through an 0.45 μ porosity filter. Metal which passes this porosity is often termed soluble. The present author believes that a significant percentage of particulate metal passes this filter porosity size.

A good deal of controversy exists on what type of bottle should be used for storage of water samples. Choices range through various glasses and plastics. The present author has found that high-density linear polyethylene is a good inexpensive container for water samples. A detailed discussion of containers is presented below in the section on standard solutions.

II. METHOD OF ANALYSIS

Atomic absorption spectrometry is the most widely used technique for the determination of trace metals in waters. The equipment is generally available in most water quality laboratories, and the approach is relatively simple. For these reasons atomic absorption methods are given exclusively in this section.

In all but highly polluted waters, most trace metals are present in amounts which are not directly detectable by flame atomic absorption. A preconcentration method is thus essential prior to flame atomic absorption analysis. Direct analysis by electrothermal atomic absorption is possible for some trace metals but usually only in soft waters. In heavy matrix waters detection limits are poorer, meaning a separation/preconcentration method must be employed.

A. Historical Introduction

The phenomenon of the absorption of radiation by atoms has been used for investigations in physics since the early part of the 19th century, when Fraunhofer observed a number of dark lines in the sun's spectrum. The first analytical application of atomic absorption was to the determination of Hg by Müller.[2] It was not until 1955, when Walsh[3] discovered the general usefulness of the approach to elemental analysis, that real analytical atomic absorption spectroscopy was born. In the relatively short period of two and one half decades since this development, atomic absorption spectroscopy has become one of the most important techniques for the analysis of waters.

During the early commercialization of atomic absorption spectroscopy, extravagant claims were made concerning the general lack of interferences encountered with the technique. This is peculiar in that Walsh, in his original paper,[3] limited his claims for superiority to a simpler pattern of atomic spectral interference and a greater tolerance to thermal fluctuations in the atomizer compared to emission spectroscopy.

Between the years 1955 and 1962 there was little useful commercial instrumentation available. The first commercial offering, an unmodulated spectrophotometer, had serious deficiencies. During the early years, in response to many requests for information, Walsh and co-workers published papers describing in detail instrumentation suitable for routine atomic absorption work, e.g., Box and Walsh,[4] Figure 1. This equipment is remarkably similar in principle to modern atomic absorption units. In fact, with the exception of the development of electrothermal atomizers and devices for automatic background correction, little of fundamental importance has been introduced into commercial units to this day. Double-beam or dual-channel units, advanced electronics, and microprocessors, etc., have only resulted in equipment which is more easily used. Modern equipment, despite its push button-black box technology, operates on the same physics and chemistry as the original equipment. Therefore, it is important to stress the need for analysts to master these principles.

B. Principles of an Atomic Absorption Spectrometer

A block diagram of an atomic absorption unit is given in Figure 2. Atoms absorb radiation only at discrete wavelengths characteristic of the absorbing species. Thus, radiation from the source, produced from a vapor of the metal of interest, is absorbed at a discrete wavelength(s) by atoms of that element in the atomizer. As a result, the radiation beam intensity is attenuated by an amount which is proportional to the concentration of the element of interest in the atomizer.

FIGURE 1.   Instrumentation suitable for routine atomic absorption work as described by Box and Walsh.[4]

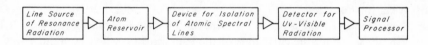

FIGURE 2.   Block diagram of an atomic absorption unit.

The function of the atomizer is to produce free atoms from the introduced sample. In atomic absorption the atomizer is usually a flame- or furnace-type electrothermal device. Radiation of a characteristic wavelength is usually produced by a hollow cathode or electrodeless discharge lamp. The desired wavelength is isolated from other absorbing or nonabosrbing lines by a monochromator. This latter device is placed after the atomizer to reduce the light flux on the detector system. The transducer (detector) is usually a photomultiplier tube.

Absorbance, the quantity usually measured in atomic absorption spectrometry, can be expressed as follows:

$$A = \log \frac{I}{Io} = a\,b\,c$$

where Io = intensity of the incident beam, I = intensity of the transmitted beam, a = a constant (characteristic of the particular system), b = path length of the optical beam and which can be kept constant, and c = concentration of the element of interest in the atomizer. This equation predicts a linear relationship between the absorbance and the concentration of the element of interest. In practice, this occurs from the detection limit over two orders of magnitude.

An atomic absorption working curve is depicted in Figure 3. An estimate of the precision obtainable is also shown in this figure. As can be seen, best precision is obtained in the middle straight line portion of the curve. At the upper end of the graph, curvature is experienced, and precision decreases markedly.

C. Applicability

Figure 4 is a periodic table showing only those elements which can be analyzed by atomic absorption. Atomic absorption finds its best and easiest application in the analysis of the transition heavy metals.

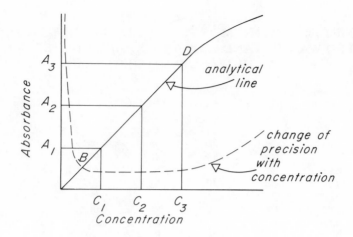

FIGURE 3. Atomic absorption working curve.

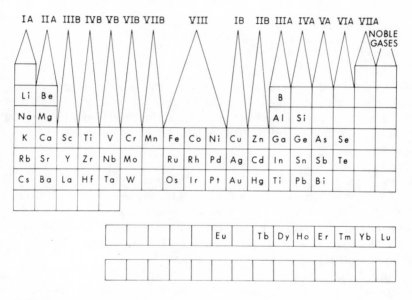

FIGURE 4. Periodic table showing only those elements which can be analyzed by atomic absorption.

D. Flame Atomizers

As a general rule, a flame atomizer should be used when applicable. In comparison with electrothermal atomizers, better precision and accuracy are obtainable. Matrix effects, both volatilization and non-specific absorption interferences, are great with electrothermal devices.

Each element, run at a given analytical line, has its characteristic atomic absorption sensitivity. Table 1 lists flame detection limits obtained by Perkin-Elmer Corporation[5] for water solutions which are very low in total dissolved salts. Detection limits, it must be remembered, are matrix dependent. Therefore, such a listing can be used in a relative sense only.

Little has been new in analytical atomic absorption spectroscopy, flame technology, since the introduction of the nitrous oxide acetylene flame by Amos and Willis.[6] There are three flames in common use today, air-hydrogen, nitrous oxide-acetylene, and air-acetylene. Of these the latter is most commonly employed. Choice of flame is basically

Table 1

FLAME DETECTION LIMITS FOR A PERKIN-ELMER® 603 USING
DISTILLED-WATER STANDARDS AND CONDITIONS RECOMMENDED
BY THE MANUFACTURER (μg/mℓ)

Element	Detection Limit	Element	Detection Limit	Element	Detection Limit	Element	Detection Limit
Ag	0.002	Co	0.01	Mg	0.001	Ru	0.07
Al[a]	0.02	Cr	0.003	Mn	0.002	Sb	0.04
As	0.2	Cu	0.002	Mo[a]	0.02	Se[c]	0.s0
Au	0.01	Fe	0.005	Na	0.002	Si[a]	0.02
B[a]	0.7	Ga[a]	0.05	Ni	0.005	Sn[a,c]	0.07
Ba[a,b]	0.008	Ge[a,c]	0.10	Os[a]	0.08	Sr[b]	0.002
Be[a]	0.001	Hg[c]	0.25	Pb[c]	0.01	Te[c]	0.03
Bi	0.025	Ir	0.6	Pd	0.02	Ti[a]	0.04
Ca	0.0005	K	0.002	Pt	0.05	Tl[c]	0.01
Cd[c]	0.001	Li	0.003	Rh	0.004	V[a]	0.04
						Zn	0.001

[a]   Nitrous oxide acetylene flame.
[b]   1000 μg/mℓ K added to suppress ionization.
[c]   Electrodeless discharge lamp.

dictated by the temperature required to atomize the analyte in the presence of potential interferences. Little use is made of the air-hydrogen flame except for elements such as As and Se, whose principal analytical wavelengths are at 1937 and 1960 Å, respectively. In the wavelength region below 2000 Å, absorption of radiation by other flames is prohibitively high.

Table 2 lists the elements together with the flame type commonly employed. Comments are given concerning interferences. The comment, "cationic interferences are minimal", refers to waters with less than few hundred μg/mℓ of the individual cations. In more concentrated solutions, any cation or anion may be considered a potential interference and may have to be added to the standard solutions. Ionization interferences can be overcome with an easily ionizable alkali metal such as K.

As can be seen, refractory elements require the high-temperature nitrous oxide-acetylene flame. Other elements, e.g., Cr and Ni, may benefit from the use of this flame in overcoming persistent interferences. On the negative side, the nitrous oxide-acetylene flame results in a high incidence of ionization interference, can yield appreciable shot-noise problems, and is generally more difficult and less desirable to work with.

Non-specific interferences due to molecular absorption and/or light scatter are much less of a problem with flame atomizers than with electrothermal devices and will be discussed under the section on the latter equipment. However, when working near the detection limit with any element, the need for background correction should be evaluated. As in the case of flames, little is new in nebulizer/pre-mix chamber technology. Ultrasonic nebulization, capable of yielding high atomization efficiencies, has yet to become a commercial success in atomic absorption.

E. Electrothermal Atomizers

L'Vov[12] introduced electrothermal atomizers into atomic absorption analysis. The term "electrothermal atomizer" will be used in this chapter to refer to electrically heated devices such as graphite furnaces and rods. The common designation "flameless atomizer" will not be employed, to avoid implications that other equipment such as cold vapor absorption tubes are being discussed in this section.

Most early commercial atomic absorption equipment was less than optimal for elec-

Table 2
FLAME TYPE TO BE USED FOR THE ELEMENTS

Element	Flame type[a]	Comments
Ag	A, l	Cation interferences minimal, avoid halides.
Al	N	Fe above 0.2% depresses, Ti enhances, H_2SO_4 depresses. Some ionization occurs.
As	H	Interferences severe; better to use hydride or direct electrothermal method.
Au	A, l	Use a *very* lean flame, cationic interferences minimal.
Bi	A, l	Significant interference from base metals, can use hydride method as alternative for complex solutions.
Cd	A, l	Cationic interferences slight.
Co	A, l	Cationic interferences slight.
Cr	A, r	Gives best sensitivity, but extensive cation interference occurs.
	A, l	Poorer sensitivity, but most cationic interference suppressed.
	N	Cationic interferences minimal.
Cu	A, l	Cationic interferences minimal.
Fe	A, l	Cationic interferences minimal.
Hg	—	Use cold absorption tube method.
Ir	A, r	Complex interference patterns, interference minimized; use a Na/Cu mixture.[7]
Mn	A, l	Cationic interferences minimal.
Mo	N, r	A variety of cations interfere, use Al to suppress these.[8]
Ni	A, l	Some interference from base cations, use very lean flame to minimize.
	N, r	Cationic interferences minimal.
Pb	A, l	Use 2833 A line. Cationic interferences minimal. Avoid SO^-_4.
Pd	A, l	Several cationic interferences, use La to suppress these.[9]
Pt	A, l	Complex cationic interferences, use La to suppress these.[10]
	N, r	Minimal cationic interferences.
Rh	A, l	Complex cationic interferences, use La to suppress.[11]
Sb	A, l	Spectral interferences occur in presence of lead using 2176 A. Some cationic interferences, use hydride or other electrothermal method in complex solutions.
Se	H	Interferences severe, use hydride or other electrothermal method.
Sn	N, r	Sensitivity best in hydrogen flame, but interferences very severe. Poor sensitivity in air acetylene flames. Use hydride or other electrothermal method if applicable.
Te	A, l	Some cationic interference, use hydride or electrothermal methods where applicable.
V	N, r	Overcome cation interferences with Al.
Zn	A, l	Minimal cationic interferences.

[a] A = Air acetylene, l = lean, N = nitrous oxide-acetylene, H = hydrogen air, r = reducing.

trothermal work. Electronic systems failed to follow the very fast transient signals obtained with electrothermal atomization. Baffling was essential in many optical systems to cut down on stray light. While rapid progress was made in solving these problems, the main drawback with the electrothermal equipment itself, i.e., lack of real control of temperature, was not adequately addressed in commercial equipment until 1977. To minimize interferences, it is often important to use the highest-power ramp available during the atomization cycle. With control of temperature, this can now be done without danger of overshooting the desired final temperature. Without temperature control capability, reproducible and accurately definable heating cycles were impossible. As a result, most of the thermal programs which now appear in the literature possess a degree of uncertainty. This will explain many of the disagreements between researchers on thermal volatilization programs and interferences. The new equipment does not accurately measure the temperature in the atomization zone, and hence disagreements, but at a much reduced level, will persist.

In spite of these negative comments the advent of electrothermal atomic absorption

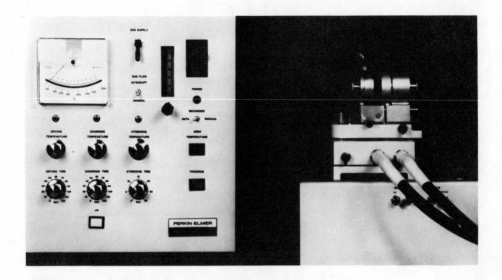

FIGURE 5.   The graphite tube furnace.

represents a milestone in the development of analytical methods for water samples. With the commercial production of devices with capability for real control of temperature, there now exists a more solid base for research into electrothermal atomization. In addition, the analyst can expect these new devices to be more reliable and useful in routine analysis.

There are two types of electrothermal device generally available today, the graphite tube furnace, e.g. Figure 5, and the carbon rod, e.g. Figure 6. Both involve samples atomized in a tube. In the case of the graphite furnace, the purge gas is injected into, and travels axially through, the center of the tube. With the carbon rod, the purge gas is injected from below the rod/tube assembly, and simply surrounds the assembly. In general, better limits of detection are possible with the furnace compared to the rod. Metallic strip atomizers were formerly commercially available. But for many elements, detection limits were poorer and interferences worse.

Electrothermal atomizers have a small sample capacity. It is difficult to manually inject the microliter volumes, applicable to these devices, reproducibly onto the atomizer. Auto samplers are now available for both the furnace and the rod and come highly recommended by the present author.

Detection limits are potentially up to three orders of magnitude better than with flames. Table 3 lists the elements often done by electrothermal atomization, and their detection limits in low salt-content solutions. Detection limits are, of course, matrix dependent. Cruz and Van Loon[13] found that in rock sample solutions the detection limits for most elements given in Table 3 were seriously degraded.

1. Interferences Using Electrothermal Atomization

Interferences in electrothermal work can be very complex. This is particularly true with heavy matrix water samples. For overcoming physical and chemical interferences, the method of standard additions can often be employed.

Physical interference problems can arise in pipetting of standards and samples when the viscosities of each are very different. With some of the highly acid or high-salt content geological sample solutions, this source of error can be significant. The method of standard additions can often be used to overcome this problem.

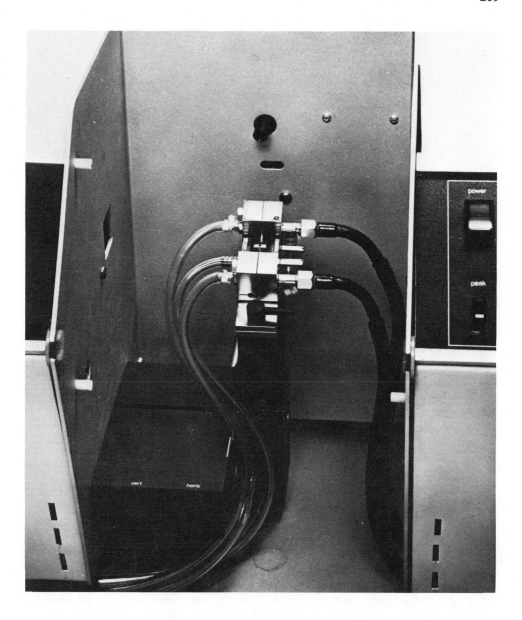

FIGURE 6.   The carbon rod.

Any phenomenon which alters the rate of volatilization of analyte from a sample compared to a standard is an interference. Coexisting cations and anions in sample solutions can alter the rate of volatilization. Interferences of this type are very poorly understood. To minimize interference it may be possible to change the composition of the matrix, either by reagent addition or by selective volatilization of matrix components prior to atomization. Signal integration or standard addition are also employed. For the latter approach to be valid, the added analyte must assume the same chemical form in the atomizer as the sample analyte. Standard addition also requires that the calibration graph be linear over the concentration range employed. Calibration graphs can have very short linear regions in electrothermal work. Of course, using standard addition is time consuming, but may be essential for the most accurate work.

With some samples, e.g., seawaters, there is a danger that during the heating cycle,

Table 3
DETECTION LIMITS (pg) FOR ANALYTE IN LOW-SALT-CONTENT WATER — PERKIN-ELMER® 603 AND HGA 2100, USING CONDITIONS RECOMMENDED BY MANUFACTURER

Element	Detection Limit	Element	Detection Limit	Element	Detection Limit
Ag	0.5	Ir	1000	Se	200
As	25	Mn	0.8	Sn	200
Au	25	Mo	15	Te	100
Bi	30	Ni	30	Ti	500
Cd	0.5	Pb	5	V	200
Co	10	Pd	30	Zn	0.2
Cr	20	Pt	500		
Cu	5	Rh	50		
Fe	10	Sb	10		

analyte can be lost prior to the atomization step. This is a particularly serious potential source of error when chloride is present. It is advisable, therefore, to convert all chlorides to a less volatile form, e.g., nitrate or sulfate, prior to the ashing step. A number of elements such as Se, Zn, Cd, and Pb are potentially volatile at relatively low temperatures. It may be possible in these cases to vaporize the analyte prior to the matrix. When this approach is not valid, recent research has shown that an element such as Ni (in the case of Se or Te) can be added to stabilize the element in the atomizer during the ashing step.

To complicate matters, analyte can undergo complex reactions after volatilization, i.e., in the gas phase. This matrix-dependent, poorly understood chemical interference, is best handled by standard addition.

By far the most important interference in using electrothermal atomizers, results from signal due to light-scatter or molecular absorption. This problem, most severe at UV wavelengths, also must be considered in the visible region of the spectrum. The effect can be so acute that the background correction capability of the instrument is exceeded. Under this circumstance much of the older equipment will still generate values without any indication of error. The consequences of such a phenomenon are obvious. If there are no warning lights indicating background correction overload, but the equipment has an energy meter, observation that the energy meter needle drops drastically toward zero can be evidence of the problem. Most commercial atomic absorption units cannot background correct when the interference produces a signal greater than an absorbance of one. Recent equipment embodies a warning light to indicate when the background correction capability has been exceeded.

Matrix modification and/or a background correction method are used to overcome nonspecific interferences. In seawater samples, compounds such as NaCl and organic materials come off at relatively low temperatures. Their generation can interfere with the determination of low-boiling-point elements (Cd, Pb, Zn, etc.) For high-boiling-point elements compounds such as NaCl and KCl can be reduced to tolerable levels by selective volatilization. The presence or formation of refractory oxides (CaO, NgO, Al_2O_3, etc.), can interfere severely with the determination of high-boiling-point metals. Selective volatilization is not possible in this latter case. It must be emphasized that the method of standard additions is not applicable to overcoming non-specific interferences.

There are a number of approaches to background correction. Most frequently an automatic, continuum source, background correction is employed. In this method the signals obtained with a continuum and line source are monitored separately. Their difference can give a background-corrected value. If narrow slit widths are used, significant absorption of the continuum radiaiton at the analyte wavelength may occur. This latter effect leads to a loss of sensitivity. Instruments which give dual-beam background correction or the equivalent are to be preferred because of drift problems inherent with continuum sources.

Non-absorbing lines can be employed for background correction. However this method, frequently used in the early years, has been largely replaced by the continuum approach. When background correction is needed in the visible region, the non-absorbing line method is still employed. A deuterium arc is usually used as a continuum source. These devices have little output in the visible region. However, some recent equipment provides a tungsten iodide continuum source for automatic correction in this region of the spectrum.

When there is direct spectral overlap between the fine structure of a molecular absorption band and the analyte line, background compensation will give erroneous results. The analyst must also realize that over- or under-correction can readily occur, using continuum source methods. Failure to ensure, by adjustment each time, that the two optical beams are coincident and fill the same fraction of the optical aperture gives "corrected values" which are in error.

III. STANDARD SOLUTIONS

A. Water

High-purity water for atomic absorption purposes can be produced by distillation or with ion-exchange resins or with a combination of both. The analyte element and the concentration at which this element must be analyzed will determine the quality of water required. Ubiquitous elements such as zinc, calcium, potassium, sodium and iron, when they must be analyzed at sub-μg/mℓ levels, dictate extreme precautions in the production and subsequent use of high-purity water. In many laboratories, including that of the present author, it is impossible to analyze these elements below about 1 to 5 ng/mℓ because of uncontrollable contamination problems.

1. Ion-Exchange Systems

Depending on the characteristics of the water supply and the purity requirements of the finished water, one or several stages are used within an ion-exchange system. Incoming water is first filtered to remove particulate material. If organic matter is a potential problem, the water can be run through an activated-charcoal column. Finally, the water is passed through one or more stages of a mixed cation-anion exchange column.

Cation and anion resins are used in H^+ and OH^- forms, respectively. Contaminant ions in the water supply exchange onto the column displacing H^+ and OH^-, which combine to form water.

2. Distillation

For many purposes one stage of distillation, using commercially available equipment, is sufficient. In general, but particularly when metal analyses are to be done, glass or quartz apparatus should be used.

A variety of organic contaminants distill over with the water. To negate this problem, alkaline potassium permanganate can be added to the distillation flask. In the presence of permanganate, most offending organic substances are oxidized to carbon dioxide and water.

When high-purity water is essential, two stages of distillation are often employed. Again, at least the final stage should be accomplished using an all-glass system. Double distillation may also be essential when the water supply is high in total dissolved solids. In this latter case a combination of ion exchange and distillation is a good approach.

For the ultimate in high-purity water, a sub-boiling distillation should be employed. As would be expected, this is extremely slow and expensive (<1 ℓ/day) and hence should be used only when ultrapure water is a necessity. For storage of this water, a Teflon® bottle minimizes contamination.

B. Preparation of Standard Solutions

Great care is, of course, required in the production of standard solutions. These will be used to prepare calibration graphs and hence will in large part determine the accuracy of an analysis. Standard reference samples must also be used to assure accuracy. These latter samples are often instrumental in pointing up errors in standard solutions.

1. Reagents

For most elements the high-purity metal is the substance of choice for preparing standard solutions. In the analysis of water a metal of at least 99.9% purity is sufficient.

The metals of some elements are not suitably stable or cannot be obtained in a sufficiently pure state. In these instances a high-purity salt may be employed. The salt chosen should not be hygroscopic and should be of known composition. High-purity metals and metal salts are available from several companies.* Table 4 contains suggestions of metals/metal salts and solvents for preparation of standard solutions. In each case, when acid is used, the final acid concentration in the diluted solution should be about 1%. Further details on standard solutions are given with the procedures.

Reagents other than metals required for standard solutions, such as acids, buffers, and releasing agents must be of high purity. A blank of these reagents should always be prepared.

There are commercially available atomic absorption standard solutions. These are satisfactory for many applications.

2. Containers

As in the case of water sample storage, there is no unanimity of opinion on the best type of container to use for the storage of standard solutions. The controversy surrounds the reported loss or gain of metals resulting from exchange and absorption on container walls.

With the possible exception of mercury the present author recommends high-density linear polyethylene for storage of standard solutions. This material, when properly washed, yields minimal metal contamination and does not suffer from the severe evaporation (breathing) problems encountered with other plastics. Teflon® containers, used by the National Bureau of Standards for reference water samples, are superior. The cost of these vessels negates their use for the storage of the large numbers of standard samples employed by most laboratories.

Plastic containers, as supplied, are often contaminated with traces of metal, particularly zinc. A rigorous regime of decontamination is thus required. Most workers recommend a detergent wash followed by an acid treatment and distilled water rinses. The present author agrees with this approach, but recommends against the use of hot concentrated acids for the acid wash. The latter may attach the surface of the plastic. Soaking 0.5 hr in a room temperature 1:1 mixture of hydrochloric or nitric acid, fol-

---

* The present author can suggest Spex Industries, Box 798, Metuchen, N.J., 08840; Johnson Mattey Chemicals, 74 Hutton Gordon, London ECIP 1AE; and Fisher Reagent Grade Chemicals, Fisher Scientific, Fairlawn, N.J.

Table 4
SUGGESTED METAL/METAL SALT AND SOLVENT FOR PREPARATION OF STANDARD SOLUTIONS

Element	Metal/Metal Salt	Solvent	Comments
Ag	silver nitrate	water	Store in amber bottle.
Al	aluminum metal	1:1 HCl	Add a drop of mercury to aid dissolution.
As	arsenous oxide	20% KOH	
Au	gold metal	aqua regia	Store in amber glass bottle.
Be	beryllium metal	1:1 HCl	Metal dust is very toxic if inhaled.
Bi	bismuth metal	1:1 HNO_3	
Ca	calcium carbonate	1:3 HCl	Add dropwise.
Cd	cadmium metal	1:1 HCl	
Co	cobalt metal	1:1 HCl	
Cr	potassium chromate	water	
Cu	copper metal	1:1 HNO_3	
Fe	iron metal	1:1 HNO_3	
Hg	mercuric oxide	1:2 HCl	
K	potassium chloride	water	
Mg	magnesium metal	1:1 HCl	
Mn	manganese metal	1:1 HNO_3	
Mo	ammonium paramolybdate	1% NH_3	The final concentration of ammonia in the dilute solution should be 1%.
Na	sodium chloride	water	
Os	ammonium chloroosmate	1% HCl	Osmium is very toxic and very volatile as the tetroxide.
Pb	lead metal	1:4 HNO_3	This is best, but an oxide coating may form on the metal making dissolution troublesome.
	lead nitrate	1% HNO_3	
Pd	palladium metal	aqua regia	
Pt	platinum metal	aqua regia	
Rh	ammonium hexachlororhodate	10% HCl	Diluted solution should be 10% in HCl.
Ru	ruthenium chloride $RuCl_3$	20%	Diluted solution should be 10% in HCl.
Sb	potassium antimony tartrate	water	
Se	selenium metal	concentrated HNO_3	Evaporate to dryness; add water and evaporate to dryness; dissolve in 10% HCl; diluted solution should be 10% in HCl.
Si	silicon dioxide	lithium metaborate fusion	Fuse with 7-fold excess of lithium metaborate and equal weight of calcium oxide in a pre-ignited graphite crucible at 925°. Pour melt into 1:24 HNO_3. Dilute with 1:24 HNO_3.
Te	tellurium dioxide	aqua regia	Dilute with 6 NHCl.
Zn	zinc metal	1:1 HNO_3	

lowed by several rinsings with the same mixture is best. Distilled water should be used for all water rinsings.

3. Procedure

The metal or metal salt should be dried at an appropriate temperature and cooled in a desiccator prior to weighing. The latter must be done to about 1 part in 1000 for most applications. Stock standard solutions are usually made to contain 1000 $\mu g/m\ell$.

The weighed product should be dissolved in a minimum of acid. Hydrochloric or nitric acid should be used when possible, bearing in mind chemical problems which might be encountered, e.g., precipitation of silver chloride in the presence of hydrochloric acid. Nitric acid is preferred when the solutions are to be used in an electrothermal atomizer. The solution must be diluted to volume in acid-washed, Class A or equivalent, volumetric flasks. This dilution and subsequent dilutions should be made so that the final acid concentration is about 1%. Maintaining a 1% acid concentration in the standard minimizes hydrolysis and ion exchange and absorption problems on container walls. For some elements, e.g., sodium, potassium, calcium, etc. it is not necessary to maintain a 1% acid content. These latter elements are not prone to hydrolysis. Suggested dissolution approaches are given in Table 4.

Preservatives may be necessary for a few elements. For example, in the case of the mercury a few µg/mℓ gold or 0.05% potassium dichromate yields more stable solutions.

IV. SOLVENT EXTRACTION

Ion-exchange chromatography and solvent extraction are the most frequently used methods for preconcentration and/or separation of metals from matrix constituents in waters. In the present author's laboratory, solvent extraction has been found to be the most satisfactory approach. Apart from yielding a satisfactory separation, solvent extraction has the advantage that the metals of interest are in an organic solvent. If the proper solvent is chosen, an enhanced signal compared to water solutions is obtained when flame atomization is employed.

Solvent extraction methods abound for concentrating trace metal ions in waters, prior to atomic absorption analysis. Unfortunately, the majority of these have been developed without regard for important theoretical data available from such sources as Stary,[14] Morrison and Freiser,[15] and Zolotov.[16] There have only been a few critical studies of solvent extraction-atomic absorption procedures. As a result, available procedures are seldom optimized with respect to pH range, buffer, ionic strength, stability, equilibration time, etc. This means it is often impossible for the analyst in a laboratory to obtain good results on a routine basis.

Proposed solvent extraction methods for the atomic absorption analysis of trace metals were examined in the present author's laboratory. The following is a summary of the important considerations which resulted.

The solvent used to extract metal complexes must show a number of desirable characteristics:

1.   extract the desired metal chelates
2.   be immiscible with the aqueous solution
3.   not tend to form emulsions
4.   have good burning characteristics
5.   enhance rather than suppress the atomic absorption sensitivity as compared to the metal in water

Work was carried out on a number of likely solvents. Benzene and xylene were eliminated because of the turbulent and unstable flames they produced. Decanol proved to have too pungent an odor. Chloroform, a solvent widely used in colorimetric work, evaporates too quickly, leaving the solid complex behind, which then clogs the vaporization chamber. Ethyl acetate, methylisobutylketone, isoamyl acetate, and *n*-butyl acetate were also considered. Of these, ethyl acetate and methylisobutylketone gave the greatest enhancements as compared to the abosrbance of the same quantity of metal in water. Ethyl acetate is too volatile for easy use. Although methylisobutylketone is fairly soluble in water, this extractant was chosen.

The use of a buffer is mandatory in routine extraction work. This fact is not recognized by many workers. It is well known that the quantity of metal extracted is strongly dependent on the pH of the solution and that chelating agents will often alter the pH of the solution to which they are added.

The choice of the buffer is very important. It must be stable, have a high buffering capacity and not participate in any reaction. A number of buffers have been studied for solvent extraction-atomic absorption work: borate, phosphate, citrate, acetate, and formate. Solutions containing the formate buffer were found to be unstable and slowly decomposed (organic droplets appeared on container walls) after several days. An acetate buffer was also unfavorable. The latter would combine with any lead or silver in solution to form stable acetates which were not readily extracted. A citrate buffer, which was found to be stable and did not interfere with the extraction process, is best for most metals. This buffer, however, does contain considerable cadmium and iron. The buffer should be purified as well as possible of trace metal contaminants by an extraction wash, using the chelating agent. In spite of this precaution a blank must be run with each set of samples. If silver and lead are not to be analyzed, an acetate buffer is recommended because of lower cadmium and iron contamination.

Two of the main considerations affecting the choice of a chelating agent are (1) it should extract the largest number of trace metals, and (2) it should extract the metals equally well over some fairly wide range of pH of the solution. Many procedures available at present require pH adjustment to within one pH unit or less, which can result in serious errors in routine applications.

The pH dependence of nine different chelating agents was studied. These chelating agents were 8-quinolinol, acetylacetone, theoxyltrifluoroacetone, 1-(2-pyridylazo)-2-naphthol, ammonium pyrrolodinedithio-carbamate, potassium ethyl zanthate, α-benzionoxime, diethylammonium diethyldithio-carbamate, and sodium diethyldithio-carbamate. Ammonium pyrrolodine dithio-carbamate extracts the greatest number of metals over the widest pH range. Diethylammonium diethyldithio-carbamate has a stabilizing effect on all metal complexes in the system, and hence its addition is recommended.

A. Solvent Extraction Method for Trace Metals in Water[17]

In the analysis of water samples, a number of ionic species may be encountered at concentration levels far greater than those of the metals being determined. Fluoride, calcium, potassium, magnesium, sodium, phosphate, silicate, and biodegradable detergent are potential interferences present in many waters. These ions were individually added to solutions of the analyte and to absorbances obtained compared to the absorbances of solutions without interferences. Most of these ions are known masking agents; however, only the biodegradable detergent appeared to affect the absorbances of the metals to any degree.

The following procedure is suitable for the analysis of sea, brackish, and fresh waters and for most effluents. Elements which can be readily determined are silver, cadmium, cobalt, copper, iron, nickel, lead, and zinc, with detection limits of 0.6, 0.8, 1.5, 0.8, 1.3, 2.5 and 0.6 μg/ℓ, respectively.

1. Equipment and Reagents

An IL® (Instrumentation Laboratories Inc.) 153 was used for all atomic absorption measurements. pH measurements were made using a Beckman® Expandomatic pH meter.

All chemicals used were of reagent grade of the highest quality available. Standard metal solutions (1000 μg/mℓ) were prepared for Ag(I), Co(II), Fe(III), Ni(II), Cu(II), Pb(II), Cd(II), and Zn(II) from high-purity metal. Buffer was prepared by mixing the appropriate amounts of citric, boric, and phosphoric acids, in water to give a solution which is 0.5 M in each acid.

Prepare a 1% w/v mixed chelating solution of ammonium pyrrolidine dithio-carbamate and diethylammonium diethyldithio-carbamate in water. Extract this twice with methyisobutylketone.

2. Procedure
a. Extractable Metals

Acidify the water to pH = 1 with nitric acid. Place the desired sample (usually 200 mℓ) in a 250-mℓ separatory funnel fitted with a Teflon® stopcock. Add 4 mℓ of the buffer. Shake to mix well. The pH should be 4.0±0.1. If the pH must be adjusted, add sufficient 20% sodium hydroxide solution to obtain this value. Add 5mℓ of 1% mixed chelating agent. Shake briefly. Add 10 to 20 mℓ (depending on concentration factor required) of MIBK. Shake vigorously for 60 sec. Allow the layers to separate. Remove the aqueous lower layer. Retain the MIBK layer in tightly capped glass bottles until samples have been made ready for analysis. Prepare standards from multielement stock solution so that the 200 mℓ of water extracted contains four concentrations within the ranges of iron 10 to 20 μg/ℓ, copper 5 to 100 μg/ℓ, nickel 5 to 100 μg/ℓ, cadmium 1 to 20 μg/ℓ, zinc 10 to 2000 μg/ℓ, and lead 10 to 2000 μg/ℓ. In this way a direct concentration relationship exists with samples. Run a reagent blank.

b. Total Metal

Add 1 mℓ of nitric acid to the desired sample. Evaporate the sample to dryness on medium heat on a hot plate. Add 2 mℓ of hydrochloric acid + 1 mℓ of nitric acids. Evaporate to dryness. Add a drop of hydrochloric acid, and dilute to 200 mℓ. Continue as for extractable metal above. Run a blank containing all reagents.

V. ION-EXCHANGE AND CHELATING RESINS

Ion-exchange and chelating resins are used for two purposes in water analysis: (1) preconcentration of the trace metals prior to the determinative step, and (2) separation of the trace metals from interfering concomitant substances. Chelating resins are particularly useful in trace metal water analysis because of the very high selectivity of polyvalent over monovalent ions.

Ion-exchange resins consist of an insoluble polymer (commonly styrene) lattice with attached functional groups. The polymer matrix is porous to water and inorganic ions. Resins are crosslinked with divinylbenzene. The porosity of a resin is determined by the hydration of the matrix. Resins are formed into beads of various mesh sizes. Styrene polymer chains are crosslinked with divinylbenzene. Resins swell when hydrated, the amount of swelling being dependent on the degree of crosslinking.

Ion-exchange resins are available with acidic or basic functional groups and may be either weakly or strongly acidic or basic depending on the nature of the group. Acidic resins exchange cations and are called cation exchange resins. Likewise, basic resins exchange anions and are termed anion exchange resins. For strong resins the order of selectivity favors polyvalent over monovalent ions. The ion exchange affinity is inversely proportional to the radius of the hydrated ion for ions of the same charge.

A composite affinity sequence for cations is $Li^+ < Na^+ < NH_4^+ < K^+ < Rb^+ < Cs^+ < Ti^+ < Ag^+ < Mg^{+2} < Ca^{+2} < Sr^{+2} < Ba^{+2} < Fe^{+2} < Co^{+2} < Ni^{+2} < Cu^{+2} < Zn^{+2} < Al^{+3} < Sc^{+3}$. For anions the composite affinity sequence is $F^- < Cl^- < Br^- < CrO_4^{-4} < MoO_4^{-2} < PO_4^{-3} < AsO_4^{-3} NO_3^- < I^- < SO_4^{-2}$. These series vary slightly depending on individual ion-exchange resins and differing conditions.

Chelating resins are also a styrene polymer crosslinked by divinylbenzene. A typical chelating group used in these resins is imminodiacetate. This reactive group chelates with trace metals giving a 5000 to 1 selectively of divalent to monovalent ions. Chelating resins, therefore, find very useful application in the trace metal analysis of sea water.

A. Chelating Resin Method for the Analysis of Cd, Co, Cu, Fe, Mn, Ni, Pb, and Zn in Seawater.[18]

The trace metals are separated from the major cations of seawater, Na, K, Ca and Mg using Chelex® 100 resin. The pH of seawater is adjusted to 5.0 to 5.5. It is passed through a column containing Chelex® 100 resin. Ammonium acetate is used to elute the alkali and alkaline earth elements, followed by elution of the trace metals using nitric acid. The trace metals are determined by electrothermal atomic absorption spectroscopy.

Prior to electrothermal atomic absorption analysis it is essential to remove the major cationic matrix constituents of seawater. Separational techniques for trace metals commonly require large amounts of chemicals, which can lead to high blanks. Of the methods available those employing Chelex® 100 give suitably low blank values without excessive purification of reagents.

A radioactive tracer procedure was used to study the separation. The recovery of Cd, Cu, Mn, Ni, and Zn was greater than 99.9%, and recovery of Co, Pb and Fe was 99.5, 98.4, and 93.1%, respectively.

1. Equipment and Reagents

High-purity water, nitric and glacial acetic acids were prepared using sub-boiling distillation at the National Bureau of Standards (NBS). All reagents used in the separation process were prepared and stored in clean FEP Teflon® bottles unless otherwise stated.

Ammonium hydroxide was prepared by bubbling filtered ammonia gas through high-purity water until room temperature saturation was achieved.

A 1.0 M ammonium acetate solution was prepared by mixing 60 g of purified glacial acetic acid and 67g of saturated NH_4OH and diluting to 1 l in a polypropylene volumetric flask. The acidity was adjusted to pH 5.0 by dropwise addition of HNO_3 and/or NH_4OH. All reagent and sample preparations were done in a class 100 clean air laboratory.

Chelex® 100 chelating resin, 200-400 mesh size, was purchased from Bio-Rad Laboratories.

All standard stock solutions were prepared from high-purity metals or salts in sub-boiling distilled NBS acids as described by Dean and Rains.[19] Working solutions were prepared as needed.

a. Seawater

The sample was collected with a submersible pump and plastic tubing permanently submerged approximately 100 m offshore from the Institute. The seawater was pumped directly into a conventional polyethylene drum which had been cleaned first with hydrochloric and then with nitric acid and purified water prior to use. After filtration through a 0.45-μm millipore filter using an all polypropylene filter apparatus, the seawater was collected in a polyethylene carboy and acidified (to 0.6 M in HNO_3) with high-purity HNO_3 to prevent bacterial growth, to stabilize the trace element concentrations and to strip any trace elements bound by colloidal particles.

b. Atomic Absorption Apparatus

The instrumental system used in this study consisted of a Perkin-Elmer® Model 603 atomic absorption spectrometer with HGA-2100 graphite furnace (GFAAS).

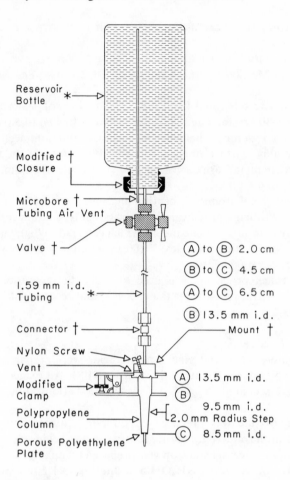

FIGURE 7.   Column separation apparatus.

c. Column Separation Apparatus (Figure 7)

The Isolab® Qs-Q polypropylene column with porous polyethylene resin support was used for 100-mℓ and 1-ℓ sample volumes. Although the same column was used for both sample volumes, the amount of resin and reservoir systems were entirely different. For the 100-mℓ sample, the QS-S 25-mℓ conventional polyethylene extension funnel was attached to the column to act as a reservoir for the sample.

For a 1-ℓ sample, the reservoir was a 1-ℓ Teflon® (FEP) bottle inverted and modified with a machined Teflon® (TFE) closure insert containing a microbore venting tube and outlet tube. The outlet was connected to a valve (TFE) by 1.59-mm (1/16 in.) I.D. Teflon® (FEP) tubing connector and linked to the reservoir with a specially machined mount (TFE) which sealed the column into the closed system. The mount contained a vent (sealed with nylon screw, allowing the removal of air from the system), as well as an inlet, and was tightly clamped to the column using the lip on the column at point B (see Figure 7). The clamp (a modified glass joint clamp) and mount provided a seal, which allowed the reservoir to be raised above the column to obtain enough pressure to control the flow rate using the pressure of the raised reservoir and the valve (Figure 7).

2. Procedure
a. Column Preparation and Purification Procedure

The column preparation procedure consisted of precleaning the columns in 1:4 HCl and then in 1:4 HNO₃ for one week in each bath, and then rinsing the columns with water after each acid wash. Load the column with a slurry of Chelex® 100 resin 200-400 mesh size (sodium form). For 100-mℓ seawater samples, use 3.2 to 3.4 mℓ of resin which covers the lower barrel of the column from point B to point C in Figure 7. For 1-ℓ samples, use 5.8 to 5.9 mℓ of resin, which fills the column from point A to point C in Figure 7. Wash the resin with 15 to 20 mℓ of 2.5 M HNO₃ (for the small and large resin volumes, respectively), in 5-mℓ portions to elute any trace metal contamination present in the resin. Then rinse two 5-mℓ volumes of water. To transform the resin to the NH⁺₄ form, add 10 to 15 mℓ of 2.0 M HN₄OH in 5-mℓ volumes. After checking the pH of the effluent to ensure basicity, rinse the column with 10 to 15 mℓ of water to remove the excess NH₄OH.

b. Column Preconcentration and Separation Procedure

For the 100-mℓ seawater sample, weigh 101.8±0.2 g directly into a clean 250-mℓ Teflon® (FEP) beaker and adjust the pH to 5.0 to 5.5 with the dropwise addition of NH₄OH. Then add 0.5 mℓ of 8 M ammonium acetate to aid in buffering the system. Any necessary agitation of the solution is done with a Teflon® stirring rod. Add a small amount of the seawater to the reservoir and column to allow the resin to undergo its natural shrinkage as it changes ionic form and pH. This shrinkage results in a resin volume of approximately one half of its original volume. After the completion of this transformation is observed (2 to 3 min), add the remaining seawater to the reservoir as needed to keep it filled; the flow rate is approximately 0.8 mℓ/min. To selectively elute Na⁺, K⁺, Ca⁺² and Mg⁺², replace them with NH⁺₄, add 40 mℓ of 1.0 M ammonium acetate to the column in 10-mℓ aliquots. At the completion of the ammonium acetate addition, add 10 mℓ of water to remove residual ammonium acetate. Elute the transition metals using 7 mℓ of 2.5 M HNO₃, and collect in clean preweighed 10-mℓ conventional polyethylene bottles. Cap the bottles with clean polyethylene-lined caps and reweigh to determine the weight of the effluent accurately.

The procedure for the 1-ℓ samples is the same as the 100-mℓ samples, with minor alterations due to apparatus (see above). Weigh the sample (1018.0±0.5 g) into a 1-ℓ Teflon® (FEP) bottle, and adjust the pH in the same manner as previously described. The bottle becomes the reservoir and is fitted with a modified closure (see Figure 7). Invert the bottle and purge the air from the system by means of the vent on the column mount. Adjust the flow rate using the valve and the height of the reservoir. Keep the flow rate to less than 0.2 mℓ/min until the shrinkage of the resin is complete. Then increase the flow rate to 1.0 mℓ/min and leave overnight to flow through the column. After passing the sample through the column, remove the valve and tubing at the connector above the column mount and replace with a smaller reservoir containing 70 mℓ of 1.0 M ammonium acetate. Adjust the flow rate to 0.5 mℓ/min until the reagent is exhausted. Wash the resin with 10 mℓ of water. Elute the transition metals with two 5-mℓ portions of 2.5 M HNO₃ into preweighed polyethylene bottles as previously described.

Introduce a 25 $\mu\ell$ sample into the graphite furnace with the autosample. Use the instrument parameters given in Table 5.

VI. DETERMINATION OF METALS IN FRESH WATER USING ELECTROTHERMAL ATOMIZATION[20]

As already stated, it is not advisable to analyze waters with high salt content (e.g., seawater, brackish water, and some effluents) directly by electrothermal atomization. Problems with nonspecific absorption are normally too great for reliable automatic

Table 5

INSTRUMENTAL PARAMETERS

Element	Wavelength (nm)	P & E 603 SI: + (nm)	P & E 603 Scale Expansion	HGA-2100 Drying (*T*-sec)	HGA-2100 Charring (*T*-sec)	HGA-2100 Atomization (*T*-sec)	HGA-2100 Gas
Cd	228.8	0.7	1	100-30	200-20	2100-7	Ar[a]
Co	240.7	0.2	2	100-30	500-30	2700-7	Ar[a]
Cu	324.7	0.7	1	100-40	700-30	2500-6	Ar[a]
Fe	248.3	0.2	2	100-30	600-30	2700-7	Ar[a]
Mn	279.5	0.7	2	100-30	300-30	2700-7	Ar[a]
Ni	232.0	0.7	5	100-30	1000-30	2700-6	Ar[a]
Pb	283.3	0.7	3	100-40	400-30	2200-7	Ar[b]
Zn	213.9	0.7	0.5	100-30	500-20	2000-7	Ar[b]

Note: *T* = temperature

[a]   Interrupt mode.
[b]   Normal mode.

background correction. The following procedure should be used for samples such as rain, snow, soft water lakes, and rivers and drinking waters.

When running samples of a moderately complex matrix, e.g., river waters, the need for the method standard additions for calibration should be determined. This can be done by comparing the results obtained by direct comparison on a typical sample with those obtained by standard addition. If the results differ, standard addition should be employed. Background correction should be used with the method of standard additions.

It is surprising to find that values obtained by electrothermal analysis are sometimes different from those obtained by extraction or ion exchange. The present author believes this to be due to the presence of fine particulate even after filtration through 0.45μ filters.

Loss of metals during ashing can occasionally be a problem. This is often due to the presence of abundant chloride. Nitric acid is, therefore, added to the samples to convert metals to the oxide during this stage. Drop by drop, addition of 50% hydrogen peroxide can also aid in the ashing and oxidation and may be used as necessary.

A. Equipment and Reagents

The following procedure is strictly for use with the HGA 2100 (HGA 74) furnace from Perkin-Elmer. Other Perkin-Elmer® equipment may require other conditions. It is almost certain that other manufacturers, furnaces, rods, etc., will require different conditions. These can usually be found experimentally, starting with those recommended by the manufacturer for standard solutions. Both Perkin-Elmer® 305B and 603 atomic absorption units have been used. A Perkin-Elmer® Model 056 recorder was employed. Background correction is essential.

Samples were injected using disposable Eppendorf® pipettes. The disposable plastic tips were treated in 3 *M* nitric acid and rinsed with water to eliminate contamination of iron and zinc. The largest single volume used was 50 $\mu\ell$.

The quality of the distilled water is crucial in this work. Detection limits are often set too high because of impurities in distilled water. Double distillation in quartz will normally give a suitable high-quality water.

Stock standard solutions (1000 $\mu g/m\ell$) are prepared by dissolving pure metals or

Table 6
TEMPERATURE PROGRAM-AAS
PARAMETERS (GENERAL CONDITIONS)

Element	Wavelength (nm)	Slit (nm)	Dry (°C)	Ash (°C)	Atomize (°C)
Cd	228.8	0.7	100	440	1800
Co	240.7	0.2	100	1000	2600
Cr	357.9	0.7	100	1000	2700
Cu	324.7	0.7	100	1000	2500
Fe	248.3	0.2	100	1200	2600
Mn	279.5	0.2	100	1000	2500
Ni	232.0	0.2	100	1200	2600
Pb	283.3	0.7	100	500	2000
Zn	213.8	0.7	100	500	2000

pure metal salts in a minimum of acid and then diluting to volume in distilled water. Nitric acid should always be used. Dilute working solutions of mixed cadmium, cobalt, chromium, copper, iron, manganese, nickel, lead, and zinc are prepared in concentrations from 0.3 to 100 $\mu g/\ell$ and are made to contain 1% nitric acid. Contamination at lower concentrations necessitates the renewal of standards at frequent intervals.

B. Procedure
1. Direct Comparison
Using an appropriate-sized pipette or auto sampler, place standards and then samples into the furnace one after the other. All waters should contain or be treated with nitric acid to prevent formation of volatile metal salts during ashing.

The atomic absorption equipment including the background corrector should be warmed up for 20 min. or until sources are stabilized. Use temperature programs and instrument parameters as outlined in Table 6. Use nitrogen purge gas in the normal flow mode.

Run 1 or 2 standards every 10 to 20 samples to check the stability of the readings obtained. Recalibrate if necessary. When going from high to low values and when using manual injections, rinse the pipette. It is a good precaution to rinse between each sample to prevent spurious contamination.

2. Method of Standard Additions
Prepare the following simple solutions for each sample; the sample alone, the sample plus a small spike of standard solution to give a concentration approximately twice the original sample and the sample similarly spiked to give a concentration three times the original sample. Run each solution in order from low to high concentrations. If a small enough volume of spiking solution is used, the slight dilution can be ignored.

VII. MERCURY AND HYDRIDE METHODS

A. Introduction
One of the major contributions in applied atomic absorption analysis was the development of the cold vapor absorption tube method for Hg by Hatch and Ott.[21] This approach, based on a colorimetric procedure published by Kimura and Miller,[22] allowed the determination of Hg in a wide variety of matrices at levels three orders of magnitude lower than atomic absorption methods available at that time. These latter methods were inadequate for the determination of the very low levels of Hg found in almost all geological and environmental samples.

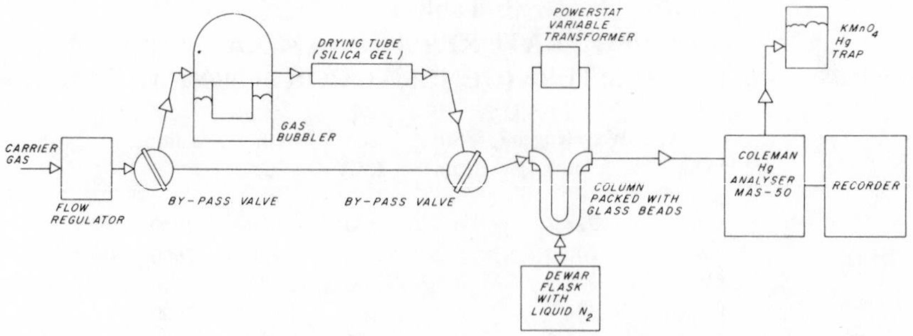

FIGURE 8.   Schematic diagram of system used to determine the presence of mercury in water.

The Marsh reaction for the generation of arsine has been known for 130 years. Another major contribution to atomic absorption analysis was the development by Holak[23] of an atomic absorption method for the analysis of As based on the production of arsine. Arsine is atomized in a flame or a low temperature furnace, the latter being generally preferred due to excessive absorption by flame gases of the useful As resonance line at 193.7 nm. Like Hg, with the advent of this method, levels of As could be determined which were three orders of magnitude lower than were possible with methods available at the time. The hydride approach has since been expanded to include the elements Sn, Ge, Sb, Bi, Te, and Se. In contrast to Hg, reliable methods have been developed using conventional electrothermal atomization, which can be used for analyzing covalent hydride elements in a number of sample types with detection limits comparable to the hydride method.

B. Determination of Mercury in Water[24]

A cold trap preconcentration step is used, in which a glass U-tube is packed with glass beads and immersed in liquid nitrogen. This avoids the use of slower amalgamation or extraction preconcentration steps. The procedure was tested on seawater, marine zooplankton, NBS orchard leaves, and oceanic manganese crustal materials. The value obtained for NBS orchard leaves was 150 ± 15 mg/g.

1. Equipment and Reagents

The mercury analyses were conducted using a Coleman Instrument® mercury analyzer (MAS 50) equipped with a Leeds & Northrup® Speedomax recorder (Model XL 601). The aqueous sample solution was contained in 250-mℓ Pyrex® (Corning Glass Works) glass bubbler placed at one end of a sampling train employing nitrogen as the purging and carrier gas. The gas-flow system also included a flow regulator, two by-pass valves, water absorber, a mercury cold-trap, a gas cell, and a gas washing bottle containing a 10% solution of potassium permanganate. A schematic diagram of the entire system is shown in Figure 8. The sparging vessel was connected directly through amber latex rubber tubing (i.d. 3 mm, length 60 cm) to a Pyrex® glass drying tube (i.d. 18 mm, length 18 cm) containing colorless silica gel (6/20 mesh) as a water absorbent. The mercury cold-trap followed this absorbing stage, and connection was made using a 75-cm length of rubber tubing. The mercury cold-trap consisted of a 50 cm Pyrex® glass (i.d. 4mm) U-tube (width 4.4 cm), which was packed with glass beads (90/100 mesh) to form a 6-cm column at the bend of the tube. The U-tube was wrapped with Nichrome® wire (diameter 0.06 mm), yielding five windings over the entire lower 15 cm of the tube. The wire-wound U-tube was placed inside an insulating Pyrex® glass covering (i.e., i.d. 10 mm, ℓ 3.8mm) such that only the nonwire-wrapped sections were exposed. This complete U-tube apparatus was designed to fit a 1-ℓ Dewar® flask containing liquid nitrogen to provide the trap and concentration step for mercury vapor. The leads from the heating wire were connected to a Powerstat® variable transformer, which allowed the column to be heated electrically upon removal from the liquid nitrogen bath. Three-way Teflon® stopcocks (4-mm bore) were placed before the bubbler and after the drying tube to permit the sparging vessel and the drying column to be bypassed during the heating and elution step. Connections were made with 3-mm-i.d. rubber latex tubing. The mercury, which is rapidly vaporized and eluted form the column during the heating step, was fed directly by polyethylene tubing (i.e., 3mm, length 70 cm) to the gas cell of the Coleman® mercury analyzer. The absorption of elemental mercury in arbitrary units was displayed on a recorder using a 25 X scale expansion. After the carrier gas had passed through the gas cell, it was directed into a 300-mℓ gas washing bottle containing 100mℓ of 10% potassium permanganate. At this stage, the elemental mercury was oxidized and removed from the gas flow.

Except where noted, all chemicals used were ACS certified reagent-grade quality supplied by the J. T. Baker Chemical Company. The nitrogen carrier gas was 99.8% pure (Chemetron Corporation). Deionized water for reagent preparation and experimental studies was produced by passing Pyrex® glass distilled water through an activated charcoal absorbent and two mixed-bed ion exchangers (Continental Dionization Service). A primary standard mercury solution was made by dissolving 1.354 g of free-dried spectrographic-grade mercuric chloride (Johnson, Matthey Chemical Ltd.) in 1000 mℓ of distilled deionized water. From this solution, 10- and 100-μg/ℓ spiking solutions were made daily. A 20% stannous chloride solution was prepared by dissolving 72 g stannous chloride in 20-mℓ concentrated hydrochloric acid and bringing to volume with 3 N sulfuric acid. The standard mercury solution, mercury spiking solutions, and the stannous chloride reagent were stored in high density polyethylene bottles. The 10% potassium permanganate oxidizing solution was prepared by dissolving 100 g potassium permanganate crystals in 1 ℓ of distilled deionized water.

The column packing material was Anaport® glass beads (80/100 mesh). In earlier studies, the column packing material employed was Analabs® Chromosor W-HP (80/100 mesh 1.5% OV 17 and 1.95% QF 1). Although the latter column was equally efficient, it must be conditioned before use at 150°C for approximately 2 hr. No column conditioning, other than normal blank measurements, was necessary with the glass bead column. Colorless silica gel (Fisher Scientific Company, 6/20 mesh) was used as the drying agent.

2. Procedure

With the U-tube column immersed in the liquid nitrogen bath, place a 100-mℓ sample solution into the gas bubbler. Insert the sparger, and switch the valves to the flow-through position. The purging rate for the nitrogen aeration is 0.5ℓ/min and 7 psi. After purging is completed (7 min), return the valves to the bypass position, and increase the flow rate of the carrier gas to 0.7 ℓ/min (7 psi). Remove the column from the liquid nitrogen bath, and heat the U-tube through the wire windings using the variable transformer. The transformer can be simply switched on by prior calibration of the voltage setting, to give a column temperature of 225° C measured on the outside wall of the U-tube after 60 sec of heating. The elemental mercury is vaporized and eluted from the column in 1 to 2 sec, and the entire operation required less than 10 sec from the time the U-tube is removed from the cold trap. During the heating and elution cycle (60 sec), the flow rate of the carrier gas decreases from 0.7 to 0.15 ℓ/min. The absorption at the mercury wavelength (253.7 nm) occurs in the gas cell 12 sec after the heating cycle is initiated. The absorption is recorded in arbitrary units and the maximum height noted. This represents the sample and system blank.

After the response has returned to the initial base line (60 sec), stop the heating and cool the column in air for 30 sec. Return the U-tube to the liquid N_2 bath. The carrier gas flow is set to 0.5 ℓ/min, and the system is ready for the stannous chloride reduction and nitrogen aeration step. The total time for this operation is 9 min.

Add 0.5 mℓ of the stannous chloride reagent to the 100-mℓ sample to reduce the mercury to its elemental state. Mix the sample solution by hand shaking for 5 sec, and then switch the Teflon® stopcocks to the flow-through position. After the latter operation, the manipulations are identical to the steps outlined in the procedure for establishing the system blank. Record the absorption peak (253.7 nm) of mercury, and note its maximum height. Add appropriate spikes of mercuric chloride standards to the sample matrix and repeat the procedure. Three spike additions are usually made. A system blank is repeated and the mercury concentration determined from an individual calibration curve for each sample.

C. Determination of Arsenic, Germanium, Antimony, Selenium, Tin, Bismuth, and Tellurium, Using a Conventional Generator[25]

The authors of this procedure have carefully studied the optimum reaction times for the various elements. The times given apply to water samples, and they caution the worker that longer reaction periods may be necessary if other sample types are analyzed.

The effect of hydride collection time (time period between adding the sodium borohydride and sweeping the generated gases into the flame) was investigated for arsenic, bismuth, germanium, antimony, selenium, tin and tellurium. Only for the analysis of bismuth and tellurium is there a need for careful monitoring of the collection time. A significant loss of sensitivity is obtained using collection times longer than 30 sec. By keeping the fourway stopcock in the sweep position, one may continuously flush the tellurium hydride into the burner. This degrades the sensitivity by about a factor of 2; however, the need for time monitoring is eliminated.

For other elements determined successfully using the sodium borohydride method (arsenic, germanium, antimony, selenium and tin), collection times of up to 2 to 3 min can be used with no loss of sensitivity. Generally, a collection time of 30 sec is sufficient, although longer collection times may be required for certain types of samples. The sodium borohydride pellet reacts vigorously when added to the acidified sample, and dissolves completely in about 20 sec. The addition of a single sodium borohydride pellet (10/32 in.), weighing approximately 200 mg was found to provide optimum results. Varying the amount of sodium borohydride from 100 to 400 mg had no noticeable effect on the sensitivity obtained for any of the elements studied. The addition of amounts larger than 500 mg caused the balloon reservoir to rupture. The use of a magnetic stirrer to agitate a sample solution of arsenic had no effect on the sensitivity obtained.

For arsenic, bismuth, antimony, and tellurium, varying the acid concentration from 1 to 6 N had no noticeable effect on the sensitivity obtained. To obtain optimum sensitivity, the acid concentration should be at least 5 N when determining selenium and 4 N or less when determining germanium. For the determination of tin, it is important that the acid concentration be 0.5 N or less, due to the very pronounced loss of sensitivity with increasing acid concentration. The suitability of other acids was not investigated.

For arsenic, the electrodeless discharge lamp (EDL) provides about a two-fold improvement of sensitivity, and for selenium the sensitivity is improved by about 30%. The sensitivity for germanium is about an order of magnitude poorer than that ob-

Table 7

ABSOLUTE SENSITIVITIES OBTAINED UTILIZING SODIUM BOROHYDRIDE REDUCTION

Element	λ nm	Spectral slit (Å)	Absolute Sensitivity (ng)	Solution detection limit (μg/ℓ)[a]	Remarks
As	193.7	7	10	0.5	Hollow cathode
			5	0.15	EDL
Bi	223.1	2	8	0.25	Collect 30 sec
			12	0.4	Continuous flow
Ge	265.1	2	270	10.0	
Sb	217.6	2	10	0.25	
Se	196.0	7	11	0.25	Hollow cathode
			9	0.15	EDL
Sn	224.6	2	7	0.2	
Te	214.3	2	14	0.25	Collect 30 sec
			27	0.75	Continuous flow

[a] Based on a 20 mℓ sample. Both electrodeless-discharge lamps and conventional hollow-cathode lamps were tested for some elements. Gains in sensitivity obtained with the former were significant.

tained for the other elements studied. The absolute sensitivities (weight of an element which gives a signal of 1% absorption) obtained for the seven elements studied are summarized in Table 7, together with solution detection limits.

1. Equipment and Reagents

All results were obtained using a Perkin-Elmer® Model 403 atomic absorption spectrophotometer equipped with a three-alot burner head, deuterium background corrector, Model 056 recorder, and Intensitron® hollow-cathode lamps. For the determination of arsenic and selenium, Perkin-Elmer® electrodeless discharge lamps were also employed. A Perkin-Elmer® arsenic/selenium sampling system was utilized for the generation and collection of the gaseous hydrides. This system utilizes a dosing stopcock for reagent introduction and a balloon reservoir for collection of the generated gases. By rotating a four-way stopcock, the argon flow can be set to bypass or flow through the generation flask (125-mℓ Erlenmeyer with a 29/42 ground glass joint). The collected hydride, plus excess hydrogen, is introduced into the burner via the auxiliary oxidant connection. The following flow settings were found to be optimum: argon 40 (13 ℓ/min) at a pressure of 20 psi, hydrogen 24 (10 ℓ/min) at a pressure of 20 psi. Reagents used where $NaBH_4$ pellets (10/32 in., available from Alfa Inorganics®), and hydrochloric acid. Standard solutions of all the elements investigated were prepared in dilute HCl.

2. Procedure

Pipette 20 mℓ of sample into the generation flask. Acidify the sample using a suitable volume of HCl, and dilute to 40 mℓ with deionized water. Connect the flask to the generation apparatus, and open the four-way stopcock for about 15 sec to admit argon, which flushes the air out of the system. After flushing, close the four-way stopcock, and add a single sodium borohydride pellet via the dosing stopcock. Continue the reaction for a time that will vary depending on the element being determined and the type of sample being analyzed. For aqueous samples, a reaction time of 30 sec is suitable. Longer reaction times may be required for some types of samples. Open the four-way stopcock which allows the auxiliary argon flow to sweep the generated gases into the burner. Obtain the absorption signal on the recorder. Close the four-way stopcock after the absorption signal has been recorded and the pen has returned to the baseline. Analyze standards, including a reagent blank, using the same procedure.

When the four-way stopcock is opened, the surge of excess hydrogen into the flame causes a sudden change in the absorption of the flame, which produces a large blank signal when operating at wavelengths below 210.0 nm. Use of the deuterium background corrector appreciably reduces this blank signal.

VIII. DETERMINATION OF CR III AND CR VI IN SEAWATER[26]

It is crucially important to be able to distinguish between the trivalent and hexavalent forms of chromium. This is because hexavalent chromium is carcinogenic at very low levels. Trivalent chromium, on the other hand, has little health effect even at ppm levels.

Chromium VI is extracted with Aliquat-336® (General Mills, Kankakee, Illinois) from weakly acidic (pH 2) sample solutions. Chromium III is recovered by extraction of pH 6 to 8 solutions containing 1 M thiocyanate. Alternatively, chromium III can be extracted after removal of chromium VI by oxidation with ammonium persulfate and extraction with Aliquat 336®. The latter method gave persistently higher results for chromium III probably because of the inclusion of chromium from highly stable complexes or colloids. Determinations are done by electrothermal atomic absorption spectroscopy.

Aliquat 336® is an ammonium salt of a high-molecular-weight tertiary amine. Tertiary amines of this type extract chromium VI in a number of chemical forms, e.g., $HCrO_4^-$, $Cr_2O_7^{-2}$, $HCr_2O_7^-$ and CrO_3Cl^- in the presence of chromium III.

The detection limits for chromium VI and III are 0.01 and 0.03 $\mu g/\ell$, respectively.

A. Equipment and Reagents

A Perkin-Elmer® 403 atomic absorption spectrometer equipped with an HGA-72 graphite furnace and number 71868 graphite tubes was used.

Standard solutions were prepared from analytical-grade chromium (III), nitrate, and potassium chromate; the chromium (III) solutions were acidified to 0.01 M- with hydrochloric acid. Aliquat-336®, a yellow viscous liquid, is a mixture of methyltri-*n*-alkylammonium chlorides with alkyl groups that consist mainly of C_8-C_{10} chains; its mean molecular weight is 475. All other chemicals were of analytical-grade quality and were tested for low chromium blanks.

Seawater samples were collected in polyethylene bottles, which were conditioned by shaking with ca. 1 M nitric acid and deionized water. In order to prevent the absorption of chromium III on the wall of the bottles, an amount of hydrochloric acid sufficient to acidify the samples to pH 2 was put into the containers prior to sampling for the chromium III analyses. With chromium VI, losses were shown to be negligible from neutral solutions. However, at pH 2, the hexavalent ion is slowly reduced to chromium III, the rate of reduction being dependent on the composition of the seawater sample. In the present investigation, chromium VI was found to be absent from all samples acidified immediately after the sampling and analyzed after a period of time of 24 hr (or more).

Solutions of chromium III and VI in deionized distilled water were used. To 500 mℓ of these solutions, 14 g of sodium chloride were added to equal the concentration in seawater.

B. Procedure

Carefully clean 1 ℓ glass separatory funnels with a (1 + 1) mixture of 4 M nitric acid and 4 M hydrochloric acid and subsequently rinse with deionized water. For the extraction of chromium VI, acidify the seawater to pH 2 with hydrochloric acid. For the extraction of chromium III, add enough potassium thiocyanate to obtain a concentration of 1 M. Neutralize the solution with sodium hydroxide to pH 6 to 8. For the oxidation of chromium III, boil 1 mℓ of a 0.1 M ammonium persulfate solution and 500 mℓ of seawater for 15 min.

After these additions of the appropriate reagents, equilibrate 500 mℓ of seawater with 5 mℓ of a 0.1 M solution of Aliquat-336® in toluene for chromium III (for chromium VI use 15 mℓ). After shaking for about 3 min, allow the phases to separate for 15 to 30 min. Determine the chromium content of the organic phase by flameless atomic absorption spectrometry at 357.9 nm. Inject aliquots (50-μl) and use the following heating program: drying at 130°C for 1 min, ashing at 1350°C for 1.5 min, and atomization at 2650°C for 15 sec.

REFERENCES

1. Gibbs, R. J., *Nature* (London), 180, 71, 1973.
2. Müller, K., *Z. Phys.*, 65, 739, 1930.
3. Walsh, A., *Spectrochin Acta*, 7, 108, 1955.
4. Box, G. F. and Walsh, A., *Spectrochin Acta*, 16, 255, 1960.
5. Analytical Methods for Atomic Absorption Spectroscopy, Perkin-Elmer Corp., Norwalk, Conn., 1976.
6. Amos, M. D. and Willis, J. B., *Spectrochin Acta*, 22, 1325, 1965.
7. Grimaldi, F. S. and Schnepfe, M. M., *Talanta*, 16, 1461, 1970.
8. Van Loon, J. C., *Atomic Absorption News*, 11, 685, 1972.
9. Sychra, V., Slevin, P. J., Matousek, J., and Bed, F., *Anal. Chim. Acta*, 52, 259, 1970.
10. Pitts, A. E., Beamish, F. E., and Van Loon, J. C., *Anal. Chim. Acta*, 50, 171, 1970.
11. Schnepfe, M. M. and Grenaldi, F. S., *Talanta*, 16, 1461, 1969.
12. L'Vov, B. V., *Inzh. Fiz. Zh.*, 2, 44, 1959.
13. Cruz, R. B. and Van Loon, J. C., *Anal. Chim. Acta*, 72, 231, 1974.
14. Stary, J., *The Solvent Extraction of Metal Chelates,* Macmillan, N.Y., 1964.
15. Morrison, G. H. and Freisen, H., *Solvent Extractions in Analytical Chemistry,* John Wiley & Sons, New York, 1957.
16. Zolotov, Y. A., *Extraction of Chelate Compounds,* Ann Arbor — Humphrey Scientific Publishers, Ann Arbor, Mich., 1970.
17. Kinrade, J. D. and Van Loon, J. C., *Anal. Chem.*, 46, 1894, 1974.
18. Kingston, M. M., Barnes, I. L., Brady, T. J., Rains, T. C., and Champ, M. A., *Anal. Chem.*, 50, 2064, 1978.
19. Dean, J. A. and Rains, T. C., Eds., *Flame Emission and Atomic Absorption Spectrometry,* Vol. 2, Marcel Dekker, N.Y., 1971.
20. Van Loon, J. C., (unpublished, 1980) Internal Laboratory Procedure.
21. Hatch, W. R. and Ott, W. L., *Anal. Chem.*, 40, 2085, 1968.
22. Kimura, Y. and Miller, V. L., *Anal. Chim. Acta*, 27, 325, 1962.
23. Holak, W., *Anal. Chem.*, 41, 1712, 1969.
24. Fitzgerald, W. F., Lyons, W. B., and Hunt, C. D., *Anal. Chem.*, 46, 1883, 1974.
25. Fernandez, F. J., *At. Absorpt. Newsl.*, 12, 93, 1973.
26. de Jong, G. J. and Brinkmann, U. A. Th., *Anal. Chim. Acta*, 98, 243, 1978.

INDEX

A

Autotrophic organisms, 135
Autotrophic phytoplankton, 15
Available phosphorus, 183

B

Background
    carbonic acid, 83
    compensation of, 221
    correction of, 216, 220, 230
    interferences in, 78, 85
Bacteria, 135
Balance of ions, 71
Barium chloride solution, 123, 127
Bathtub ring, 89
Bathythermograph (BT), 31—38
    electronic, 33—35
    expendable, 35—38
    mechanical, 31—33
Baylis turbidimeter, 21
Beer's Law, 144
Bias, 140
Bicarbonates, 138
    alkalinity of, 112
Biological activity, 141
Biomass, 134
Bismuth, 234—236
Bivalent ion-selective electrodes, 88
Blank reagents, 174, 222
Blue Black R, 73
Boiler scales, 89
    formation of, 55
Borax, 190
Boric acid, 190
Boron, 190, 201—202
    accuracy in determination of, 203, 206
    automated methods for, 202—206
    calculations for, 205
    carminic acid for analysis of, 202—203
    colorimetric method for, 201
    curcumin as colorimetric reagent for, 201—202
    fluorometric method for, 203—206
    nitrate interference in analysis of, 202
    pollution by, 190
    precision in determination of, 203, 206
    standard solution, 201, 202, 204
    toxicity of, 190
Boron manifold, 204
Bottle standards comparison, 21
Brij-35, 91
Bromocresol green, 101
BT, see Bathythermograph
Buffer, 42, 106, 119, 142, 159
    glycine-potassium hydroxide, 85
Buffered EDTA solution, 127
Buffer solution, 44, 46, 47, 72, 84, 127, 164

C

Cadmium, 225
    in seawater, 227—229

Calcium, 70, 71—83
Calcon carbonic acid, 73, 75, 76
Calibration, 166
    curves for, 95, 118, 155, 205
    of pH meter, 44—46
    of turbidimeter, 26
Calmagite indicator solution, 91
Canada Centre for Inland Waters (CCIW), 35
Canadian Department of National Health and
        Welfare, 15, 54
Carbon, 133, 135, 137
    colloidal, 136
    dissolved, 136
    forms of, 136
    inorganic, 135, 136, 154, 155
    organic, 135, 136, 153—158
    particulate organic, see Particulate organic
        carbon
    soluble, 136
    suspended, 136
    total, 136
Carbonate alkalinity, 103, 109, 112
Carbonate alkalinity manifold, 110
Carbonate hardness, 88
Carbonates, 138
    inorganic, 153
Carbon cycle, 135—136
Carbon dioxide, 135, 138
    evolution method for, 109, 111
Carbonic acid background, 83
Carbon rod, 218
Carminic acid for boron analysis, 202—203
Cation-exchange resin, 82, 221
Cations, 70, 71, 91
    hardness-causing, 90
CCIW, see Canada Centre for Inland Waters
Cell constant, 49, 50
Chelating resins, 226—229
Chloramine-T, 193—196
Chloranilic acid, 72
Chloride, 70, 113—120
    IC method for, 120
    interference of in iodide determination, 207
    ISE method for, 119—120
    standard solution, 115
4'-Chloro-2-hydroxy-4-methoxybenzophenone
        (CHMB), 203
Chromatography, 82, 224
Chromium III, 236—237
Chromium IV, 236—237
Cobalt, 225
    in seawater, 227—229
Coefficient of variation, 13
Coking plants, 209
Cold vapor absorption tube, 231
Colloidal carbon, 136
Color, 3, 10—15
    apparent, 11
    standards for, 12
    true, 11
Color chart, 40
Colorimeter, 160, 202
Colorimetric reagents for boron, 201—202